Clinical Chemistry, Immunology and Laboratory Quality Control

Clinical Chemistry, Immunology and Laboratory Quality Control

A Comprehensive Review for Board Preparation, Certification and Clinical Practice

Amitava Dasgupta, PhD, DABCC

Professor of Pathology and Laboratory Medicine,
University of Texas Medical School at Houston

Amer Wahed, MD

Assistant Professor of Pathology and Laboratory Medicine,
University of Texas Medical School at Houston

ELSEVIER

AMSTERDAM • BOSTON • HEIDELBERG • LONDON
NEW YORK • OXFORD • PARIS •SAN DIEGO
SAN FRANCISCO • SINGAPORE • SYDNEY • TOKYO

Elsevier
525 B Street, Suite 1900, San Diego, CA 92101-4495, USA
32 Jamestown Road, London NW1 7BY, UK
225 Wyman Street, Waltham, MA 02451, USA

Notice
No responsibility is assumed by the publisher for any injury and/or damage to persons, or property as a matter of products liability, negligence or otherwise, or from any use or operation of any methods, products, instructions or ideas contained in the material herein. Because of rapid advances in the medical sciences, in particular, independent verification of diagnoses and drug dosages should be made.

Medicine is an ever-changing field. Standard safety precautions must be followed, but as new research and clinical experience broaden our knowledge, changes in treatment and drug therapy may become necessary or appropriate. Readers are advised to check the most current product information provided by the manufacturer of each drug to be administered to verify the recommended dose, the method and duration of administrations, and contraindications. It is the responsibility of the treating physician, relying on experience and knowledge of the patient, to determine dosages and the best treatment for each individual patient. Neither the publisher nor the authors assume any liability for any injury and/or damage to persons or property arising from this publication.

British Library Cataloguing-in-Publication Data
A catalogue record for this book is available from the British Library

Library of Congress Cataloging-in-Publication Data
A catalog record for this book is available from the Library of Congress

ISBN: 978-0-12-407821-5

For information on all Academic Press publications
visit our website at elsevierdirect.com

Printed and bound in the United States of America

Transferred to Digital Printing, 2014

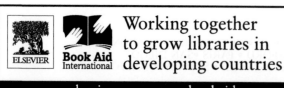

Dedication

Dedicated to our wives, Alice and Tanya.

Contents

Preface

There are excellent clinical chemistry textbooks, so the question may arise: Why this book? From our many years of teaching experience, we have noticed that few pathology residents are fond of clinical chemistry or will eventually choose a career in chemical pathology. However, learning clinical chemistry, immunology, and laboratory statistics is important for not only passing the American Board of Pathology, but also for a subsequent career as a pathologist. If, after a fellowship, a pathology resident chooses an academic career, he or she may be able to consult with a M.D. or Ph.D. level clinical chemist colleague for laboratory issues involving quality control, but in private practice a good knowledge of laboratory statistics and quality control is essential because a smaller hospital may not have a dedicated clinical chemist on staff. These professionals can use this book as a comprehensive review of pertinent topics.

We have been using our resources for teaching our residents and students, and many of them have provided positive feedback after taking the boards. As clinical chemistry topics are relatively new to a typical resident, these resources provided a smooth transition into the field. This motivated us to refine our resources into book form. Hopefully this book will help junior residents get a good command of the subject before pursuing a more advanced understanding of clinical chemistry by studying a textbook in clinical chemistry or a laboratory medicine textbook. In addition, a first year Ph.D. fellow in clinical chemistry may also find this book helpful to become familiar with this field before undertaking more advanced studies in clinical chemistry. We decided to add hemoglobinopathy to this book because in our residency program we train residents both in serum protein electrophoresis and hemoglobinopathy during their clinical chemistry/immunology rotation, although in other institutions a resident may be exposed to hemoglobinopathy interpretation during the hematology rotation. Ph.D. clinical chemistry fellows also require exposure to this topic. We hope this book will successfully help pathology residents to have a better understanding of the subject as well as to be comfortable with their preparation for the board exam. Moreover, this book should also help individuals taking the National Registry of Certified Chemists (NRCC) clinical chemistry certification examination. We have included a detailed Key Points section at the end of each chapter, which should serve as a good resource for final review for the board. This book is not a substitute for any of the well recognized textbooks in clinical chemistry.

We would like to thank our pathology residents, especially Jennifer Dierksen, Erica Syklawer, Richard Poe Huang, Maria Gonzalez, and Angelica Padilla, for critically reading the manuscript and making helpful suggestions. In addition, special thanks to Professor Stephen R. Master, Perelman School of Medicine, University of Pennsylvania, for providing two figures for use in this book. Dr. Buddha Dev Paul also kindly provided a figure for the book. Last, but not least, we would like to thank our resident Andres Quesada for drawing several figures for this book. If our readers find this book helpful, our hard work will be duly rewarded.

Amitava Dasgupta

Amer Wahed

Houston, Texas

CHAPTER 1

Instrumentation and Analytical Methods

1.1 INTRODUCTION

Various analytical methods are used in clinical laboratories (Table 1.1). Spectrophotometric detections are probably the most common method of analysis. In this method an analyte is detected and quantified using a visible (400–800 nm) or ultraviolet wavelength (below 380 nm). Atomic absorption and emission, as well as fluorescence spectroscopy, also fall under this broad category of spectrophotometric detection. Chemical sensors such as ion-selective electrodes and pH meters are also widely used in clinical laboratories. Ion-selective electrodes are the method of choice for detecting various ions such as sodium, potassium, and related electrolytes in serum or plasma. In blood gas machines chemical sensors are used that are capable of detecting hydrogen ions (pH meter) as well as the partial pressure of oxygen during blood gas measurements. Another analytical method used in clinical laboratories is chromatography, but this method is utilized less frequently than other methods such as immunoassays, enzymatic assays, and colorimetric assays that can be easily adopted on automated chemistry analyzers.

1.2 SPECTROPHOTOMETRY AND RELATED TECHNIQUES

Spectroscopic methods utilize measurement of a signal at a particular wavelength or a series of wavelengths. Spectrophotometric detections are used in many assays (including atomic absorption, colorimetric assays, enzymatic assays, and immunoassays) as well as for detecting elution of the analyte of interest from a column during high-performance liquid chromatography (HPLC).

Colorimetry was developed in the 19th century. The principle is based on measuring the intensity of color after a chemical reaction so that the

CONTENTS

A. Dasgupta and A. Wahed: Clinical Chemistry, Immunology and Laboratory Quality Control
DOI: http://dx.doi.org/10.1016/B978-0-12-407821-5.00001-2

Table 1.1 Assay Principles and Instrumentation in the Clinical Chemistry Laboratory

Detection Method	Various Assays/Analytical Instrument
Spectrophotometric detection	Colorimetric assays
	Atomic absorption
	Enzymatic assays
	Various immunoassays
	High-performance liquid chromatography with ultraviolet (HPLC-UV) or fluorescence detection
Chemical sensors	Various ion-selective electrodes and oxygen sensors
Flame ionization detection	Gas chromatography
Mass spectrometric detection	Gas chromatography/mass spectrometry (GC/MS), high-performance liquid chromatography (HPLC)/mass spectrometry (LC/MS) or tandem mass spectrometry (LC/MS/MS)
	Inductively coupled plasma mass spectrometry (ICP-MS)

concentration of an analyte could be determined using the absorption of the colored compound. Use of the Trinder reagent to measure salicylate level in serum is an example of a colorimetric assay. In this assay, salicylate reacts with ferric nitrate to form a purple complex that is measured in the visible wavelength. Due to interferences from endogenous compounds such as bilirubin, this assay has been mostly replaced by more specific immunoassays [1]. Please see Chapter 2 for an in-depth discussion on immunoassays.

Spectrophotometric measurements are based on Beer's Law (sometimes referred to as the Beer–Lambert Law). When a monochromatic light beam (light with a particular wavelength) is passed through a cell containing a specimen in a solution, part of the light is absorbed and the rest is passed through the cell and reaches the detector. If Io is the intensity of the light beam going through the cell and Is the intensity of the light beam coming out of the cell (transmitted light), then Is should be less than Io. However, part of the light may be scattered by the cell or absorbed by the solvent in which the analyte is dissolved, or even absorbed by the material of the cell. To correct this, one light beam of the same intensity is passed through a reference cell containing solvent only and another through the cell containing the analyte of interest. If Ir is the intensity of the light beam coming out of the reference cell, its intensity should be close to Io. Transmittance (T) is defined as Is/Io. Therefore, correcting for scattered light and other non-specific absorption, we can assume transmittance of the analyte in solution should be Is/Ir. In spectrophotometry, transmittance is often measured as absorption (A) because there is a linear relationship between absorbance and concentration of the analyte in the solution (Equation 1.1):

$$A = -\log T = -\log Is/Ir = \log Ir/Is \qquad (1.1)$$

Transmittance is usually expressed as a percentage. For example, if 90% of the light is absorbed, then only 10% of the light is being transmitted, where Ir is 100 (this assumes no light was absorbed when the beam passed through the reference cell, i.e. Io is equal to Ir) and Is is 10. Therefore (Equation 1.2):

$$A = \log 100/10 = \log 10 = 1 \qquad (1.2)$$

If only 1% of the light is transmitted, then Ir is 100 and Is is 1 and the value of absorbance is as follows (Equation 1.3):

$$A = \log 100/1 = \log 100 = 2 \qquad (1.3)$$

Therefore, the scale of absorbance is from 0 to 2, where a zero value means no absorbance.

Absorption of light also depends on the concentration of the analyte in the solvent as well as on the length of the cell path (Equation 1.4):

$$A = \log Ir/Is = a.b.c \qquad (1.4)$$

In this equation, "a" is a proportionality constant termed "absorptivity," "b" is the length of the cell path, and "c" is the concentration. Therefore, if "b" is 1 cm and the concentration of the analyte is expressed as moles/L, then "a" is "molar absorptivity" (often designated as epsilon, "ε"). The value of "ε" is a constant for a particular compound and wavelength under prescribed conditions of pH, solvent, and temperature (Equation 1.5):

$$A = \varepsilon bc, \quad \text{or} \quad \varepsilon = A/bc \qquad (1.5)$$

For example, if "b" is 1 cm and the concentration of the compounds is 1 mole/L, then $A = \varepsilon$. Therefore, from the measured absorbance value, concentration of the analyte can be easily calculated from the measured absorbance value, known molar absorptivity, and length of the cell (Equation 1.6):

$$A = \varepsilon bc, \quad \text{or} \quad \text{concentration } "c" = A/\varepsilon b \qquad (1.6)$$

1.3 ATOMIC ABSORPTION

Atomic absorption spectrophotometric techniques are widely used in clinical chemistry laboratories for analysis of various metals, although this technique is capable of analyzing many elements (both metals and non-metals), including trace elements that can be transformed into atomic form after vaporization. Although many elements can be measured by atomic absorption, in clinical laboratories, lead, zinc, copper, and trace elements are the most commonly measured in blood. The following steps are followed in atomic absorption spectrophotometry:

- The sample is applied (whole blood, serum, urine, etc.) to the sample cup.
- Liquid solvent is evaporated and the dry sample is vaporized to a gas or droplets.
- Components of the gaseous sample are converted into free atoms; this can be achieved in either a flame or flameless manner using a graphite chamber that can be heated after application of the sample.
- A hollow cathode lamp containing an inert gas like argon or neon at a very low pressure is used as a light source. Inside the lamp is a metal cathode that contains the same metal as the analyte of analysis. For example, for copper analysis a hollow copper cathode lamp is needed. For analysis of lead, a hollow lead cathode lamp is required.
- Atoms in the ground state then absorb a part of the light emitted by the hollow cathode lamp and are boosted into the excited state. Therefore, a part of the light beam is absorbed and results in a net decrease in the intensity of the beam that arrives at the detector. By application of the principles of Beer's Law, the concentration of the analyte of interest can be measured.
- Zimmerman correction is often applied in flameless atomic absorption spectrophotometry in order to correct for background noise; this produces more accurate results.

Because atoms for most elements are not in the vapor state at room temperature, flame or heat must be applied to the sample to produce droplets or vapor, and the molecular bonds must be broken to produce atoms of the element for further analysis. An exception is mercury because mercury vapor can be formed at room temperature. Therefore, only "cold vapor atomic absorption" can be used for analysis of mercury.

Inductively coupled plasma mass spectrometry (ICP-MS) is not a spectrophotometric method, but is a mass spectrometric method that is used for analysis of elements, especially trace elements found in minute quantities in biological specimens. This technique has much higher sensitivity than atomic absorption methods, and is capable of analyzing elements present in parts per trillion in a specimen. In addition, this method can be used to analyze most elements (both metals and non-metals) found in the periodic table. In ICP-MS, samples are introduced into argon plasma as aerosol droplets where singly charged ions are formed that can then be directed to a mass filtering device (mass spectrometry). Usually a quadrupole mass spectrometer is used in an ICP-MS analyzer where only a singly charged ion can pass through the mass filter at a certain time. ICP-MS technology is also capable of accurately measuring isotopes of an element by using an isotope dilution technique. Sometimes an additional separation method such as high-performance liquid chromatography can be coupled with ICP-MS [2].

1.4 ENZYMATIC ASSAYS

Enzymatic assays often use spectrophotometric detection of a signal at a particular wavelength. For example, an enzymatic assay of ethyl alcohol (alcohol) utilizes alcohol dehydrogenase enzyme to oxidize ethyl alcohol into acetaldehyde. In this process co-factor NAD (nicotinamide adenine dinucleotide) is converted into NADH. While NAD does not absorb light at 340 nm, NADH does. Therefore, absorption of light is proportional to alcohol concentration in serum or plasma (see Chapter 18). Another example of an enzymatic assay is the determination of blood lactate. Lactate in the blood is converted into pyruvate by the enzyme lactate dehydrogenase, and in this process NAD is converted into NADH and measured spectrophotometrically at 340 nm. Various enzymes, especially liver enzymes such as aminotransferases (AST and ALT), can be measured by coupled enzymatic reactions. For example, AST converts 2-oxoglutarate into L-glutamate and at the same time converts L-aspartate into oxaloacetate. Then the generated oxaloacetate can be converted into L-malate by malate dehydrogenase; in this process NADH is converted into NAD. The disappearance of the signal (NADH absorbs at 340 nm, but NAD does not) is measured and can be correlated to AST concentration. However, enzyme activities can also be measured by utilizing their abilities to convert their substrates into products that have absorbance in the visible or UV range. For example, gamma glutamyl transferase (GGT) activity can be measured by its ability to convert gamma-glutamyl p-nitroanilide into p-nitroaniline (which absorbs at 405 nm). Enzymatic activity is expressed as U/L, which is equivalent to IU/L (international unit/L).

Cholesterol, high-density lipoprotein cholesterol (HDL-C), and triglycerides are often measured using enzymatic assays, where end point signals are measured using the spectrophotometric principles of Beer's Law. Cholesterol exists in blood mostly as cholesterol ester (approximately 85%). Therefore, it is important to convert cholesterol ester into free cholesterol prior to assay.

$$\text{Cholesterol esters} \xrightarrow{\text{Cholesterol Ester Hydrolase}} \text{Cholesterol} + \text{Fatty Acids}$$

$$\text{Cholesterol} + \text{Oxygen} \xrightarrow{\text{Cholesterol Oxidase}} \text{Cholest-4-en-3-one} + \text{Hydrogen Peroxide}$$

Hydrogen peroxide (H_2O_2) is then measured in a peroxidase-catalyzed reaction that forms a colored dye, absorption of which can be measured spectrophotometrically in the visible region. From this, concentration of cholesterol can be calculated.

$$H_2O_2 + Phenol + 4\text{-aminoantipyrine} \longrightarrow Quinoneimine\ dye + water$$

1.5 IMMUNOASSAYS

Immunoassays are based on the principle of antigen−antibody reactions; there are various formats for such immunoassays. In many immunoassays, the final signal generated (UV absorption, fluorescence, chemiluminescence, turbidimetry) is measured using spectrophotometric principles via a suitable spectrophotometer. This topic is discussed in detail in Chapter 2.

1.6 NEPHELOMETRY AND TURBIDIMETRY

Turbidity results in a decrease of intensity of the light beam that passes though a turbid solution due to light scattering, reflectance, and absorption. Measurement of this decreased intensity of light is measured in turbidimetric assays. However, in nephelometry, light scattering is measured. In common nephelometry, scattered light is measured at a right angle to the scattered light. Antigen−antibody reactions may cause turbidity, and either turbidimetry or nephelometry can be used in an immunoassay for quantification of an analyte. Therefore, both nephelometry and turbidimetry are spectroscopic techniques. Although nephelometry can be used for analysis of small molecules, it is more commonly used for analysis of relatively big molecules such as immunoglobulin, rheumatoid factor, etc.

1.7 CHEMICAL SENSORS

Chemical sensors are capable of detecting specific chemical species present in the biological matrix. More recently, biosensors have been developed for measuring a particular analyte. However, in a clinical chemistry laboratory, chemical sensors are various types of ion-selective electrodes capable of detecting a variety of ions, including hydrogen ions (pH meter). Chemical sensors capable of detecting selective ions can be classified under three broad categories:

- Ion-selective electrodes.
- Redox electrodes.
- Carbon dioxide-sensing electrodes.

Ion-selective electrodes selectively interact with a particular ion and measure its concentration by measuring the potential produced at the membrane–sample interface, which is proportional to the logarithm of the concentration (activity) of the ion. This is based on the Nernst equation (Equation 1.7):

$$E = Eo - \frac{RT}{nF} \ln \frac{\text{Reduced ions}}{\text{Oxidized ions}} \quad (1.7)$$

E is the measured electrode potential, Eo is the electrode potential under standard conditions (values are published), R is the universal gas constant (8.3 Joules per Kelvin per mole), n is the number of electrons involved, and F is Faraday's constant (96485 Coulombs per mole). Inserting these values we can transform this into Equation 1.8:

$$E = Eo - \frac{0.0592V}{n} \log \frac{\text{Reduced ions}}{\text{Oxidized ions}} \quad (1.8)$$

In ion-selective electrodes, a specific membrane is used so that only ions of interest can filter through the membrane and can reach the electrode to create the membrane potential. Polymer membrane electrodes are used to determine concentrations of electrolytes such as sodium, potassium, chloride, calcium, lithium, magnesium, as well as bicarbonate ions. Glass membrane electrodes are used for measuring pH and sodium, and are also a part of the carbon dioxide sensor.

- Valinomycin can be incorporated in a potassium selective electrode.
- Partial pressure of oxygen is measured in a blood gas machine using an amperometric oxygen sensor.
- Optical oxygen sensors or enzymatic biosensors can also be used to measure partial pressure of oxygen in blood.

1.8 BASIC PRINCIPLES OF CHROMATOGRAPHIC ANALYSIS

Chromatography is a separation method that was developed in the 19th century. The first method developed was column chromatography, where a mixture is applied at the top of a silica column (solid phase) and a non-polar solvent such as hexane is passed through the column (mobile phase). Due to differential interactions of various components present in the mixture with the solid and mobile phases, each component can be separated based on its polarity. For example, if "A" (most polar), "B" (medium polarity), and "C" (non-polar) are applied as a mixture to a silica column (followed by hexane), then "A" (being polar) should have the highest interaction with silica and "C" should have the least interaction. In addition, compound "C" (being non-polar) should be more soluble in hexane, which is a non-polar solvent and should elute from the column first. Compound "A" should be least soluble in hexane, and, due to the higher affinity for silica, should elute last, and compound "B" should elute after "C" but before "A." The differential interaction of a component in the mixture with the solid phase and mobile phase (partition coefficient) is the basis of chromatographic analysis. There are two major forms of chromatography used in clinical laboratories:

- Gas chromatography, also known as gas liquid chromatography.
- Liquid chromatography, especially high-performance liquid chromatography.

In addition, thin-layer chromatography (TLC) is sometimes used in a toxicological laboratory to screen for illicit drugs in urine. In TLC separation, migration of the compound on a specific absorbent under specific developing solvent(s) is determined by the characteristic of the compound. This is expressed by comparing the migration of the compound to that of the solvent front, and is called the retardation factor (R_f). Typically, compounds are spotted at the edge of a paper strip and a mixture of polar solvents is allowed to migrate through the paper as the mobile phase.

Compounds are separated based on the principle of partition chromatography. Various detection techniques can be used for detecting compounds of interest after separation. UV (ultraviolet) detection is a very popular method due to its simplicity. The TLC method lacks specificity for compound identification and is rarely used in therapeutic drug monitoring, although the ToxiLab technique (a type of paper chromatography) is used as a screening technique for qualitative analysis of drugs of abuse in urine specimens in some clinical laboratories.

In 1941, Martin and Synge first predicted the use of a gas instead of a liquid as the mobile phase in a chromatographic process. Later, in 1952, James and Martin systematically separated volatile compounds (fatty acids) using gas chromatography (GC). The bases of this separation are a difference in vapor pressure of the solutes and Raoult's Law [3]. Originally, GC columns started with wide-bore coiled columns packed with an inert support of high surface area. Currently, capillary columns are used for better resolution of compounds in GC, and columns are coated with liquid phases such as methyl, methyl–phenyl, propylnitrile, and other functional groups chemically bonded to the silica support. The effectiveness of the GC column is based on the number of theoretical plates (n), as defined by Equation 1.9:

$$n = 16(t_r/w_b)^2 \qquad (1.9)$$

Here, t_r is retention time of the analyte and w_b is the width of the peak at the baseline.

Major features of GC include the following:

- GC can be used for separation of relatively volatile small molecules. Because GC separations are based on differences in vapor pressures (boiling points), compounds with higher vapor pressures (low boiling points) will elute faster than compounds with lower vapor pressures (high boiling points).
- Generally, boiling point increases with increasing polarity.
- Sometimes for GC analysis, a relatively non-volatile compound (e.g. a relatively polar drug metabolite) can be converted into a non-polar compound by chemically modifying a polar functional group into a non-polar group. For example, a polar amino group ($-NH2$) can be converted into a non-polar group ($-NH-CO-CH3$) by reaction with acetic acid and acetic anhydride. This process is called derivatization.
- Compounds are typically identified by the retention time (RT) or travel time needed to pass through the GC column. Retention times depend on flow rate of gas (helium or an inert gas) through the column, the nature of the column, and the boiling points of the analytes.
- After separation by GC, compounds can be detected by a flame-ionization detector (FID), electron-capture detector (ECD), nitrogen-phosphorus detector (NPD), or other type of electrochemical detector.
- Mass spectrometer (MS) is a specific detector for GC because mass spectral fragmentation patterns are specific for compounds (except optical isomers). Gas chromatography combined with mass spectrometry (GC-MS) is widely used in clinical laboratories for analysis of drugs of abuse.

Gas chromatography is used in toxicology laboratories for analysis of volatiles (methanol, ethanol, propanol, ethyl glycol, and propylene glycol), various drugs of abuse, and selected drugs such as pentobarbital. One major limitation of GC is that only small molecules capable of existing in the vapor (gaseous) state without decomposition can be analyzed by this method. Therefore, polar molecules and molecules with higher molecular weight (e.g. the immunosuppressant cyclosporine) cannot be analyzed by GC. On the other hand, liquid chromatography can be used for analysis of both polar and non-polar molecules.

High-performance liquid chromatography (also called high-pressure liquid chromatography) is usually used in clinical laboratories in order to achieve better separation; the solid stationary phase is composed of tiny particles (approximately 5 microns). In order for the mobile phase to move through the column a high pressure must be created. This is achieved by using a high-performance pump. The elution of analytes from the column is monitored by a detection method, and a computer can be used for data acquisition and analysis. Major features of liquid chromatography include:

- Normal-phase chromatography. For separation of polar compounds a polar stationary phase such as silica is used; the mobile phase (solvent passing through the column) should be a non-polar solvent such as hexane, carbon tetrachloride, etc.
- Reverse-phase chromatography. For separation of relatively non-polar molecules, a non-polar stationary phase such as derivatized silica is used; the mobile phase is a polar solvent such as methanol or acetonitrile. Commonly used derivatized silica in chromatographic columns includes C-18 (an 18-carbon fatty acid chain linked to the silica molecule), C-8, and C-6.

Elution of a compound from a liquid chromatography column can be monitored by the following methods:

- Ultraviolet–visible (UV–Vis) spectrophotometry. Of note: UV detection is more common because many analytes absorb wavelengths in the UV region.
- Refractive index detection. In this method the change in refractive index of the mobile phase (solvent) due to elution of a peak from the column is measured. This method is far less sensitive than UV detection and is not used in clinical chemistry laboratories.
- Fluorescence detection. This is a very sensitive technique that is in general more sensitive than UV.
- Mass spectrometric detection. This method uses either one or two mass spectrometers (tandem mass spectrometry) as a very powerful detection system. High-performance liquid chromatography combined with tandem mass spectrometry (LC/MS/MS) is the most sensitive and robust method available in a clinical laboratory.

When only solvent (mobile phase) is coming out of a column, a baseline response is observed. For example, if methanol is eluted from a column and the UV detector is set at 254 nm to measure tricyclic antidepressant drugs, then no absorption should be recorded because methanol does not absorb at 254 nm. On the other hand, when amitriptyline or another tricyclic antidepressant is eluted from the column, a peak should be observed because tricyclic antidepressants absorb UV light at 254 nm (Figure 1.1). Similarly, if any other detector type is used, a response is observed in the form of a peak when an analyte elutes from the column. The time it takes for an analyte to elute from the column after injection is called "retention time," and depends on the partition coefficient (differential interaction of the analyte with the stationary and mobile phases). Retention time is usually expressed in

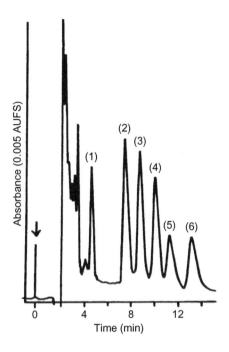

FIGURE 1.1

Chromatogram of a serum extract containing various tricyclic antidepressants and an internal standard:
(1) beta-naphthylamine, the internal standard, (2) doxepin, (3) desipramine, (4) nortriptyline,
(5) imipramine, and (6) amitriptyline. Absorbance to monitor elution of peaks was measured at 254 nm
at the UV region. Mobile phase composition was methanol/acetonitrile/phosphate buffer (0.1 mol/L). Final
pH of the mobile phase was 6.5 and a C-18 reverse-phase column was used to achieve
chromatographic separation. The 0 time (indicated as an arrow) is the injection point [4]. (© *American
Association for Clinical Chemistry. Reprinted with permission.)*

minutes. When analytes of interest are separated from each other completely, it is called baseline separation. Basic principles of retention time of a compound include:

- An increase in flow rate decreases retention time of a compound. For example, if the retention time of A is 5 min, the retention time of B is 7 min, but the retention time of C is 15 min, and initial flow rate of the mobile phase through the column is 1 mL/min, then after elution of B at 7 min, the flow rate can be increased to 3 mL/min to shorten the retention time of C in order to reduce the run time.
- If compounds A and B have the same or very similar partition coefficients for a particular stationary phase and mobile phase combination, then compounds A and B cannot be separated by chromatography using the same stationary phase and mobile phase composition. A different stationary phase, mobile phase, or both

stationary and mobile phase may be needed to separate compound A from B.

- Sometimes more than one solvent is used to compose the mobile phase by mixing predetermined amounts of two solvents. This is called the "gradient," but if only one solvent is used in the mobile phase it is called an "isocratic condition." Using more than one solvent in the mobile phase may improve the chromatographic separation.
- Sometimes heating the column to 40−60°C can improve separation between peaks. This is often used for chromatographic analysis of immunosuppressants.

1.9 MASS SPECTROMETRY COUPLED WITH CHROMATOGRAPHY

Mass spectrometry, as mentioned earlier, is a very powerful detection method that can be coupled with a gas chromatography or a high-performance liquid chromatography analyzer. Mass spectrometric analysis takes place at very low pressure, except for the recently developed atmospheric pressure chemical ionization mass spectrometry. During mass spectrometric analysis, analyte molecules in the gaseous phase are bombarded with high-energy electrons (electron ionization) or a charged chemical compound with low molecular weight such as charged ammonia ions (chemical ionization). During collision, analyte molecules lose an electron to form a positively charged ion that may also undergo further decomposition (fragmentation) into smaller charged ions. If the analyte molecule loses one electron and retains its identity, it forms a molecular ion (m/z) where m is the molecular weight of the analyte and z is the charge (usually a value of 1). The fragmentation pattern depends on the molecular structure, including the presence of various functional groups in the molecule. Therefore, the fragmentation pattern is like a fingerprint of the molecule and only optical isomers produce identical fragmentation patterns. The mass spectrometric detector can detect ions with various molecular mass and construct a chromatogram which is usually m/z in the "x" axis, with the intensity of the signal (ion strength) at the "y" axis. Although positive ions are more commonly produced during a mass spectrometric fragmentation pattern, negative ions are also generated, especially during chemical ionization mass spectrometry. Therefore, negative ions can also be monitored, although this is done less often than positive ion mass spectrometry in clinical toxicology laboratories. Major features to remember in coupling a mass spectrometer with a chromatography set-up include:

- Because mass spectrometry occurs in a vacuum, after elution of an analyte with the carrier gas from the column, the carrier gas must be removed quickly in order to have volatile analyte entering the mass spectrometer. This is achieved with a high-performance turbo pump at the interface of the gas chromatograph and mass spectrometer.

- Most commonly, an electron ionization mass spectrometer is coupled with a gas chromatograph. However, gas chromatography combined with chemical ionization mass spectrometry is gaining more traction in toxicology laboratories.
- One advantage of chemical ionization mass spectrometry is that it is a soft ionization method, and usually a good molecular ion peak as adduct ($M + H^+$, molecular ion adduct with hydrogen; or $M + NH_4^+$, molecular ion adduct with ammonia) can be observed. In contrast, an $M +$ molecular ion peak in the electron ionization method can be a very weak peak for certain analytes.
- A quadrupole detector is usually used in the mass spectrometer.
- Combining a high-performance liquid chromatography apparatus with a mass spectrometer is a big challenge because a liquid is eluted from the column. Therefore, an interface must be used to remove the liquid mobile phase quickly prior to mass spectrometric analysis. However, with the discovery of electrospray ionization, and more recently atmospheric pressure chemical ionization mass spectrometry, this problem has been circumvented.
- Electrospray ionization is the most common mass spectrometric method used in liquid chromatography combined with the mass spectrometric method (LC/MS).
- Sometimes instead of one mass spectrometer, two mass spectrometers are used so that parent ions can undergo further fragmentation in a second mass spectrometer to produce a very specific parent ion/daughter ion pattern. This improves both sensitivity and specificity of the analysis. This method is called liquid chromatography combined with tandem mass spectrometry (LC/MS/MS).

1.10 EXAMPLES OF THE APPLICATION OF CHROMATOGRAPHIC TECHNIQUES IN CLINICAL TOXICOLOGY LABORATORIES

Chromatographic methods are used in the toxicology laboratory in the following situations:

- Therapeutic drug monitoring where there is no commercially available immunoassay for the drug.
- Immunoassays are commercially available but have poor specificity. Good examples are immunoassays for immunosuppressants (cyclosporine, tacrolimus, sirolimus, everolimus, and mycophenolic acid) where metabolite cross-reactivity may produce a 20−50% positive bias as

compared to a specific chromatographic method. For therapeutic drug monitoring of immunosuppressants, LC/MS or LC/MS/MS is the gold standard and preferred method of analysis.

- Legal blood alcohol determination (GC is the gold standard).
- GC/MS or LC/MS is needed for confirmation of drugs of abuse for legal drug testing.

Subramanian et al. described LC/MS analysis of nine anticonvulsants: zonisamide, lamotrigine, topiramate, phenobarbital, phenytoin, carbamazepine, carbamazepine-10,11-diol, 10-hydroxycarbamazepine, and carbamazepine-10,11-epoxide. Sample preparation included solid-phase extraction for all anticonvulsants. HPLC separation was achieved by a reverse-phase C-18 column (4.6×50 mm, 2.2 µm particle size) with a gradient mobile phase of acetate buffer, methanol, acetonitrile, and tetrahydrofuran. Four internal standards were used. Detection of peaks was achieved by atmospheric pressure chemical ionization mass spectrometry in selected ion monitoring mode with constant polarity switching [5]. Verbesselt *et al.* described a rapid HPLC assay with solid-phase extraction for analysis of 12 antiarrhythmic drugs in plasma: amiodarone, aprindine, disopyramide, flecainide, lidocaine, lorcainide, mexiletine, procainamide, propafenone, sotalol, tocainide, and verapamil [6]. Concentrations of encainide and its metabolites can be determined in human plasma by HPLC [7].

The presence of benzoylecgonine, the inactive major metabolite of cocaine, must be confirmed by GC/MS in legal drug testing (such as pre-employment drug testing) if the initial immunoassay screen is positive. The carboxylic acid in benzoylecgonine must be derivatized prior to GC/MS analysis. A representative spectrum of the propyl ester of benzoylecgonine is shown in Figure 1.2. Molecular ion and fragment ions from the side chain are the major ions. Fragment ion m/z 82 is unique to the core structure of the compound. The ion at m/z 331 is the molecular ion.

1.11 AUTOMATION IN THE CLINICAL LABORATORY

Automated analyzers are widely used in clinical laboratories for speed, ease of operation, and because they allow a technologist to load a batch of samples for analysis, program the instrument, and walk away. The analyzer then automatically pipets small amounts of specimen from the sample cup, mixes it with reagent, records the signal, and, finally, produces the result. Therefore, the automation sequence follows similar steps to analysis via a manual laboratory technique, except that each step here is mechanized. The most common configuration of automated analyzers is "random access analyzers,"

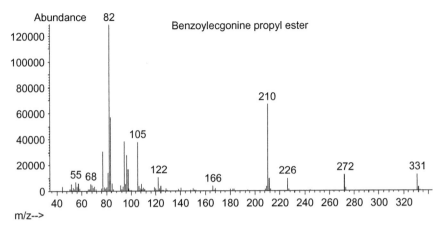

FIGURE 1.2

Mass spectrum of benzoylecgonine propyl ester. *(Courtesy of Dr. Buddha Dev Paul.)*

where multiple specimens can be analyzed for a different selection of tests. More recently, manufacturers have introduced modular analyzers that provide improved operational efficiency. Automated analyzers can be broadly classified under two categories:

- Open systems, where a technologist is capable of programming parameters for a test using reagents prepared in-house or from a different vendor than the manufacturer.
- Closed systems, where the analyzer requires that the reagent be in a unique container or format that is usually marketed by the manufacturer of the instrument or a vendor authorized by the manufacturer. Usually such proprietary reagents are more expensive than reagents available from multiple vendors that can be only be adapted to an open system analyzer.

Most automated analyzers have bar code readers so that the instrument can identify a patient's specimen from the bar code. Moreover, many automated analyzers can be interfaced to the laboratory information system (LIS) so that after verification by the technologist and subsequent release of the result, it is automatically transmitted to the patient record; this eliminates the need for manual entry of the result in the computer. This is not only time-efficient, but is also useful for preventing transcription errors during manual entry of the result in the LIS.

More recently, total automation systems are available where, after receiving the specimen, the automated system can process the specimen, including automated centrifugation, aliquoting, and delivery of the aliquot to the analyzer. Robotic arms make this total automation in a clinical laboratory feasible.

1.12 ELECTROPHORESIS (INCLUDING CAPILLARY ELECTROPHORESIS)

Electrophoresis is a technique that utilizes migration of charged solutes or analytes in a liquid medium under the influence of an applied electrical field. This is a very powerful technique for analysis of proteins in serum or urine, as well as analysis of various hemoglobin variants. Please see Chapter 22 for an in-depth discussion on this topic.

KEY POINTS

- Major analytical methods used in the clinical chemistry laboratory include spectrophotometry, chemical sensors, gas chromatography with various detectors, gas chromatography combined with mass spectrometry, high-performance liquid chromatography, and liquid chromatography combined with mass spectrometry or tandem mass spectrometry.

- Spectrophotometric measurements are based on Beer's Law (sometimes referred to as the Beer–Lambert Law). In spectrophotometry, transmittance is often measured as absorption ("A") because there is a linear relationship between absorbance and concentration of the analyte in the solution. $A = -\log T = -\log Is/Ir = \log Ir/Is$, where Ir is the intensity of the light beam transmitted through the reference cell (containing only solvent) and Is is the intensity of the transmitted light through the cell containing the analyte of interest dissolved in the same solvent as the reference cell. The scale of absorbance is from 0 to 2, where a zero value indicates "no absorbance."

- Absorption of light also depends on the concentration of the analyte in the solvent as well as on the length of the cell path. Therefore, $A = \log Ir/Is = a.b.c$, where "a" is a proportionality constant termed "absorptivity," "b" is the length of the cell path, and "c" is the concentration. If "b" is 1 cm and the concentration of the analyte is expressed as moles/L, then "a" is the "molar absorptivity," often designated as epsilon ("ε"). The value of "ε" is a constant for a particular compound and wavelength under prescribed conditions of pH, solvent, and temperature.

- In atomic absorption spectrophotometry (used for analysis of various elements, including heavy metals), components of gaseous samples are converted into free atoms. This can be achieved in a flame or flameless manner using a graphite chamber that can be heated after application of the sample. In atomic absorption spectrophotometry, a hollow cathode lamp containing an inert gas like argon or neon at a very low pressure is used as a light source. The metal cathode contains the analyte of interest; for example, for copper analysis, the cathode is made of copper. Atoms in the ground state then absorb a part of the light emitted by the hollow cathode lamp to boost them into the excited state. Therefore, a part of the light beam is absorbed and results in a net decrease in the intensity of the beam that arrives at the detector. Applying the principles of Beer's Law, the concentration of the analyte of interest can be measured. Zimmerman's correction is often applied in flameless atomic absorption spectrophotometry in order to correct for background noise in order to produce more accurate results. Mercury is vaporized at room temperature. Therefore, "cold vapor atomic absorption" can be used only for analysis of mercury.

- Inductively coupled plasma mass spectrometry (ICP-MS) is not a spectrophotometric method, but is a mass spectrometric method that is used for analysis of elements, especially trace elements found in small quantities in biological specimens.

- Chemical sensors are capable of detecting various chemical species present in the biological matrix. Chemical sensors capable of detecting selective ions can be classified under three broad categories: ion-selective electrodes, redox electrodes, and carbon dioxide-sensing electrodes.

- Valinomycin can be incorporated into a potassium-selective electrode.

- Gas chromatography can be used for separation of relatively volatile small molecules where compounds with higher vapor pressures (low boiling points) will elute faster than compounds with lower vapor pressures (high boiling points). Compounds are typically identified by the retention time (RT), or travel time, needed to pass through the GC column. Retention times depend on the flow rate of gas (helium or an inert gas) through the column, nature of the column, and boiling points of analytes. After separation by GC, compounds can be detected by a flame-ionization detector (FID), electron-capture detector (ECD), or nitrogen-phosphorus detector (NPD). However, the mass spectrometer is the most specific detector for gas chromatography.

- Although gas chromatography can be applied only for analysis of relatively volatile compounds or compounds that can be converted into volatile compounds using chemical modification of the structure (derivatization), high-performance liquid chromatography (HPLC) is capable of analyzing both polar and non-polar compounds. Common detectors used in HPLC systems include ultraviolet (UV) detectors, fluorescence detectors, or electrochemical detectors. However, liquid chromatography combined with mass spectrometry is a superior technique and a very specific analytical tool. Electrospray ionization is commonly used in liquid chromatography and combined with mass spectrometry or tandem mass spectrometry (MS/MS).

- Automated analyzers can be broadly classified under two categories: open systems where a technologist is capable of programming parameters for a test using reagents prepared in-house or obtained from a different vendor than the manufacturer of the analyzer, and closed systems where the analyzer requires that the reagent be in a unique container or format that is usually marketed by the manufacturer of the instrument or a vendor authorized by the manufacturer.

REFERENCES

[1] Dasgupta A, Zaidi S, Johnson M, Chow L, Wells A. Use of fluorescence polarization immunoassay for salicylate to avoid positive/negative interference by bilirubin in the Trinder salicylate assay. Ann Clin Biochem 2003;40:684–8.

[2] Profrock D, Prange A. Inductively couples plasma-mass spectrometry (ICP-MS) for quantitative analysis in environmental and life sciences: a review of challenges, solutions and trends. Appl Spectrosc 2012;66:843–68.

[3] James AT, Martin AJP. Gas-liquid partition chromatography: the separation and microestimation of volatile fatty acids from formic acid to dodecanoic acid. Biochem J 1952;50:679–90.

[4] Proeless HF, Lohmann HJ, Miles DG. High performance liquid-chromatographic determination of commonly used tricyclic antidepressants. Clin Chem 1978;24:1948–53.

[5] Subramanian M, Birnbaum AK, Remmel RP. High-speed simultaneous determination of nine antiepileptic drugs using liquid chromatography–mass spectrometry. Ther Drug Monit 2008;30:347–56.

[6] Verbesselt R, Tjandramaga TB, de Schepper PJ. High-performance liquid chromatographic determination of 12 antiarrhythmic drugs in plasma using solid phase extraction. Ther Drug Monit 1991;13:157–65.

[7] Dasgupta A, Rosenzweig IB, Turgeon J, Raisys VA. Encainide and metabolites analysis in serum or plasma using a reversed-phase high-performance liquid chromatographic technique. J Chromatogr 1990;526:260–5.

CHAPTER 2

Immunoassay Platform and Designs

2.1 APPLICATION OF IMMUNOASSAYS FOR VARIOUS ANALYTES

Immunoassays are available for analysis of over 100 different analytes. Most immunoassay methods use specimens without any pretreatment and the assays can be run on fully automated, continuous, high-throughput, random access systems. These assays use very small sample volumes ($10\,\mu L - 50\,\mu L$), reagents can be stored in the analyzer, most have stored calibration curves on the automated analyzer system, they are often stable for $1-2$ months, and results can be reported in $10-30$ minutes. Immunoassays offer fast throughput, automated reruns, auto-flagging (to alert for poor specimen quality such as hemolysis, high bilirubin, and lipemic specimens that may affect test result), high sensitivity and specificity, and results can be reported directly into the laboratory information system (LIS). However, immunoassays do suffer from interferences from both endogenous and exogenous factors.

2.2 IMMUNOASSAY DESIGN AND PRINCIPLE

Immunoassay design can be classified under two broad categories:

- Competition immunoassay: This design uses only one antibody specific for the analyte molecule and is widely used for detecting small analyte molecules such as various therapeutic drugs and drugs of abuse.
- Immunometric or non-competitive (sandwich) immunoassay: This design uses two analyte-specific antibodies that recognize different parts of the analyte molecule, and is used for analysis of large molecules such as proteins and polypeptides.

CONTENTS

A. Dasgupta and A. Wahed: Clinical Chemistry, Immunology and Laboratory Quality Control
DOI: http://dx.doi.org/10.1016/B978-0-12-407821-5.00002-4

Depending on the need of the separation between the bound labels (labeled antigen–antibody complex) versus free labels, the immunoassays may be further sub-classified into homogenous or heterogenous formats.

- Homogenous immunoassay format: After incubation, no separation between bound and free label is necessary.
- Heterogenous immunoassay format: Bound label must be separated from the free label before measuring the signal.

In competitive immunoassays, predetermined amounts of labeled antigen and antibody are added to the specimen followed by incubation. In the basic design of a competitive immunoassay, analyte molecules present in the specimen compete with analyte molecules labeled with a tag and are added to the sample in a predetermined amount for a limited number of binding sites in the antibody molecules (also added to the specimen in a predetermined amount). After incubation, the signal is measured with (heterogenous format) or without (homogenous format) separating labeled antigen molecules bound to antibody molecules from labeled antigen molecules (which are free in solution). Let's take the hypothetical scenario presented in Figure 2.1. In Scenario 1, four labeled antigen molecules and two antigen molecules present in the specimen are competing for three binding antibodies, while in Scenario 2 more antigen molecules (analyte) are present. As expected in the

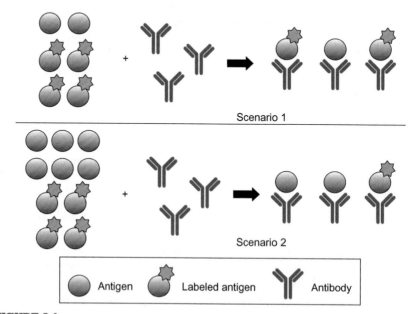

Scenario 1

Scenario 2

Antigen   Labeled antigen   Antibody

FIGURE 2.1

Competitive immunoassay. This figure is reproduced in color in the color plate section. *(Courtesy of Stephen R. Master, MD, PhD, Perelman School of Medicine, University of Pennsylvania.)*

competitive assay format in Scenario 1, more labeled antigen molecules would bind with the antibody than in Scenario 2. If a signal is produced when a labeled antigen is bound with an antibody molecule, as with the FPIA assay (fluorescence polarization immunoassay), then more signals will be generated in Scenario 1 than Scenario 2. Therefore, the general conclusions are as follows:

- If the signal is generated when a labeled antigen binds with an antibody molecule, then the signal is inversely proportional to analyte concentration in the specimen (e.g. FPIA assay design).
- If the signal is generated by an unbound labeled antigen, then the assay signal is directly proportional to the analyte concentration (e.g. enzyme multiplied immunoassay technique, EMIT).

In the non-competitive (sandwich) assay (Figure 2.2), captured antibodies specific to the analyte are immobilized on a solid support (microparticle bead, microtiter plate, etc.). After the specimen is added, a predetermined time is allowed for incubation of the analyte with the antibody and then liquid reagent containing the second antibody conjugated to a molecule for generating the signal (e.g. an enzyme) is added. Alternatively, after adding patient serum, liquid reagent may be added followed by single incubation. Then a sandwich is formed. After incubation, excess antibody may be washed off by a washing step and a substrate for the enzyme can be added for generating a signal that can be measured. Analyte concentration is directly proportional to the intensity of the signal.

Antibodies used in immunoassays can be either monoclonal or polyclonal. Polyclonal antibodies can be raised using animals such as rabbits, sheep, or goats by injecting analyte (as antigen) along with an adjuvant. An analyte with a small molecular weight (such as therapeutic drugs or drugs of abuse)

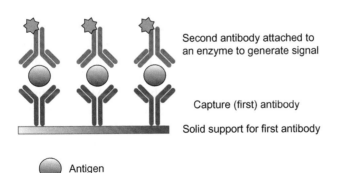

Second antibody attached to an enzyme to generate signal

Capture (first) antibody

Solid support for first antibody

Antigen

FIGURE 2.2
Sandwich immunoassay. This figure is reproduced in color in the color plate section. *(Courtesy of Stephen R. Master, MD, PhD, Perelman School of Medicine, University of Pennsylvania.)*

is most commonly injected as the conjugate to a large protein. Appearance of analyte-specific antibodies in the animal's sera is monitored, and when a sufficient concentration of the antibody is reached, the animal is bled. Then serum antibodies are purified from serum and used in an immunoassay. Since there are many clones of the antibodies specific for the analyte, these antibodies are called polyclonal. In newer technologies, a plasma cell of the animal can be selected as producing the optimum antibody, and then it can be fused to an immortal cell. The resulting tumor cell grows uncontrollably and produces only the single clone of the desired antibody. Such antibodies, called monoclonal antibodies, may be grown in live animals or in cell culture. Sometimes instead of using the whole antibody, fragments of the antibody, generated by digestion of the antibody with peptidases (e.g. Fab, Fab', or their dimeric complexes), are also used as reagents.

2.3 VARIOUS COMMERCIALLY AVAILABLE IMMUNOASSAYS

Many immunoassays are commercially available for analysis of a variety of analytes. These assays use different labels and different methods for generating and measuring signals, but the basic principles are the same as described in the immunoassay design section. FPIA, EMIT, CEDIA, KIMS, and LOCI assays are examples of homogenous competitive immunoassay designs. Common commercial assays are summarized in Table 2.1.

- In the fluorescent polarization immunoassay (FPIA), the free label (which is a relatively small molecule) attached to the analyte molecule has different Brownian motion than when the label is complexed to a large antibody. FPIA is a homogenous competitive assay where, after incubation, the fluorescence polarization signal is measured without separation of bound labels from free labels. If the labeled antigen is bound to the antibody molecule, then the signal is generated, and when the labeled antigen is free in the solution, no signal is produced. Therefore, signal intensity is inversely proportional to the analyte concentration. Abbott Laboratories first introduced this assay design [1].
- Enzyme multiplied immunoassay technique (EMIT) was first introduced by the Syva Company; it is a homogenous competitive immunoassay. In this immunoassay design, the antigen is labeled with glucose 6-phosphate dehydrogenase enzyme. The active enzyme reduces nicotinamide adenine dinucleotide (NAD, no signal at 340 nm) to NADH (absorbs at 340 nm), and the absorbance is monitored at 340 nm. When labeled antigen binds with the antibody molecule, the enzyme becomes inactive. Therefore, the signal is produced by the free label, and signal intensity is proportional to the analyte concentration.

Table 2.1 Examples of Various Types of Commercially Available Immunoassays

Immunoassay Types	Example	Format	Assay Sinal
Competition (small molecules: ≤1000 Dalton)	FPIA (Abbott) —Therapeutic drugs —Abused drugs	Homogenous	Fluorescence polarization
	EMIT (Syva) —Therapeutic drugs —Abused drugs	Homogenous	Absorbance at 340 nm (enzyme modulation)
	CEDIA (Thermo Fisher: Microgenics) —Therapeutic drugs —Abused drugs	Homogenous	Colorimetry (enzyme modulation)
	KIMS® (Roche) —Abused drugs	Homogenous	Optical detection
	LOCI (Siemens)* —Various analytes	Homogenous	Chemiluminescence
Sandwich (analytes, MW > 1000 D)	TIA* (Siemens, Roche) —Serum proteins	Homogenous	Turbidimetry
	CLIA (Multiple^) —Hormones, proteins	Heterogenous	Chemiluminescence
	CLIA (Roche) —Hormones, proteins	Heterogeneous	Electrochemiluminescence

^LOCI assays are available in both competition and sandwich format for analysis of both small and large molecules
*Multiple manufacturers (Abbott, Beckman, Siemens etc.) use this heterogenous sandwich format for manufacturing commercially available immunoassays for analysis of large molecules such as proteins.

- The cloned enzyme donor immunoassay (CEDIA) method is based on recombinant DNA technology to produce a unique homogenous enzyme immunoassay system. The assay principle is based on the bacterial enzyme beta-galactosidase, which has been genetically engineered into two inactive fragments. The small fragment is called the enzyme donor (ED), which can freely associate in the solution with the larger part called the enzyme acceptor (EA) to produce an active enzyme that is capable of cleaving a substrate that generates a color change in the medium that can be measured spectrophotometrically. In this assay, drug molecules in the specimen compete for limited antibody binding sites with drug molecules conjugated with the ED fragment. If drug molecules are present in the specimen, then they bind to the antibody binding sites and leave drug molecules conjugated with ED free to form active enzyme by binding with EA; a signal is generated and the intensity of the signal is proportional to the analyte concentration. Many therapeutic drugs and drugs of abuse manufactured by Microgenic Corporation use the CEDIA format, although other commercial assays also use this format [2].

- Kinetic interaction of microparticle in solution (KIMS): In this assay, in the absence of antigen (analyte) molecules, free antibodies bind to drug microparticle conjugates to form particle aggregates that result in an increase in absorption, which is optically measured at various visible wavelengths (500−650 nm). When antigen molecules are present in the specimen, antigen molecules bind with free antibody molecules and prevent the formation of particle aggregates; this results in diminished absorbance in proportion to the drug concentration. The On-Line Drugs of Abuse Testings immunoassays marketed by Roche Diagnostics (Indianapolis, IN) are based on the KIMS format.
- Luminescent oxygen channeling immunoassay (LOCI) is a homogenous competitive immunoassay where the reaction mixture is irradiated with light to generate singlet oxygen molecules; this results in the formation of a chemiluminescent signal. This technology is used in the Siemens Dimension Vista® automated assay system [3].

2.4 HETEROGENOUS IMMUNOASSAYS

In heterogenous immunoassays the bound label is physically separated from the unbound label prior to measuring the signal. The separation is often done magnetically using paramagnetic particles, and after separation of bound from free using a washing step, the bound label is reacted with other reagents to generate the signal. This is the mechanism in many chemiluminescent immunoassays (CLIA) where the label may be a small molecule that generates a chemiluminescent signal. Examples of immunoassay systems where the chemiluminescent labels generate signals by chemical reaction are the ADVIA Centaur® from Siemens and the Architect® from Abbott [4]. An example where the small label is activated electrochemically is the ELECSYS® automated immunoassay system from Roche Diagnostics [5]. The label may also be an enzyme (enzyme-linked immunosorbent assay, ELISA) that generates chemiluminescent, fluorometric, or colorimetric signals depending on the enzyme substrates used. Examples of commercial automated assay systems using ELISA technology and chemiluminescent labels are Immulite® (Siemens) and ACCESS® from Beckman-Coulter [6,7]. Another type of heterogenous immunoassay uses polystyrene particles. If these are particles are micro-sized, that type of assay is called micro-particle enhanced immunoassay (MEIA) [8]. If the immunoassay format utilizes a radioactive label, the assay is called a radioimmunoassay (RIA). Today, RIA is rarely used due to safety and waste disposal issues involving radioactive materials.

2.5 CALIBRATION OF IMMUNOASSAYS

Like all quantitative assays, immunoassays also require calibration. Calibration is a process of analyzing samples containing analytes of known concentrations (calibrators) and then fitting the data into a calibration curve so that concentration of the analyte in an unknown specimen can be calculated by linking the signal to a particular value on the calibration curve. For calibration purposes, known amounts of the analyte are added to a matrix similar to the serum matrix to prepare a series of calibrators with concentrations varying from zero calibrator (contains no analyte) to a calibrator containing the highest targeted concentration of the analyte (which is also the upper limit of analytical measurement range, AMR). The minimum number of calibrators needed to calibrate an assay is two (one zero calibrator and another calibrator representing the upper limit of AMR), and many immunoassays are based on a two-calibration system. However, in some immunoassays, five or six calibrators may be used with one zero calibrator, one representing the upper end of AMR, and the other calibrators in between concentrations.

The calibration curve can be a straight line or a curved line fitting to a polynomial function or logit function. Regardless of the curve-fitting method, the signal generated during analysis of an unknown patient sample can be extrapolated to determine the concentration of the analyte using the calibration curve. For example, the LOCI myoglobin assay on the Dimension Vista analyzers (Siemens Diagnostics) is a homogenous sandwich chemiluminescent immunoassay based on LOCI technology that uses six levels of calibrators for construction of the calibration curve. Level A (myoglobin concentration zero), Level B (110 ng/mL), and Level C (1100 ng/mL) calibrators are supplied by the manufacturer, and during calibration the instrument auto-dilutes Level B and Level C calibrators to produce calibrators with intermediate myoglobin concentrations. The chemiluminescence signal is measured at 612 nm and the intensity of the signal is proportional to the concentration of myoglobin in the specimen; the calibration curve fits to a linear equation (Figure 2.3).

2.6 VARIOUS SOURCES OF INTERFERENCE IN IMMUNOASSAYS

Even though immunoassays are widely used in the clinical laboratory, they suffer from the following types of interferences, which render false positive or false negative results:

- Endogenous components (e.g. bilirubin, hemoglobin, lipids, and paraproteins) may interfere with immunoassays.
- Interferences from the other endogenous and exogenous components.
- System- or method-related errors (e.g. pipetting probe contamination and carry-over). Most modern instruments have various ways to eliminate carry-over issues, typically by using disposable probes or a washing protocol between analyses.

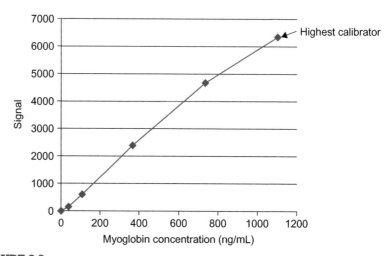

FIGURE 2.3
Calibration curve of myoglobin using Vista 1500 analyzer (Siemens Diagnostics).

- Heterophilic interference is caused by endogenous human antibodies in the sample.
- Interferences from macro-analytes (endogenous conjugates of analyte and antibody), macro-enzymes, and rheumatoid factors.
- Prozone (or "hook") effect: If a very high amount of analyte is present in the specimen, observed values may be much lower than the true analyte concentration (false negative result).

2.7 INTERFERENCES FROM BILIRUBIN, HEMOLYSIS, AND HIGH LIPID CONTENT

Bilirubin is derived from the hemoglobin of aged or damaged red blood cells. Bilirubin does not contain iron, but is rather a derivative of the heme group. Some part of serum bilirubin is conjugated as glucuronides ("direct" bilirubin) and the unconjugated bilirubin is referred to as indirect bilirubin. In normal adults, total bilirubin concentrations in serum are from 0.3 to 1.2 mg/dL. In different forms of jaundice, total bilirubin may increase to as high as 20 mg/dL. Major issues of bilirubin interference are as follows:

- Usually, a total bilirubin concentration below 20 mg/dL does not cause interference but concentrations over 20 mg/dL may cause problems.
- The interference of bilirubin in assays is mainly caused by bilirubin absorbance at 454 or 461 nm.
- Bilirubin may also interfere with an assay by chemically reacting with a component of the reagent.

Hemoglobin is mainly released by hemolysis of red blood cells (RBC). Hemolysis can occur in vivo, during venipuncture and blood collection, or during sample processing. Hemoglobin interference depends on its concentration in the sample. Serum appears hemolyzed when the hemoglobin concentration exceeds 20 mg/dL. The absorbance maxima of the heme moiety in hemoglobin are at 540 to 580 nm wavelengths. However, hemoglobin begins to absorb around 340 nm and then absorbance increases at 400−430 nm as well. Interference of hemoglobin (if the specimen is grossly hemolyzed) is due to interference with the optical detection system of the assay.

All lipids in plasma exist as complexed with proteins that are called lipoproteins, and particle size varies from 10 nm to 1000 nm (the higher the percentage of the lipid, the lower the density of the resulting lipoprotein and the larger the particle size). The lipoprotein particles with high lipid content are micellar and are the main source of assay interference. Unlike bilirubin and hemoglobin, lipids normally do not participate in chemical reactions and mostly cause interference in assays due to their turbidity and capability of scattering light, as in nephelometric assays.

2.8 INTERFERENCES FROM ENDOGENOUS AND EXOGENOUS COMPONENTS

Immunoassays are affected by a variety of endogenous and exogenous compounds, including heterophilic antibodies. The key points regarding immunoassay interferences include:

- Endogenous factors such as digoxin-like immunoreactive factors only affect digoxin immunoassays. Please see Chapter 15 for a more detailed discussion.
- Structurally similar molecules are capable of cross-reacting with the antibody to cause falsely elevated (positive interference) or falsely lowered results (negative interference). Negative interference occurs less frequently than positive interference, but may be clinically more dangerous. For example, if the result of a therapeutic drug is falsely elevated compared to the previous measurement, the clinician may question the result, but if the value is falsely lower, the clinician may simply increase the dose without realizing that the value was falsely lower due to interference. That can cause drug toxicity in the patient.
- Interference from drug metabolites is the most common form of interference, although other structurally similar drugs may also be the cause of interference. See also Chapter 15.

2.9 INTERFERENCES OF HETEROPHILIC ANTIBODIES IN IMMUNOASSAYS

Heterophilic antibodies are human antibodies that interact with assay antibody interferences. Features of heterophilic antibody interference in immunoassays include:

- Heterophilic antibodies may arise in a patient in response to exposure to certain animals or animal products or due to infection by bacterial or viral agents, or non-specifically.
- Among heterophilic antibodies, the most common are human anti-mouse antibodies (HAMA) because of wide use of murine monoclonal antibody products in therapy or imaging. However, other anti-animal antibodies in humans have also been described that can interfere with an immunoassay.
- If a patient is exposed to animals or animal products, or suffers from an autoimmune disease, the patient may have heterophilic antibodies in circulation.
- Heterophilic antibodies interfere most commonly with sandwich assays that are used for measuring large molecules, but rarely interfere with competitive assays. Most common interferences of heterophilic antibodies are observed with the measurement of various tumor markers.
- In the sandwich-type immunoassays, heterophilic antibodies can form the "sandwich complex" even in the absence of the target antigen; this generates mostly false positive results. False negative results due to the interference of heterophilic antibodies are rarely observed.
- Heterophilic antibodies are absent in urine. Therefore, if a serum specimen is positive for an analyte, for example, human chorionic gonadotropin (hCG), but beta-hCG cannot be detected in the urine specimen, it indicates interference from heterophilic antibodies in the serum hCG measurement.
- Another way to investigate heterophilic antibody interference is serial dilution of a specimen. If serial dilution produces a non-linear result, it indicates interference in the assay.
- Interference from heterophilic antibodies may also be blocked by adding any commercially available heterophilic antibody blocking agent in the specimen prior to analysis.
- For analytes that are also present in the protein-free ultrafiltrate (relatively small molecules), analysis of the analyte in the protein-free ultrafiltrate can eliminate interference from heterophilic antibodies because, due to large molecular weights, heterophilic antibodies are absent in protein-free ultrafiltrates.

Heterophilic antibodies are more commonly found in sick and hospitalized patients with reported prevalences of 0.2%−15%. In addition, rheumatoid factors that are IgM type antibodies may be present in the serum of patients suffering from rheumatoid arthritis and certain autoimmune diseases. Rheumatoid factors may interfere with sandwich assays and the mechanism of interference is similar to the interference caused by heterophilic antibodies. Commercially available rheumatoid factor blocking agent may be used to eliminate such interferences.

CASE REPORT

A 58-year-old man without any familial risk for prostate cancer visited his primary care physician and his prostate-specific antigen (PSA) level was 83 ng/mL (0–4 ng/mL is normal). He was referred to a urologist and his digital rectal examination was normal. In addition, a prostate biopsy, abdominal tomodensitometry, whole body scan, and prostatic MRI were performed, but no significant abnormality was observed. However, due to his very high PSA level (indicative of advance stage prostate cancer) he was treated with androgen deprivation therapy with goserelin acetate and bicalutamide. After 3 months he still had no symptoms, his prostate was atrophic on digital rectal examination, and he had suppressed testosterone levels as expected. However, his PSA level was still highly elevated (122 ng/mL) despite no radiographic evidence of advanced cancer. At that point his serum PSA was analyzed by a different assay (Immulite PSA, Cirrus Diagnostics, Los Angeles) and the PSA level was < 0.3 ng/mL. The treating physician therefore suspected a false positive PSA by the original Access Hybritech PSA assay (Hybritech, San Diego, CA), and interference of heterophilic antibodies was established by treating specimens with heterophilic antibody blocking agent. Re-analysis of the high PSA specimen showed a level below the detection limit. This patient received unnecessary therapy for his falsely elevated PSA level due to the interference of heterophilic antibody [9].

CASE REPORT

A 64-year-old male during a routine visit to his physician was diagnosed with hypothyroidism based on elevated TSH (thyroid stimulating hormone) levels, and his clinician initiated therapy with levothyroxine (250 microgram per day). Despite therapy, there were still increased levels of TSH (33 mIU/L) and his FT4 level was also elevated. The endocrinologist at that point suspected that TSH levels measured by the Unicel Dxi analyzer (Beckman Coulter) were falsely elevated due to interference. Serial dilution of the specimen showed non-linearity, an indication of interference. When the specimen was analyzed using a different TSH assay (immunoradiometric assay (IRMA), also available from Beckman Coulter), the TSH value was 1.22 mIU/L, further confirming the interference with the initial TSH measurement. The patient had a high concentration of rheumatoid factor (2700 U/mL) and the authors speculated that his falsely elevated TSH was due to interference from rheumatoid factors [10].

2.10 INTERFERENCES FROM AUTOANTIBODIES AND MACRO-ANALYTES

Autoantibodies (immunoglobulin molecules) are formed by the immune system of an individual capable of recognizing an antigen on that person's own tissues. Several mechanisms may trigger the production of autoantibodies, for example, an antigen formed during fetal development and then sequestered may be released as a result of infection, chemical exposure, or trauma, as occurs in autoimmune thyroiditis. The autoantibody may bind to the analyte-label conjugate in a competition-type immunoassay to produce a false positive or false negative result. Circulating cardiac troponin I autoantibodies may be present in patients suffering from acute cardiac myocardial infarction where troponin I elevation is an indication of such an episode. Unfortunately, the presence of circulating cardiac troponin I autoantibodies may falsely lower cardiac troponin I concentration (negative interference) using commercial immunoassays, thus complicating the diagnosis of acute myocardial infarction [11]. However, falsely elevated results due to the presence of autoantibodies are more common than false negative results. Verhoye *et al.* found three patients with false positive thyrotropin results that were caused by interference from an autoantibody against thyrotropin. The interfering substance in the affected specimens was identified as an autoantibody by gel-filtration chromatography and polyethylene glycol precipitation [12].

Often the analyte can conjugate with immunoglobin or other antibodies to generate macro-analytes, which can falsely elevate the true value of the analyte. For example, macroamylasemia and macro-prolactinemia can produce falsely elevated results in amylase and prolactin assays, respectively. In macro-prolactinemia, the hormone prolactin conjugates with itself and/or with its autoantibody to create macro-prolactin in the patient's circulation. The macro-analyte is physiologically inactive, but often interferes with many prolactin immunoassays to generate false positive prolactin results [13]. Such interference can be removed by polyethylene glycol precipitation.

CASE REPORT

A 17-year-old girl was referred to a University hospital for having a persistent elevated level of aspartate aminotransferase (AST). One year earlier, her AST level was 88 U/L as detected during her annual school health check, but she had no medical complaints. She was not on any medication and had a regular menstrual cycle. Her physical examination at the University hospital was unremarkable. All laboratory test results were normal, but her AST level was further elevated to 152 U/L. All serological tests for hepatitis were negative.

On further follow-up her AST level was found to have increased to 259 U/L. At that point it was speculated that her elevated AST was due to interference, and further study by gel-filtration showed a species with a molecular weight of 250 kilodaltons. This was further characterized by immunoelectrophoresis and immunoprecipitation to be an immunoglobulin (IgG kappa-lambda globulin) complexed AST that was causing the elevated AST level in this girl. These complexes are benign [14].

2.11 PROZONE (OR "HOOK") EFFECT

The Prozone or hook effect is observed when a very high amount of an analyte is present in the sample but the observed value is falsely lowered. This type of interference is observed more commonly in sandwich assays. The mechanism of this significant negative interference is the capability of a high level of an analyte (antigen) to reduce the concentrations of "sandwich" (antibody 1:antigen:antibody 2) complexes that are responsible for generating the signal by forming mostly single antibody:antigen complexes. The hook effect has been reported with assays of a variety of analytes, such as β-hCG, prolactin, calcitonin, aldosterone, cancer markers (CA 125, PSA), etc. The best way to eliminate the hook effect is serial dilution. For example, if the hook effect is present and the original value of an analyte (e.g. prolactin) was 120 ng/mL, then 1:1 dilution of the specimen should produce a value of 60 ng/mL; but if the observed value was 90 ng/mL (which was significantly higher than the expected value), the hook effect should be suspected. In order to eliminate the hook effect, a 1:10, 1:100, or even a 1:1000 dilution may be necessary so that the true analyte concentration will fall within the analytical measurement range (AMR) of the assay..

CASE REPORT

A 16-year-old girl presented to the emergency department with a 2-week history of nausea, vomiting, vaginal spotting, and lower leg edema. On physical examination, a lower abdomen palpable mass was found. The patient admitted sexual activity, but denied having any sexually transmitted disease. Molar pregnancy was suspected, and the quantitative β-subunit of human chorionic gonadotropin (β-hCG) concentration was 746.2 IU/L; however, the urine qualitative level was negative. Repeat of the urinalysis by a senior technologist also produced a negative result. At that point the authors suspected the hook effect and dilution of the serum specimen (1:1) produced a non-linear value (455.2 IU/L), which further confirmed the hook effect. After a 1:10 dilution, the urine test for β-hCG became positive, and finally, by using a 1:10,000 dilution of the specimen, the original serum β-hCG concentration was determined to be 3,835,000 IU/L. Usually the hook effect is observed with a molar β-hCG level in serum because high amounts of β-hCG are produced by molar pregnancy [15].

KEY POINTS

- Immunoassays can be competitive or immunometric (non-competitive, also known as sandwich). In competitive immunoassays only one antibody is used. This format is common for assays of small molecules such as a therapeutic drugs or drugs of abuse. In the sandwich format two antibodies are used and this format is more commonly used for assays of relative large molecules.

- Homogenous immunoassay format: After incubation, no separation between bound and free label is necessary.
- Heterogenous immunoassay format: The bound label must be separated from the free label before measuring the signal.
- Commercially available immunoassays use various formats, including FPIA, EMIT, CEDIA, KIMS, and LOCI. In the fluorescent polarization immunoassay (FPIA), the free label (a relatively small molecule) attached to the analyte (antigen) molecule has different Brownian motion than when the label is complexed to a large antibody (140,000 or more Daltons). FPIA is a homogenous competitive assay where after incubation the fluorescence polarization signal is measured; this signal is only produced if the labeled antigen is bound to the antibody molecule. Therefore, intensity of the signal is inversely proportional to the analyte concentration.
- EMIT (enzyme multiplied immunoassay technique) is a homogenous competitive immunoassay where the antigen is labeled with glucose 6-phosphate dehydrogenase, an enzyme that reduces nicotinamide adenine dinucleotide (NAD, no signal at 340 nm) to NADH (absorbs at 340 nm), and the absorbance is monitored at 340 nm. When a labeled antigen binds with the antibody molecule, the enzyme label becomes inactive and no signal is generated. Therefore, signal intensity is proportional to analyte concentration.
- The Cloned Enzyme Donor Immunoassay (CEDIA) method is based on recombinant DNA technology where bacterial enzyme beta-galactosidase is genetically engineered into two inactive fragments. When both fragments combine, a signal is produced that is proportional to the analyte concentration.
- Kinetic interaction of microparticle in solution (KIMS): In the absence of antigen molecules free antibodies bind to drug microparticle conjugates to form particle aggregates that result in an increase in absorption that is optically measured at various visible wavelengths (500–650 nm).
- Luminescent oxygen channeling immunoassays (LOCI): The immunoassay reaction is irradiated with light to generate singlet oxygen molecules in microbeads ("Sensibead") coupled to the analyte. When bound to the respective antibody molecule, also coupled to another type of bead, it reacts with singlet oxygen and chemiluminescence signals are generated that are proportional to the concentration of the analyte–antibody complex.
- Usually total bilirubin concentration below 20 mg/dL does not cause interferences, but concentrations over 20 mg/dL may cause problems. The interference of bilirubin is mainly caused by its absorbance at 454 or 461 nm.
- Various structurally related drugs or drug metabolites can interfere with immunoassays.

- Heterophilic antibodies may arise in a patient in response to exposure to certain animals or animal products, due to infection by bacterial or viral agents, or use of murine monoclonal antibody products in therapy or imaging. Heterophilic antibodies interfere most commonly with sandwich assays used for measuring large molecules, but rarely with competitive assays, causing mostly false positive results.
- Heterophilic antibodies are absent in urine. Therefore, if a serum specimen is positive for an analyte (e.g. human chorionic gonadotropin, hCG), but beta-hCG cannot be detected in the urine specimen, it indicates interference from a heterophilic antibody in the serum hCG measurement. Another way to investigate heterophilic antibody interference is serial dilution of a specimen. If serial dilution produces a non-linear result, it indicates interference in the assay. Interference from heterophilic antibodies can also be blocked by adding commercially available heterophilic antibody blocking agents to the specimen prior to analysis.
- Autoantibodies are formed by the immune system of a person that recognizes an antigen on that person's own tissues, and may interfere with an immunoassay to produce false positive results (and less frequently, false negative results). Often the endogenous analyte of interest will conjugate with immunoglobin or other antibodies to generate macro-analytes, which can falsely elevate a result. For example, macroamylasemia and macro-prolactinemia can produce falsely elevated results in amylase and prolactin assays, respectively. Such interference can be removed by polyethylene glycol precipitation.
- Prozone ("hook") effect: Very high levels of antigen can reduce the concentrations of "sandwich" (antibody 1:antigen:antibody 2) complexes responsible for generating the signal by forming mostly single antibody:antigen complexes. This effect, known as the prozone or hook effect (excess antigen), mostly causes negative interference (falsely lower results). The best way to eliminate the hook effect is serial dilution.

REFERENCES

[1] Jolley ME, Stroupe SD, Schwenzer KS, Wang CJ, et al. Fluorescence polarization immunoassay III. An automated system for therapeutic drug determination. Clin Chem 1981;27:1575−9.

[2] Jeon SI, Yang X, Andrade JD. Modeling of homogeneous cloned enzyme donor immunoassay. Anal Biochem 2004;333:136−47.

[3] Snyder JT, Benson CM, Briggs C, et al. Development of NT-proBNP, Troponin, TSH, and FT4 LOCI(R) assays on the new Dimension (R) EXL with LM clinical chemistry system. Clin Chem 2008;54:A92 [Abstract #B135].

[4] Dai JL, Sokoll LJ, Chan DW. Automated chemiluminescent immunoassay analyzers. J Clin Ligand Assay 1998;21:377−85.

[5] Forest J-C, Masse J, Lane A. Evaluation of the analytical performance of the Boehringer Mannheim Elecsys® 2010 Immunoanalyzer. Clin Biochem 1998;31:81−8.

[6] Babson AL, Olsen DR, Palmieri T, Ross AF, et al. THE IMMULITE assay tube: a new approach to heterogeneous ligand assay. Clin Chem 1991;37:1521−2.

[7] Christenson RH, Apple FS, Morgan DL. Cardiac troponin I measurement with the ACCESS® immunoassay system: analytical and clinical performance characteristics. Clin Chem 1998;44:52−60.

[8] Montagne P, Varcin P, Cuilliere ML, Duheille J. Microparticle-enhanced nephelometric immunoassay with microsphere-antigen conjugate. Bioconjugate Chem 1992;3:187−93.

[9] Henry N, Sebe P, Cussenot O. Inappropriate treatment of prostate cancer caused by heterophilic antibody interference. Nat Clin Pract Urol 2009;6:164−7.

[10] Georges A, Charrie A, Raynaud S, Lombard C, et al. Thyroxin overdose due to rheumatoid factor interferences in thyroid-stimulating hormone assays. Clin Chem Lab Med 2011;49:873−5.

[11] Tang G, Wu Y, Zhao W, Shen Q. Multiple immunoassays systems are negatively interfered by circulating cardiac troponin I autoantibodies. Clin Exp Med 2012;12:47−53.

[12] Verhoye E, Bruel A, Delanghe JR, Debruyne E, et al. Spuriously high thyrotropin values due to anti-thyrotropin antibody in adult patients. Clin Chem Lab Med 2009;47:604−6.

[13] Kavanagh L, McKenna TJ, Fahie-Wilson MN, et al. Specificity and clinical utility of methods for determination of macro-prolactin. Clin Chem 2006;52:1366−72.

[14] Matama S, Ito H, Tanabe S, Shibuya A, et al. Immunoglobulin complexed aspartate aminotransferase. Intern Med 1993;32:156−9.

[15] Er TK, Jong YJ, Tsai EM, Huang CL, et al. False positive pregnancy in hydatidiform mole. Clin Chem 2006;52:1616−8.

CHAPTER 3

Pre-Analytical Variables

3.1 LABORATORY ERRORS IN PRE-ANALYTICAL, ANALYTICAL, AND POST-ANALYTICAL STAGES

Accurate clinical laboratory test results are important for proper diagnosis and treatment of patients. Factors that are important to obtaining accurate laboratory test results include:

- Patient Identification: The right patient is identified prior to specimen collection by matching at least two criteria.
- Collection Protocol: The correct technique and blood collection tube have been used for sample collection to avoid tissue damage, prolonged venous stasis, or hemolysis.
- Labeling: After collection, the specimen was labeled properly with correct patient information; specimen misidentification is a major source of pre-analytical error.
- Specimen Handling: Proper centrifugation (in the case of serum or plasma specimen analysis) and proper transportation of specimens to the laboratory.
- Storage Protocol: Maintaining proper storage of specimens prior to analysis in order to avoid artifactual changes in analyte; for example, storing blood gas specimens in ice if the analysis cannot be completed within 30 min of specimen collection.
- Interference Avoidance: Proper analytical steps to obtain the correct result and avoid interferences.
- LIS Reports: Correctly reporting the result to the laboratory information system (LIS) if the analyzer is not interfaced with the LIS.
- Clinician Reports: The report reaching the clinician must contain the right result, together with interpretative information, such as a reference range and other comments that aid clinicians in the decision-making process.

35

A. Dasgupta and A. Wahed: Clinical Chemistry, Immunology and Laboratory Quality Control
DOI: http://dx.doi.org/10.1016/B978-0-12-407821-5.00003-6

Table 3.1 Common Laboratory Errors

Type of Error
Pre-Analytical Errors
Tube filling error
Patient identification error
Inappropriate container
Empty tube
Order not entered in laboratory information system
Specimen collected wrongly from an infusion line
Specimen stored improperly
Contamination of culture tube
Analytical Errors
Inaccurate result due to interference
Random error caused by the instrument
Post-Analytical Errors
Result communication error
Excessive turnaround time due to instrument downtime

Failure at any of these steps can result in an erroneous or misleading laboratory result, sometimes with adverse outcomes. The analytical part of the analysis involves measurement of the concentration of the analyte corresponding to its "true" level (as compared to a "gold standard" measurement) within a clinically acceptable margin of error (the total acceptable analytical error, TAAE). Errors can occur at any stage of analysis (pre-analytical, analytical, and post-analytical). It has been estimated that pre-analytical errors account for more than two-thirds of all laboratory errors, while errors in the analytical and post-analytical phases account for only one-third of all laboratory errors. Carraro and Plebani reported that, among 51,746 clinical laboratory analyses performed in a three-month period in the author's laboratory (7,615 laboratory orders, 17,514 blood collection tubes), clinicians contacted the laboratory regarding 393 questionable results out of which 160 results were confirmed to be due to laboratory errors. Of the 160 confirmed laboratory errors, 61.9% were determined to be pre-analytical errors, 15% were analytical errors, while 23.1% were post-analytical errors [1]. Types of laboratory errors (pre-analytical, analytical, and post-analytical) are summarized in Table 3.1.

In order to avoid pre-analytical errors, several approaches can be taken, including:

- The use of hand-held devices connected to the LIS that can objectively identify the patient by scanning a patient attached barcode, typically a wrist band.

- Retrieval of current laboratory orders from the LIS.
- Barcoded labels are printed at the patient's side, thus minimizing the possibility of misplacing the labels on the wrong patient samples.

When classifying sources of error, it is important to distinguish between *cognitive errors* (mistakes), which are due to poor knowledge or judgment, and *non-cognitive errors* (commonly known as slips and lapses), which are due to interruptions in a process during even routine analysis involving automated analyzers. Cognitive errors can be prevented by increased training, competency evaluation, and process aids (such as checklists); non-cognitive errors can be reduced by improving the work environment (e.g. re-engineering to minimize distractions and fatigue). The vast majority of errors are non-cognitive slips and lapses performed by the personnel directly involved in the process. These can be easily avoided.

The worst pre-analytical error is incorrect patient identification where a physician may act on test results from the wrong patient. Another common error is blood collection from an intravenous line that may falsely increase test results for glucose, electrolytes, or a therapeutic drug due to contamination with infusion fluid.

CASE REPORT

A 59-year-old woman was admitted to the hospital due to transient ischemic heart attack. During the first day of hospitalization she experienced generalized tonic-clonic seizure and a 1000 mg intravenous phenytoin-loading dose was administered followed by an oral dose of 100 mg of phenytoin every three hours for a total of three doses. For the next five days, the patient received 100 mg phenytoin intravenously or orally every 8 hours. On the evening of Day 5 she received two additional 300 mg doses of phenytoin intravenously. Beginning with Day 7 the dose was 100 mg intravenously every 6 hours. On Day 5, phenytoin concentration was 17.0 µg/mL and on Day 7 phenytoin concentration was 13.4 µg/mL. Surprisingly on Day 8, phenytoin concentration was at life-threatening level of 80.7 µg/mL, although the patient did not show any symptom of phenytoin toxicity. Another sample drawn 7 hours later showed a phenytoin level of 12.4 µg/mL. It was suspected that a falsely elevated serum phenytoin level was due to drawing of the specimen from the same line through which the intravenous phenytoin was administered [2].

3.2 ORDER OF DRAW OF BLOOD COLLECTION TUBES

The correct order of draw for blood specimens is as follows:

- Microbiological blood culture tubes (yellow top).
- Royal blue tube (no additive); trace metal analysis if desired.
- Citrate tube (light blue).
- Serum tube (red top) or tube with gel separator/clot activator (gold top or tiger top).

- Heparin tube (green top).
- EDTA tube (ethylenediamine tetraacetic acid; purple/lavender top).
- Oxalate-fluoride tube (gray top).

Tubes with additives must be thoroughly mixed by gentle inversion as per manufacturer-recommended protocols. Erroneous test results may be obtained when the blood is not thoroughly mixed with the additive. When trace metal testing on serum is ordered, it is advisable to use trace element tubes. Royal-Blue Monoject® Trace Element Blood Collection Tubes are available for this purpose. These tubes are free from trace and heavy metals.

3.3 ERRORS WITH PATIENT PREPARATION

There are certain important issues regarding patient preparation for obtaining meaningful clinical laboratory test results. For example, glucose testing and lipid panel must be done after the patient fasts overnight. Although cholesterol concentration is not affected significantly by meals, after meals chylomicrons are present in serum that can significantly increase the triglyceride level.

Physiologically, blood distribution differs significantly in relation to body posture. Gravity pulls the blood into various parts of the body when recumbent, and the blood moves back into the circulation, away from tissues, when standing or ambulatory. Blood volume of an adult in an upright position is 600−700 mL less than when the person is lying on a bed, and this shift directly affects certain analytes due to dilution effects. Therefore, concentrations of proteins, enzymes, and protein-bound analytes (thyroid-stimulating hormone (TSH), cholesterol, T4, and medications like warfarin) are affected by posture; most affected are factors directly influencing hemostasis, including renin, aldosterone, and catecholamines. It is vital for laboratory requisitions to specify the need for supine samples when these analytes are requested. Several analytes show diurnal variations, most importantly cortisol and TSH (Table 3.2). Therefore, the time of specimen collection may affect test results.

3.4 ERRORS WITH PATIENT IDENTIFICATION AND RELATED ERRORS

Accurate patient and specimen identification is required for providing ordering clinicians with correct results. Regulatory agencies like The Joint Commission (TJC) have made it a top priority in order to ensure patient

Table 3.2 Common Analytes that show Diurnal Variation

Analyte	Comment
Cortisol	Much higher concentration in the morning than afternoon
Renin	Maximum activity early morning, minimum in the afternoon
Iron	Higher levels in the morning than afternoon
TSH	Maximum level 2 AM–4 AM while minimum level 6 PM–10 PM
Insulin	Higher in the morning than later part of the day
Phosphate	Lowest in the morning, highest in early afternoon
ALT	Higher level in the afternoon than morning

Abbreviations: TSH, Thyroid stimulating hormone; ALT, Alanine aminotransferase.

safety. Patient and specimen misidentification occurs mostly during the pre-analytical phase:

- Accurate identification of a patient requires verification of at least two unique identifiers from the patient and ensuring that those match the patient's prior records.
- If a patient is unable to provide identifiers (i.e. neonate or a critically ill patient) a family member or nurse should verify the identity of the patient.
- Information on laboratory requisitions or electronic orders must also match patient information in their chart or electronic medical record. Specimens should not be collected unless all identification discrepancies have been resolved.

The specimens should be collected and labeled in front of the patient and then sent to the laboratory with the test request. Non-barcoded specimens should be accessioned, labeled with a barcode (or re-labeled, if necessary), processed (either manually or on an automated line), and sent for analysis. Identification of the specimen should be carefully maintained during centrifugation, aliquoting, and analysis. Most laboratories use barcoded labeling systems to preserve sample identification. Patient misidentification can have a serious adverse outcome on a patient, especially if the wrong blood is transfused to a patient due to misidentification of the blood specimen sent to the laboratory for cross-matching. In this case a patient could die from receiving the wrong blood group.

Although errors in patient identification occur mostly in the pre-analytical phase, errors can also occur during the analytical and even post-analytical phases. Results from automated analyzers are electronically transferred to the LIS through an interface, but if direct transfer of the result from a particular instrument is not available, errors can occur during manual transfer of the results. Dunn and Morga reported that, out of 182 specimen misidentifications they studied, 132 misidentifications occurred in the pre-analytical stage. These misidentifications were due to wrist bands labeled for wrong patient, laboratory tests ordered for the wrong patient, selection of the wrong medical record from a menu of similar names and social security numbers, specimen mislabeling during collection associated with batching of specimens and printed labels, misinformation from manual entry of laboratory forms, failure of two-source patient identification for clinical laboratory specimens, and failure of two-person verification of patient identity for blood bank specimens. In addition, 37 misidentification errors during the analytical phase were associated with mislabeled specimen containers, tissue cassettes, or microscopic slides. Only 13 events of misidentification occurred in the post-analytical stage; this was due to reporting of results into the wrong medical record and incompatible blood transfusions due to failure of two-person verification of blood products [3].

CASE REPORT

A 68-year-old male presented to the hospital with sharp abdominal pain. The patient underwent an appendectomy and received one unit of type A blood. The patient developed disseminated intravascular coagulation and died 24 hours after receiving the transfusion. Postmortem analysis of the patient's blood revealed that he was actually type O. The patient had been sharing a room with another patient whose blood was type A. The specimen sent to the blood bank had been inappropriately labeled [4].

Delta checks are a simple way to detect mislabels. A delta check is a process of comparing a patient's result to his or her previous result for any one analyte over a specified period of time. The difference or "delta," if outside pre-established rules, may indicate a specimen mislabel or other pre-analytical error.

3.5 ERROR OF COLLECTING BLOOD IN WRONG TUBES: EFFECT OF ANTICOAGULANTS

Blood specimens must be collected in the right tube in order to get accurate test results. It is important to have the correct anticoagulant in the tube (different anticoagulant tubes have different colored tops). Anticoagulants are used to prevent coagulation of blood or blood proteins to obtain plasma or whole blood specimens. The most routinely used anticoagulants are ethylenediamine tetraacetic acid (EDTA), heparin (sodium, ammonium, or lithium salts), and citrates (trisodium and acid citrate dextrose). In the optimal anticoagulant, blood ratio is essential to preserve analytes and prevent clot or fibrin formation via various mechanisms. Proper anticoagulants for various tests are as follows:

- Potassium ethylenediamine tetraacetic acid (EDTA; purple top tube) is the anticoagulant of choice for complete blood count (CBC).
- EDTA is also used for blood bank pre-transfusion testing, flow cytometry, hemoglobin A1C, and most common immunosuppressive drugs such as cyclosporine, tacrolimus, sirolimus, and everolimus; another immunosuppressant, mycophenolic acid, is measured in serum or plasma instead of whole blood.
- Heparin (green top tube) is the only anticoagulant recommended for the determination of pH blood gases, electrolytes, and ionized calcium. Lithium heparin is commonly used instead of sodium heparin for general chemistry tests. Heparin is not recommended for protein electrophoresis and cryoglobulin testing because of the presence of fibrinogen, which co-migrates with beta-2 monoclonal proteins.
- For coagulation testing, citrate (light blue top) is the appropriate anticoagulant.
- Potassium oxalate is used in combination with sodium fluoride and sodium iodoacetate to inhibit enzymes involved in the glycolytic pathway. Therefore, the oxalate/fluoride (gray top) tube should be used for collecting specimens for measuring glucose levels.

Although lithium heparin tubes are widely used for blood collection for analysis of many analytes in the chemistry section of a clinical laboratory, a common mistake is to collect specimens for lithium analysis in a lithium heparin tube. This can cause clinically significant falsely elevated lithium values that may confuse the ordering physician.

CASE REPORT

A healthy 15-month-old female was brought in by her mother after ingesting an unknown amount of nortriptyline and lithium carbonate at an undetermined time. The mother reported that the patient had vomited after ingestion. Vital signs were normal. The patient was lethargic but easily aroused, and the physical examination was unremarkable. Initial ECG was also normal for age. The initial lithium level in the serum was 1.4 mEq/L, and a nortriptyline concentration of 36 ng/mL indicated that none of the drug level was in a toxic region. The patient was treated with activated charcoal, but 13 hours after admission her serum lithium concentration was elevated to 3.1 mEq/L. The patient was given 1 mg/kg oral sodium polystyrene sulfonate, the rate of IV fluids was doubled, and the patient was started on an IV dopamine infusion. However, at 15 h her serum lithium level was 1.6 mEq/L. Review of her records revealed that the specimen was collected in a lithium heparin tube. A 19-hour serum lithium concentration was 0.6 mEq/L, and the patient was discharged within 24 h after admission without further incident [5].

3.6 ISSUES WITH URINE SPECIMEN COLLECTION

Urinalysis remains one of the key diagnostic tests in the modern clinical laboratory, and, as such, proper timing and collection techniques are important. Urine is essentially an ultrafiltrate of blood. Examination of urine may take several forms: microscopic, chemical (including immunochemical), and electrophoresis. Three different timings of collection are commonly encountered. The most common is the random or "spot" urine collection. However, if it would not unduly delay diagnosis, the first voided urine in the morning is generally the best sample. This is because the first voided urine is generally the most concentrated and contains the highest concentration of sediment. The third timing of collection is the 12- or 24-hour collection. This is the preferred technique for quantitative measurements, such as for creatinine, electrolytes, steroids, and total protein. The usefulness of these collections is limited, however, by poor patient compliance.

For most urine testing, a clean catch specimen is optimal, with a goal of collecting a "midstream" sample for testing. In situations where the patient cannot provide a clean catch specimen, catheterization is another option, but must be performed only by trained personnel. Urine collection from infants and young children prior to toilet training can be facilitated through the use of disposable plastic bags with adhesive surrounding the opening.

For point of care urinalysis (e.g. urine dipstick and pregnancy testing) any clean and dry container is acceptable. Disposable sterile plastic cups and even clean waxed paper cups are often employed. If the sample is to be sent for culture, the specimen should be collected in a sterile container. For routine urinalysis and culture, the containers should not contain preservative. For specific analyses, some preservatives are acceptable. The exception to this is for timed collections where hydrochloric acid, boric acid, or glacial acetic acid is used as a preservative.

Storage of urine specimens at room temperature is generally acceptable for up to two hours. After this time the degradation of cellular and some chemical elements becomes a concern. Likewise bacterial overgrowth of both pathologic as well as contaminating bacteria may occur with prolonged storage at room temperature. Therefore, if more than two hours will elapse between collection and testing of the urine specimen, it must be refrigerated. Refrigerated storage for up to 12 hours is acceptable for urine samples destined for bacterial culture. Again, proper patient identification and specimen labeling is important to avoid errors in reported results.

3.7 ISSUES WITH SPECIMEN PROCESSING AND TRANSPORTATION

After collection, specimens require transportation to the clinical laboratory. If specimens are collected in the outpatient clinic of the hospital and analyzed in the hospital laboratory, transportation time may not be a factor. However, if specimens are transported to the clinical laboratory or a reference laboratory, care must taken in shipping specimens. Ice packs or cold packs are especially useful for preserving specimens at lower temperatures because analytes are more stable at lower temperature. Turbulence during transportation, such as transporting specimens in a van to the main laboratory, can even affect concentrations of certain analytes.

Many clinical laboratory tests are performed on either serum or plasma. Due to the instability of certain analytes in unprocessed serum or plasma, separation of serum or plasma from blood components must be performed as soon as possible, and definitely within two hours of collection. Appropriate preparation of specimens prior to centrifugation is required to ensure accurate laboratory results. Serum specimens must be allowed ample time to clot prior to centrifugation. Tubes with clot activators require sufficient mixing and at least 30 minutes of clotting time, Plasma specimens must be mixed gently according to manufacturer's instructions to ensure efficient release of additive/anticoagulant.

3.8 SPECIAL ISSUES: BLOOD GAS AND IONIZED CALCIUM ANALYSIS

Specimens collected for blood gas determinations require special care, as the analytes are very sensitive to time, temperature, and handling. In standing whole blood samples, pH falls at a rate of 0.04−0.08/hour at 37°C, 0.02−0.03/hour at 22°C, and <0.01/hour at 4°C. This drop in pH is concordant with decreased glucose and increased lactate. In addition, pCO_2 increases around 5.0 mmHg/hour at 37°C, 1.0 mmHg/hour at 22°C, and 0.5 mmHg/hour at 4°C. At 37°C, pO_2 decreases by 5−10 mmHg/hour, but only 2 mmHg/hour at 22°C. Ideally, all blood gas specimens should be measured immediately and never stored. A plastic syringe transported at room temperature is recommended if analysis will occur within 30 minutes of collection, but a glass syringe should be used if more than 30 minutes are needed prior to analysis and specimens are stored in ice. Bubbles must be completely expelled from the specimen prior to transport, as the pO_2 will be significantly increased and pCO_2 decreased within 2 minutes [6].

Blood gas analyzers re-heat samples to 37°C for analysis to recapitulate physiological temperature. However, for patients with abnormal body temperature, either hyperthermia due to fever, or induced hypothermia in patients undergoing cardiopulmonary bypass, a temperature correction should be made to determine accurate pH, pO_2, and pCO_2 results.

Ionized calcium is often measured with ion-sensitive electrodes in blood gas analyzers. Ionized calcium is inversely related to pH: decreasing pH decreases albumin binding to calcium, thereby increasing free, ionized calcium.

Therefore, specimens sent to the lab for ionized calcium determinations should be handled with the same caution as other blood gas samples since pre-analytical errors in pH will impact ionized calcium results [7].

KEY POINTS

- Errors in the clinical laboratory can occur in pre-analytical, analytical, or post-analytical steps. Most errors (almost two-thirds of all errors) occur in pre-analytical steps.
- During specimen collection, a patient must be identified by matching at least two criteria. Blood should be collected in the correct tube following the correct order of draw.
- Correct order of drawing blood: (1) microbiological blood culture tubes (yellow top), (2) royal blue tube (no additive) if trace metal analysis is desired, (3) citrate tube (light blue), (4) serum tube (red top) or tube with gel separator/clot activator (gold top or tiger top), (5) heparin tube (green top), (6) EDTA tube (purple/lavender top), and (7) oxalate-fluoride tube (gray top).
- Proper centrifugation (in the case of analyzing serum or plasma specimens) and proper transportation of the specimen to the laboratory are required, as well as maintaining proper storage of the specimen prior to analysis in order to avoid artifactual changes in the analyte.
- EDTA (purple top tube) is the anticoagulant of choice for the complete blood count (CBC). The EDTA tube is also used for blood bank pre-transfusion testing, flow cytometry, hemoglobin A1C, and most common immunosuppressive drugs such as cyclosporine, tacrolimus, sirolimus, and everolimus; another immunosuppressant, mycophenolic acid, is measured in serum or plasma instead of whole blood.
- Heparin (green top tube) is the only anticoagulant recommended for the determination of pH blood gases, electrolytes, and ionized calcium. Lithium heparin is commonly used instead of sodium heparin for general chemistry tests. Heparin is not recommended for protein electrophoresis and cryoglobulin testing because of the presence of fibrinogen, which co-migrates with beta-2 monoclonal proteins.
- For coagulation testing, citrate (light blue top) is the appropriate anticoagulant.
- Potassium oxalate is used in combination with sodium fluoride and sodium iodoacetate to inhibit enzymes involved in the glycolytic pathway. Therefore the oxalate/fluoride (gray top) tube should be used for collecting specimens for measuring glucose level.
- Ideally, all blood gas specimens should be measured immediately and never stored. A plastic syringe, transported at room temperature, is recommended if analysis will occur within 30 minutes of collection. Otherwise, a specimen must be stored in ice. Glass syringes are recommended for delayed analysis because glass

does not allow the diffusion of oxygen or carbon dioxide. Bubbles must be completely expelled from the specimen prior to transport, as the pO_2 will be significantly increased and pCO_2 decreased within 2 minutes.

REFERENCES

[1] Carraro P, Plebani M. Errors in STAT laboratory; types and frequency 10 years later. Clin Chem 2007;53:1338−42.

[2] Murphy JE, Ward ES. Elevated phenytoin concentration caused by sampling through the drug-administered line. Pharmacotherapy 1991;11:348−350.

[3] Dunn EJ, Morga PJ. Patient misidentification in laboratory medicine: a qualitative analysis of 227 root cause analysis reports in the Veteran Administration. Arch Pathol Lab Med 2010;134:244−55.

[4] Aleccia J. Patients still stuck with bill for medical errors. 2008 2/29/2008 8:26:51 AM ET [cited 2012 06/28/2012]; Available from: <http://www.msnbc.msn.com/id/23341360/ns/health-health_care/t/patients-still-stuck-bill-medical-errors/#.T-yk5vVibJs>.

[5] Lee DC, Klachko MN. Falsely elevated lithium levels in plasma samples obtained in lithium containing tubes. J Toxicol Clin Toxicol 1996;34:467−9.

[6] Knowles TP, Mullin RA, Hunter JA, Douce FH. Effects of syringe material, sample storage time, and temperature on blood gases and oxygen saturation in arterialized human blood samples. Respir Care 2006;51:732−6.

[7] Toffaletti J, Blosser N, Kirvan K. Effects of storage temperature and time before centrifugation on ionized calcium in blood collected in plain vacutainer tubes and silicone-separator (SST) tubes. Clin Chem 1984;30(4):553−6.

CHAPTER 4

Laboratory Statistics and Quality Control

4.1 MEAN, STANDARD DEVIATION, AND COEFFICIENT OF VARIATION

In an ideal situation, when measuring a value of the analyte in a specimen, the same value should be produced over and over again. However, in reality, the same value is not produced by the instrument, but a similar value is observed. Therefore, the most basic statistical operation is to calculate the mean and standard deviation, and then to determine the coefficient of variation (CV). Mean value is defined as Equation 4.1:

$$\text{Mean } (\overline{X}) = \frac{X_1 + X_2 + X_3 + \cdots\cdots + X_n}{n} \qquad (4.1)$$

Here, X_1, X_2, X_3, etc., are individual values and "n" is the number of values.

After calculation of the mean value, standard deviation (SD) of the sample can be easily determined using the following formula (Equation 4.2):

$$\text{SD} = \sqrt{\frac{\sum (x_1 - \overline{x})^2}{n-1}} \qquad (4.2)$$

Here, X_1 is the individual value from the sample and n is again the number of observations.

Standard deviation represents the average deviation of an individual value from the mean value. The smaller the standard deviation, the better the precision of the measurement. Standard deviation is the square root of variance. Variance indicates deviation of a sample observation from the mean of all values and is expressed as sigma. Therefore (Equation 4.3):

$$\sigma = \sqrt{\text{SD}} \qquad (4.3)$$

CONTENTS

A. Dasgupta and A. Wahed: Clinical Chemistry, Immunology and Laboratory Quality Control
DOI: http://dx.doi.org/10.1016/B978-0-12-407821-5.00004-8

Coefficient of variation is also a very important parameter because CV can be easily expressed as a percent value; the lower the CV, the better the precision for the measurement. The advantage of CV is that one number can be used to express precision instead of stating both mean value and standard deviation. CV can be easily calculated with Equation 4.4:

$$CV = SD/Mean \times 100 \tag{4.4}$$

Sometimes standard error of mean is also calculated (Equation 4.5).

$$Standard\ error\ of\ mean = SD/\sqrt{n} \tag{4.5}$$

Here, n is the number of data points in the set.

4.2 PRECISION AND ACCURACY

Precision is a measure of how reproducible values are in a series of measurements, while accuracy indicates how close a determined value is to the target values. Accuracy can be determined for a particular test by analysis of an assayed control where the target value is known. This is typically provided by the manufacturer or made in-house by accurately measuring a predetermined amount of analyte and then dissolving it in a predetermined amount of a solvent matrix where the matrix is similar to plasma. An ideal assay has both excellent precision and accuracy, but good precision of an assay may not always guarantee good accuracy.

4.3 GAUSSIAN DISTRIBUTION AND REFERENCE RANGE

Gaussian distribution (also known as normal distribution) is a bell-shaped curve, and it is assumed that during any measurement values will follow a normal distribution with an equal number of measurements above and below the mean value. In order to understand normal distribution, it is important to know the definitions of "mean," "median," and "mode." The "mean" is the calculated average of all values, the "median" is the value at the center point (mid-point) of the distribution, while the "mode" is the value that was observed most frequently during the measurement. If a distribution is normal, then the values of the mean, median, and mode are the same. However, the value of the mean, median, and mode may be different if the distribution is skewed (not

Gaussian distribution). Other characteristics of Gaussian distributions are as follows:

- Mean ± 1 SD contain 68.2% of all values.
- Mean ± 2 SD contain 95.5% of all values.
- Mean ± 3 SD contain 99.7% of all values.

A Gaussian distribution is shown in Figure 4.1. Usually, reference range is determined by measuring the value of an analyte in a large number of normal subjects (at least 100 normal healthy people, but preferably 200–300 healthy individuals). Then the mean and standard deviations are determined.

The reference range is the mean value −2 SD to the mean value +2 SD. This incorporates 95% of all values. The rationale for reference range to be the mean ± 2 SD is based on the fact that the lower end of abnormal values and upper end of normal values may often overlap. Therefore, mean ± 2 SD is a conservative estimate of the reference range based on measurement of the analytes in a healthy population. Important points for reference range include:

- Reference range may be the same between males and females for many analytes, but reference range may differ significantly between males and females for certain analytes such as sex hormones.
- Reference range of an analyte in an adult population may be different from infants or elderly patients.
- Although less common, reference range of certain analytes may be different between different ethnic populations.
- For certain analytes such as glucose, cholesterol, triglycerides, high-density and low-density cholesterol, etc., there is no reference range but

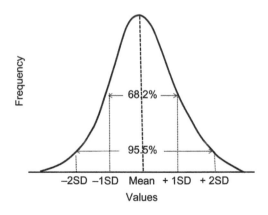

FIGURE 4.1

A Gaussian distribution showing percentage of values within a certain standard deviation from the mean. *(Courtesy of Andres Quesda, M.D., Department of Pathology and Laboratory Medicine, University of Texas-Houston Medical School.)*

there are desirable ranges which are based on the study of a large population and risk factors associated with certain values of analytes (e.g. various lipid parameters and risk of cardiovascular diseases).

Although many analytes in the normal population when measured follow normal distribution, not all analytes follow that pattern (e.g. cholesterol and triglycerides). In this case distribution is skewed and, as expected, mean, median, and mode values are different.

4.4 SENSITIVITY, SPECIFICITY, AND PREDICTIVE VALUE

An assay cannot be 100% sensitive or specific because there is some overlap between values of a particular biochemical parameter observed in normal individuals and patients with a particular disease (Figure 4.2). Therefore, during measurement of any analyte there is a gray area where few abnormal values are generated from analysis of specimens from healthy people (false positive) and few normal results are generated from patients (false negative).

- The gray area depends on the width of normal distribution as well as the reference range of the analyte.

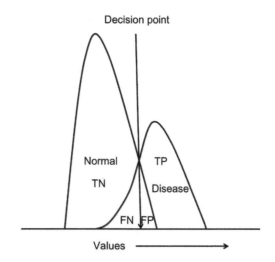

FIGURE 4.2

Distribution of values in normal and diseased states where TN: true negative values; TP: true positive values; FN: false negative values; and FP: false positive values. *(Courtesy of Andres Quesda, M.D., Department of Pathology and Laboratory Medicine, University of Texas-Houston Medical School.)*

- False positive results may mislead the clinician and lead to unnecessary investigation and diagnostic tests as well as increased anxiety of the patient.
- A false negative result is more dangerous than a false positive result because diagnosis of a disease may be missed or delayed, which can cause serious problems.
- For a test, as clinical sensitivity increases, specificity decreases. For calculating clinical sensitivity, specificity, and predictive value of a test, the following formulas can be used:
  - TP = True positive (result correctly identifies a disease)
  - FP = False positive (result falsely identifies a disease)
  - TN = True negative (result correctly excludes a disease when the disease is not present in an individual)
  - FN = False negative (result incorrectly excludes a disease when the disease is present in an individual).

Therefore, when assay results are positive, results are a combination of TP and FP, and when assay results are negative, results are combination of TN and FN (Equations 4.6–4.8).

Sensitivity (individuals with disease who show positive test results)

$$= \frac{TP}{TP + FN} \times 100 \tag{4.6}$$

Specificity (individuals without disease who show negative test results)

$$= \frac{TN}{TN + FP} \times 100 \tag{4.7}$$

$$\text{Positive predictive value} = \frac{TP}{TP + FP} \times 100 \tag{4.8}$$

A positive predictive value is the proportion of individuals with disease who showed a positive value compared to all individuals tested. Let us consider an example where a particular analyte was measured in 100 normal individuals and 100 individuals with disease. The following observations were made: TP = 95, FP = 5, TN = 95, and FN = 5. Therefore, sensitivity = 95/(95 + 5) × 100 = 95%, and specificity = 95/(95 + 5) × 100 = 95%.

4.5 RANDOM AND SYSTEMATIC ERRORS IN MEASUREMENTS

Random errors and systematic errors are important issues in the laboratory quality control process. Random errors are unavoidable and occur due to imprecision of an analytical method. On the other hand, systematic errors have certain characteristics and are often due to errors in measurement using a particular assay. Because random errors cannot be eliminated or controlled, the goal of quality control in a clinical laboratory is to avoid or minimize systematic errors. Usually recalibration of the assay is the first step taken by a clinical laboratory technologist to correct systematic error, but more serious problems such as instrument malfunction may also be responsible for systematic errors.

4.6 LABORATORY QUALITY CONTROL: INTERNAL AND EXTERNAL

Good quality control is the heart of a good laboratory operation. Because the value of an analyte in a patient's specimen is unknown, clinical laboratory professionals rely on producing accurate results using controls for an assay. Controls can be purchased from a commercial source or can be made in-house. A control is defined as a material that contains the analyte of interest with a known concentration. It is important that the control material has a similar matrix to serum or plasma. Different types of controls used in clinical laboratories are listed below:

- Assayed Control: The value of the analyte is predetermined. Most commercially available controls have predetermined values of various analytes. The target value must be verified before use.
- Un-Assayed Control: The target value is not predetermined. This control must be fully validated (run at least 20 times in a single run and then run once a day for 20 consecutive days to establish a target value).
- Homemade Control: If the assayed control material is not easily commercially available (e.g. for an esoteric test), the control material may be prepared by the laboratory staff by dissolving correctly weighed pure material in an aqueous-based solvent or in serum or whole blood (for an analyte not present in humans, e.g. a drug).

Commercially available control materials may be obtained as a ready-to-use liquid control or as a lyophilized powder. If control material is available in the form of lyophilized powder, it must be reconstituted prior to use by strictly following the manufacturer's recommended protocol. Control materials must be stored in a refrigerator following manufacturer's recommendations and the expiration date of the control must be clearly visible so that an expired control is not used by mistake. Usually low, medium, and high controls of an analyte are used to indicate analyte concentrations both in a normal physiological state and a disease state. At least two controls must be used for each analyte (high and low controls). Control materials must be run along with patient samples or at least once in each shift (a minimum of three times in a 24 h period) depending on the assay.

Quality control in the laboratory may be both internal and external. Internal quality control is essential and results are plotted in a Levey–Jennings chart as discussed below. The most common example of external quality control is analysis of CAP (College of American Pathologists) proficiency samples for most tests offered by a clinical laboratory. Proficiency samples may not be available for a few esoteric tests. CLIA 88 (Clinical Laboratory Improvement Act) requires all clinical laboratories to register with the government and to disclose all tests these laboratories offer. The test may be "waived tests" or "non-waived tests:"

- "Waived tests" are ones where laboratories can perform such tests as long as they follow manufacturer protocol. Enrolling in an external proficiency-testing program such as a CAP survey is not required for waived tests.
- "Non-waived tests" are moderately complex or complex tests. Laboratories performing such tests are subjected to all CLIA regulations and must be inspected by CLIA inspectors every two years or by inspectors from non-government organizations such as CAP or Joint Commission on Accreditation of Healthcare Organization (JCAHO). In addition, a laboratory must participate in an external proficiency program (most commonly CAP proficiency surveys) and must successfully pass proficiency testing in order to operate legally. A laboratory must produce correct results for four of five external proficiency specimens for each analyte, and must have at least an 80% score for three consecutive challenges.
- Since April 2003, clinical laboratories must perform method validation for each new test, even if such test already has FDA approval.

Currently, most common external proficiency testing samples are offered by CAP, and there are proficiency specimens for 580 analytes. The major features of CAP external proficiency testing include:

- CAP proficiency samples are mailed to participating laboratories three times a year and there are at least five samples for each analyte during this period.
- CAP proficiency samples have matrix similar to patient specimens and such specimens must be analyzed just like a regular patient specimen. For example, a CAP specimen cannot be analyzed in duplicate or only on the day shift; such practice to pass CAP proficiency testing is a violation of established practice guidelines.
- CAP proficiency testing results must be reported to CAP and later graded or ungraded results must arrive at the laboratory for evaluation by laboratory professionals. A laboratory director or designee must sign results of a CAP survey and must act if the laboratory fails a survey.

- CAP proficiency test results are graded based on performance of all participating laboratories. There are various criteria for acceptability of a result. Results must be within ± 2 SD of the peer group mean (calculated by taking into account all values reported by participating laboratories) or a fixed percentage of a target value (i.e. within 10% of target value) or the result must be within a fixed deviation from the target value (e.g. within ± 4 mol/L of the target value).
- The best way to evaluate CAP proficiency testing results of an individual clinical laboratory is to use the e-lab solution available from the CAP for downloading.
- If CAP proficiency testing is not available, then the laboratory must validate the test every six months by comparing values obtained by the test with values obtained by a reference laboratory or another laboratory offering the test (using split samples). Alternatively, if proficiency samples are available from another source, for example, AACC (American Association for Clinical Chemistry), passing such proficiency testing is also acceptable.
- In addition to the CAP external proficiency-testing program, a laboratory may participate in other proficiency testing programs. However, for laboratory accreditation by CAP, it is required that the laboratory must participate in a CAP proficiency survey, provided that the proficiency specimen is available from the CAP.

There are a number of publications that indicate that participating in external proficiency surveys such as offered by CAP is useful in improving the quality of a clinical laboratory operation [1−3].

4.7 LEVEY−JENNINGS CHART AND WESTGARD RULES

In addition to participating in the CAP program, clinical laboratories must run control specimens every shift, at least three times in a 24 h cycle. Also, instruments must be calibrated as needed in order to maintain good laboratory practice. Calibration is needed for all assays that a clinical laboratory offers. Calibration of immunoassays is discussed in Chapter 2. However, other assays are calibrated using calibrators that are either commercially available or homemade:

- Calibrators are defined as materials that contain known amounts of the analyte of interest. For a single assay, at least two calibrators are needed for calibration, a zero calibrator (contains no analyte) and a high calibrator containing the amount of the analyte that represents the upper end of the analytical measurement range. However, five to six calibrators are commonly used for calibration. One calibrator must be a zero calibrator and the highest calibrator must contain a concentration of the analyte at the upper end of the analytical measurement range. Other calibrators usually have concentrations in between the zero calibrator and the highest calibrator, and represent normal values of the analyte as well as values expected in a disease state (for drugs, values below therapeutic range, between therapeutic ranges, and then toxic range).

- Controls are materials that contain a known amount of the analyte. The matrix of the control must be similar to the matrix of the patient's sample; for example, matrix of the control must resemble serum for assays conducted in serum or plasma.

A Levey–Jennings chart is commonly used for recording observed values of controls during daily operation of a clinical laboratory. A Levey–Jennings chart is a graphical representation of all control values for an assay during an extended period of laboratory operation. In this graphical representation, values are plotted with respect to the calculated mean and standard deviation, and if all controls are within the mean and ± 2 SD, then all control values are within acceptable limits and all runs during that period will have acceptable performance (Figure 4.3). In this figure, all glucose low controls were within acceptable limits for the entire month. The Levey–Jennings chart must be constructed for each control (low and high control or low, medium, and high control) for each assay the laboratory offers. For example, if the laboratory runs two controls (low and high) for each test and offers 100 tests, then there will be 100×2, or 200 Levey–Jennings charts each month. Usually a Levey–Jennings chart is constructed for one control for one month. The laboratory director or designee must review all Levey–Jennings charts each month and sign them for compliance with an accrediting agency.

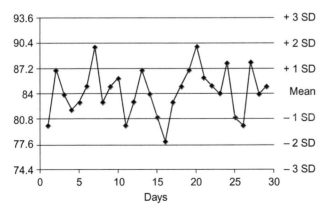

FIGURE 4.3
Levey–Jennings chart with no violation.

Table 4.1 Westgard Rules

Violation	Comments	Accept/Reject Run	Error Type
1_{2s}	One control value is outside ± 2 SD limit, but other control within ± 2 SD limit	Accept run	Random
1_{3s}	One control exceeds ± 3 SD	Reject run	Random
2_{2s}	Both controls outside ± 2 SD limit, or two consecutive controls outside limit	Reject run	Systematic
R_{4s}	One control $+2$ SD and another -2 SD	Reject run	Random
4_{1S}	Four consecutive control exceeding $+1$ SD or -1 SD	Reject run*	Systematic
$10\times$	Ten consecutive control values falling on one side of the mean	Reject run*	Systematic

*Although these are rejection rules, a laboratory may consider these violations as warnings and may accept the runs and take steps to correct such systematic errors.

However, if technologists review results of the control during a run and accept the run if the value of the control is within an acceptable range established by the laboratory (usually a mean of ± 2 SD), then the laboratory supervisor can review all control data on a daily basis; usually the supervisor reviews all control data weekly.

Usually Westgard rules are used for interpreting a Levey–Jennings chart, and for certain violations a run must be rejected and the problem resolved prior to resuming testing of a patient's samples. Various errors can occur in Levey–Jennings charts, including shift, trend, and other violations (Table 4.1). The basic principle is that control values must fall within ± 2 SD of the mean, but there are some situations when violation of Westgard rules occurs despite control values that are within the ± 2 SD limits of the mean. Usually 1_{2s} is a warning rule and occurs due to random error (Figure 4.4), and other rules are rejection rules. In addition, shift (Figure 4.5) and trend (Figure 4.6) may be observed in Levey–Jennings charts, indicating systematic errors where corrective actions must be taken. When 10 or more consecutive control values are falling on one side of the mean, a shift is observed ($10\times$ rule). In addition, when a $10\times$ violation is observed, it may also indicate a trend when control values indicate an upward trend.

4.8 DELTA CHECKS

Delta checks are an additional quality control measure adopted by the computer of an automated analyzer or the laboratory information system (LIS) where a value is flagged if the value deviates more than a predetermined limit from the previous value in the same patient. The limit of deviation for

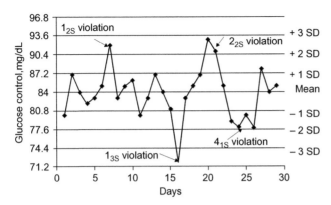

FIGURE 4.4

Levey–Jennings chart showing certain violations.

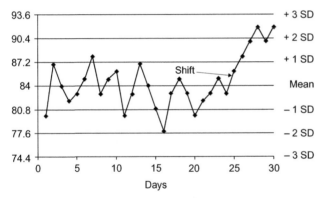

FIGURE 4.5

Levey–Jennings chart showing shift of control values.

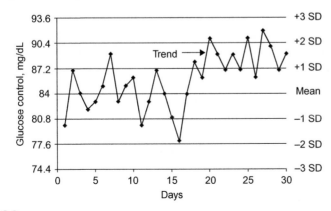

FIGURE 4.6

Levey–Jennings chart showing trend.

each analyte is set by laboratory professionals. The basis of the delta check is that the value of an analyte in a patient should not deviate significantly from the previous value unless certain intervention is done; for example, a high glucose value may decrease significantly following administration of insulin. If a value is flagged as a failed delta check, then a further investigation should be made. A phone call to the nurse may address issues such as erroneous results due to collection of a specimen from an IV line or collection of the wrong specimen. Quality control of the assay must also be addressed to ensure that the erroneous result is not due to instrument malfunction.

The value of a delta check is usually based on one of the following criteria:

- Delta difference: current value−previous value should be within a predetermined limit.
- Delta percent change: delta difference/current value.
- Rate difference: delta difference/delta interval × 100.
- Rate percent change: delta percentage change/delta interval.

4.9 METHOD VALIDATION/EVALUATION OF A NEW METHOD

Since April 2003, clinical laboratories must perform method validation for each new test implemented in the laboratory even though such tests have FDA approval. The following are steps for method validation as well as implementation of a new method in the clinical laboratory:

- Within-run assay precision must be validated by running low, medium, and high controls, or low and high controls 20 times each in a single run. Then mean, standard deviation, and CV must be calculated individually for low, medium, and high control.
- Between-run assay precision must be established by running low, medium, and high control, or low and high control once daily for 20 consecutive days. Then mean, standard deviation, and CV must be calculated.
- Although assay linearity is provided by the manufacturer, it must be validated in the clinical laboratory prior to running patient specimens. Linearity is essentially the calibration range of the assay (also called "analytical measurement range"). In order to validate the linearity, a high-end calibrator or standard can be selected and then diluted to produce at least four to five dilutions that cover the entire analytical measurement range. Then, if the observed value matches the expected value, the assay can be considered linear over the stated range.
- The detection limit should be traditionally determined by running a zero calibrator or blank specimen 20 times and then determining the mean

and standard deviation. The detection limit (also called the lower limit of detection) is the mean $+2$ SD value. However, the guidelines of the Clinical Laboratory Standard Institute (CLSI, E17 protocol) advise that a specimen with no analyte (blank specimen) should be run; then the Limit of Blank (LoB) $=$ Mean $+ 1.654$ SD. This should be established by running blank specimens 60 times, but if a company already established a guideline, then 20 runs are enough. Limit of Quantification is usually defined as a concentration where CV is 20% or less [4].

■ Comparison of a new method with an existing method is a very important step in method validation. For this purpose, at least 100 patient specimens must be run in the laboratory at the same time with both the existing method and the new method. It is advisable to batch patient samples and then run these specimens by both methods on the same day, and, if possible, at the same time (by splitting specimens). Then results obtained by the existing method should be plotted in the x-axis (reference method) and corresponding values obtained by the new method should be plotted in the y-axis. Linear regression is the simplest way of comparing results obtained by the existing method in the laboratory and the new method. The linear regression equation is the line of best fit with all data points. A computer can produce the linear regression line as well as an equation called a linear regression equation, which is the equation representing a straight line (regression line), Equation 4.9:

$$y = mx + b \qquad\qquad (4.9)$$

■ Here, "m" is called the slope of the line and "b" is the intercept. The computer calculates the equation of the regression line using a least squares approach. The software also calculates "r," the correlation coefficient, using a complicated formula.

4.10 HOW TO INTERPRET THE REGRESSION EQUATION?

The regression equation ($y = mx + b$) provides a lot of important information regarding how the new method (y) compares with the reference method (x). Interpretations of a linear regression equation include:

■ Ideal value: $m = 1$, $b = 0$, and $y = x$. In reality this never happens.
■ If the value of m is less than 1.0, then the method shows negative bias compared to the reference method. Bias can be calculated as $1 - m$; for example, if the value of "m" is 0.95, then the negative bias is $1 - 0.95 = 0.05$, or $0.05 \times 100 = 5\%$.

■ If the value of m is over 1.0, it indicates positive bias in the new method. For example, if m is 1.07, then positive bias in the new method is $1.07 - 1 = 0.07$, or $0.07 \times 100 = 7\%$.

■ The intercept "b" can be a positive or negative value and must be a relatively small number.

■ An ideal value of "r" (correlation coefficient) is 1, but any value above 0.95 is considered good, and a value of 0.99 is considered excellent. The correlation coefficient indicates how well the new method compares with the existing method, but cannot tell anything about any inherent bias in the new method. Therefore, slope must be taken into account to determine bias.

In our laboratory, we evaluated a new immunoassay for mycophenolic acid, an immunosuppressant, with a HPLC-UV method, the current method in our laboratory, using specimens from 60 transplant recipients after de-identifying specimens [5]. The regression equation was as follows (Equation 4.10):

$$y = 1.1204 \times + 0.0881 \ (r = 0.98) \tag{4.10}$$

This equation indicated that there was an average 12.04% positive bias with the new immunoassay method compared to the reference HPLC-UV method in determining mycophenolic acid concentration. This was most likely due to cross-reactivity of mycophenolic acid acyl glucuronide with the mycophenolic acid assay antibody because metabolite does not interfere with mycophenolic acid determination using HPLC-UV. However, the correlation coefficient of 0.98 indicates good agreement between both methods.

4.11 BLAND–ALTMAN PLOT

Although linear regression analysis is useful for method comparison, such analysis is affected by extreme values (where one or a series of "x" values differs widely from the corresponding "y" values) because equal weights are given to all points. A Bland–Altman plot compares two methods by plotting the difference between the two measurements on the y-axis, and the average of the two measurements on the x-axis. The difference between two methods can be expressed as a percentage difference between two methods or a fixed difference such as 1 SD or 2 SD or a fixed number. It is easier to see bias between two methods using a Bland–Altman plot.

4.12 RECEIVER–OPERATOR CURVE

A receiver–operator curve (ROC) is often used to make an optimal decision for a test. ROC plots the true positive rate of a test (sensitivity) either as a

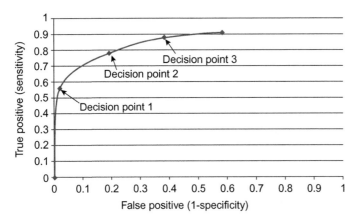

FIGURE 4.7
Receiver—operator curve (ROC) showing various decision points.

scale of $0-1$ (1 is highest sensitivity) or as a percent on the y-axis versus a false positive rate (1-specificity). As sensitivity increases, the specificity decreases. In Figure 4.7, a hypothetical ROC curve is given. If decision point 1 is selected for the test value then sensitivity of the test is 0.57 or 57% but specificity is very high (99%, in the scale 1-specificity: 0.01). On the other hand if a higher value of the test is selected for a decision point (decision point 3), the sensitivity has been increased to nearly 90% but specificity was decreased to 42% (in the scale 1-specificity: 0.58) (Figure 4.7). Therefore, a decision point can be made which can be used for making a clinical decision. In general, the closer the decision point is to the y-axis, the better the specificity.

4.13 WHAT IS SIX SIGMA?

Six sigma originated from Motorola Corporation's approach for total quality management during manufacturing with an objective to reduce defects in manufacturing. Although six sigma was originally developed for a manufacturing process, the principles can be applied to total quality improvement of any operation, including a clinical laboratory operation. The goal of six sigma is to achieve an error rate of 3.4 out of one million for a process or an error rate of only 0.00034%. An error rate of 0.001% is considered a 5.8 sigma. The goal of a clinical laboratory operation is to reduce the error rate to at least 0.1% (4.6 sigma), but preferably 0.01% (5.2 sigma) or higher. Improvement can be made during any process of the laboratory operation (pre-analytical, analytical, or post-analytical) with an overall goal of reducing laboratory errors.

4.14 ERRORS ASSOCIATED WITH REFERENCE RANGE

Reference ranges are given with patients' values to help clinicians interpret laboratory test results. However, most reference ranges include values in the range of mean ± 2 SD as observed with the normal population. Therefore, reference range only accounts for 95% of the values observed in healthy individuals for the particular tests, and statistically 5% of the values of the normal population should fall outside the reference range. If more than one test is used, then a greater percentage of the values should fall outside the reference range. The likelihood of "n" test results falling within the reference range can be calculated with Equation 4.11:

$$\% \text{ Results falling within normal range} = 0.95^n \times 100 \qquad (4.11)$$

The percent of results falling outside the reference range in normal people is shown in Equation 4.12:

$$(1 - 0.95^n) \times 100 \qquad (4.12)$$

For example, if five tests are ordered for health screening of a healthy person, then Equation 4.13 holds true:

$$
\begin{aligned}
\% \text{ Results falling outside normal range} &= (1 - 0.95^5) \times 100 \\
&= (1 - 0.773) \times 100 = 22.7\%
\end{aligned}
\qquad (4.13)
$$

In Table 4.2, examples of a number of tests falling within and outside the reference range are given.

Table 4.2 Testing and Reference Range*		
Number of Tests	**Results within Reference Range**	**Outside Reference Range**
1	95%	5%
2	90%	10%
3	85.7%	14.3%
4	81.4%	18.6%
5	77.3%	22.7%
6	73.5%	26.5%
10	59.8%	40.2%

*For multiple tests ordered in a healthy subject, chances of the number of tests falling within the reference range and the number of tests falling outside the reference range.

4.15 BASIC STATISTICAL ANALYSIS: STUDENT t-TEST AND RELATED TESTS

A new method can be validated against an existing method by using regression analysis as stated earlier in the chapter. Bias can be calculated based on the analysis of the slope or Bland—Altman plot. However, in some instances, bias between the two methods can be significant and in this case a laboratory professional needs to know if values on an analyte determined by the reference method are significantly different from the values determined by the new method. This can be calculated by the mean of two sets of values and the standard deviation using Student t-test:

- The Student t-test is useful to determine if one set of values is different from another set of values based on the difference between mean values and standard deviations. This statistical test is also useful in clinical research to see if values of an analyte in the normal state are significantly different from the values observed in a disease state.
- The Student t-test is only applicable if both distributions of values are normal (Gaussian).
- If the "t" value is significant based on the degrees of freedom ($n_1 + n_2 - 1$, where n_1 and n_2 represent the number of values in set 1 and set 2 distributions), then the null hypothesis (there is no difference between two sets of values) is rejected and it is assumed that values in the set 1 distribution are statistically different from values in the set 2 distribution. The value of t can be easily obtained from published tables.
- The F-test is a measure of differences in variances and can also be used to see if one set of data is different from another set of data. The F-test can be used for analysis of multiple sets of data, when it is called ANOVA (analysis of variance).
- If the distribution of data is non-Gaussian, then neither the t-test nor the F-test can be used. In this case, the Wilcoxon rank sum test (also known as the Mann—Whitney U test) should be used.

The formulas for the t-test and Mann—Whitney U test can be found in any textbook on statistics. However, a detailed discussion on these statistical methods is beyond the scope of this book.

KEY POINTS

- The formula for coefficient of variation (CV): $CV = SD/mean \times 100$.
- Standard error of mean $= SD/\sqrt{n}$, where n is the number of data points in the set.
- If a distribution is normal, the value of the mean, median, and mode is the same. However, the value of the mean, median, and mode may be different if the distribution is skewed (not a Gaussian distribution).

- In Gaussian distributions, the mean ± 1 SD contains 68.2% of all values, the mean ± 2 SD contains 95.5% of all values, and the mean ± 3 SD contains 99.7% of all values in the distribution.
- The reference range when determined by measuring an analyte in at least 100 healthy people and the distribution of values in a normal Gaussian distribution is calculated as mean ± 2 SD.
- For calculating sensitivity, specificity, and predictive value of a test, the following formulas can be used, where TP = true positive, FP = False positive, TN = True negative, and FN = False negative: (a) Sensitivity (individuals with disease who show positive test results) = (TP/(TP + FN)) × 100; (b) Specificity (individuals without disease who show negative test results) = (TN/(TN + FP)) × 100; and (c) Positive predictive value = (TP/(TP + FP)) × 100.
- In a clinical laboratory, three types of control materials are used: assayed control where the value of the analyte is predetermined, un-assayed control where the target value is not predetermined, and homemade control where the control material is not easily commercially available (e.g. an esoteric test).
- Quality control in the laboratory may be both internal and external. Internal quality control is essential and results are plotted in a Levey–Jennings chart; the most common example of external quality control is analysis of CAP (College of American Pathologists) proficiency samples.
- "Waived tests" are not complex and laboratories can perform such tests as long as they follow manufacturer's protocol. Enrolling in an external proficiency-testing program such as a CAP survey is not required for waived tests.
- "Non-waived tests" are moderately complex or complex tests and laboratories performing such tests are subjected to all CLIA regulations and must be inspected by CLIA inspectors every two years or by inspectors from non-government organizations such as CAP or Joint-Commission on Accreditation of Healthcare Organization (JCAHO). In addition, for all non-waived tests laboratories must participate in an external proficiency program, most commonly CAP proficiency surveys, and must successfully pass proficiency testing in order to operate legally. A laboratory must produce correct results for four of five external proficiency specimens for each analyte, and must have at least an 80% score for three consecutive challenges.
- Since April 2003, clinical laboratories must perform method validation for each new test, even if such test is already FDA approved.
- A Levey–Jennings chart is a graphical representation of all control values for an assay during an extended period of laboratory operation. In this graphical representation, values are plotted with respect to the calculated mean and standard deviation. If all controls are within the mean and ± 2 SD, then all control values were within acceptable limits and all runs during that period have acceptable performance. A Levey–Jennings chart must be constructed for each control (low and high control, or low, medium, and high control) for each assay the laboratory offers. The laboratory director or designee must review all

Levey–Jennings charts each month and sign them for compliance with an accrediting agency.

- Usually Westgard rules are used for interpreting Levey–Jennings charts, and for certain violations, a run must be rejected and the problem must be resolved prior to resuming testing of patients' samples. Various errors can occur in Levey–Jennings charts, including shift, trend, and other violations. Usually 1_{2s} is a warning rule and occurs due to random error; other rules are rejection rules (see Table 4.1).

- A delta check is important to identify laboratory errors and can be based on any of the criteria, including delta difference, delta percent change (delta difference/current value), rate difference (delta difference/delta interval $\times$ 100), or rate percent change (delta percentage change/delta interval). Usually within and between runs precision is expressed as CV. Then linearity of the assay is revalidated. Detection limits should be determined by running a zero calibrator or blank specimen 20 times and then determining the mean and standard deviation. The detection limit (also called the lower limit of detection) is considered as a mean + 2 SD value, but more sophisticated methods of calculating limit of detection have also been described.

- Comparison of a new method with an existing method is a very important step in method validation. For this purpose, at least 100 patient specimens must be run with the existing method in the laboratory at the same time as the new method. Then values are plotted and a linear regression equation determines the line of best fit as expressed by the equation $y = mx + b$, where "m" is the slope of the line and "b" is the intercept. The computer calculates the equation of the regression line using a least squares approach. The software also calculates "r," the correlation coefficient, by using a complicated formula. The ideal value of m is 1, while the ideal value of b is zero. In reality, if slope is less than 1.0, it indicates negative bias with the new method compared to the old method, and if the slope is over 1.0, it indicates positive bias.

- A receiver–operator curve (ROC) is often used to make an optimal decision level for a test. ROC plots the true positive rate of a test (sensitivity) either as a scale of 0–1 (1 is highest sensitivity) or as percent on the y-axis versus a false positive rate (1-specificity).

- Six sigma goal is achieved if the error rate is only 3.4 out of one million processes, or error rate is only 0.00034%.

- The likelihood of "n" test results falling within the reference range can be calculated from the formula % results falling within normal range $= 0.95^n \times 100$. Therefore % results falling outside the reference range in normal people is $(1 - 0.95^n) \times 100$.

- The Student t-test is useful for determining if one set of values is different from another set of values based on the difference between mean values and standard deviations. This statistical test is also useful in clinical research to see if values of an analyte in the normal state are significantly different from the values observed in a disease state.

REFERENCES

[1] Jenny RW, Jackson KY. Proficiency test performance as a predictor of accuracy of routine patient testing for theophylline. Clin Chem 1993;39:76−81.

[2] Theolen D, Lawson NS, Cohen T, Gilmore B. Proficiency test performance and experience with College of American Pathologist's programs. Arch Pathol Lab Med 1995;119:307−11.

[3] Boone DJ. Literature review of research related to the Clinical Laboratory Improvement Amendments of 1988. Arch Pathol Lab Med 1992;116:681−93.

[4] Armbuster DA, Pry T. Limit of blank, limit of detection and limit of quantification. Clin Biochem Rev 2008;29(Suppl. 1):S49−51.

[5] Dasgupta A, Tso G, Chow L. Comparison of mycophenolic acid concentrations determined by a new PETINIA assay on the Dimension EXL analyzer and a HPLC-UV method. Clin Biochem 2013;46:685−7.

CHAPTER 5

Water, Homeostasis, Electrolytes, and Acid–Base Balance

CONTENTS

5.1 DISTRIBUTION OF WATER AND ELECTROLYTES IN THE HUMAN BODY

Water is a major constituent of the human body that represents approximately 60% of body weight in men and 55% of body weight in women. Two-thirds of the water in the human body is associated with intracellular fluid and one-third is found in extracellular fluid. Extracellular fluid is composed mostly of plasma (containing 92% water) and interstitial fluid. A major extracellular electrolyte is sodium. The human body contains approximately 4,000 mmol of sodium out of which 70% is present in an exchangeable form; the rest is found in bone. The intracellular concentration of sodium is 4–10 mmol/L. The normal sodium level in human serum is 135–145 mmol/L. Potassium is the major intracellular electrolyte with an intracellular concentration of approximately 150 mmol/L. The normal potassium level in serum is usually considered to be 3.5–5.1 mmol/L. The balance between intracellular and extracellular electrolytes is maintained by a sodium–potassium ATPase pump present in cell membranes.

Along with sodium and potassium, other major electrolytes of the human body are chloride and bicarbonate. Electrolytes are classified either as positively charged ions known as cations (sodium, potassium, calcium, and magnesium, etc.) or negatively charged ions known as anions (chloride, bicarbonate, phosphate, sulfate, etc.). Four major electrolytes of the human body (sodium, potassium, chloride, and bicarbonate) play important roles in human physiology, including:

- Maintaining water homeostasis of the body.
- Maintaining proper pH of the body (7.35 to 7.45).
- Maintaining optimal function of the heart.
- Participating in various physiological reactions.
- Co-factors for some enzymes.

A. Dasgupta and A. Wahed: Clinical Chemistry, Immunology and Laboratory Quality Control
DOI: http://dx.doi.org/10.1016/B978-0-12-407821-5.00005-X

It is important to drink plenty of water and take in adequate salt on a daily basis to maintain proper health. Healthy adults (age 19–50) should consume 1.5 g of sodium and 2.3 g of chloride each day or 3.8 g of salt each day to replace lost salt. The tolerable upper limit of daily salt intake is 5.8 g (5800 mg), but many Americans exceed this limit. The average daily sodium intake is 3.5–6 g (3,500–6,000 mg) per day. Processed foods contain high amounts of sodium because manufacturers add it for food preservation. For example, a can of tomato juice may contain up to 1,000 mg of sodium. Adults should consume 4.7 g of potassium each day, but many Americans do not meet this recommended potassium requirement. Potassium-rich foods include bananas, mushrooms, spinach, almonds, and a variety of other fruits and vegetables. High sodium intake may cause hypertension. The Dietary Approaches to Stopping Hypertension (DASH) eating plan recommends not more than a daily intake of 1,600 mg (1.6 g) of sodium. In general, high sodium intake increases blood pressure; replacing a high sodium diet with a diet low in sodium and high in potassium can decrease blood pressure. Sodium and potassium are freely absorbed from the gastrointestinal tract, and excess sodium is excreted by the kidneys. Potassium filtered through glomerular filtration in the kidneys is almost completely reabsorbed in the proximal tubule and is secreted in the distal tubules in exchange for sodium under the influence of aldosterone. Interestingly, African Americans excrete less urinary potassium than Caucasians even while consuming similar diets in the DASH trail. However, consuming a diet low in sodium may reduce this difference [1].

5.2 PLASMA AND URINE OSMOLALITY

Plasma osmolality is a way to measure the electrolyte balance of the body. Osmolality (measured by an osmometer in a clinical laboratory) is technically different than osmolarity, which can be calculated based on the measured sodium, urea, and glucose concentration of the plasma. Osmolality is a measure of osmoles of solutes per kilogram of a solution where osmolarity is a measure of osmoles per liter of solvent. Because one kilogram of plasma is almost one liter in volume, osmolality and osmolarity of plasma can be considered as the same for all practical purposes. Normal plasma osmolality is 275–300 milliosmoles/kg (mOsm/kg) of water while urine osmolality is 50–1,200 mOsm/kg of water. Although plasma and urine osmolality can be measured by using an osmometer, it is also calculated by the following formula (Equation 5.1):

$$\text{Plasma osmolality} = 2 \times \text{Sodium} + \text{Glucose} + \text{Urea (all concentrations in mmol/L)} \qquad (5.1)$$

Although the sodium value is expressed as mmol/L, in clinical laboratories concentrations of glucose and urea are expressed as mg/dL. Therefore the formula can be modified as follows to calculate osmolality (Equation 5.2):

$$\text{Plasma osmolality} = 2 \times [\text{Sodium in mmol/L}] \\ + [\text{Glucose in mg/dL}]/18 + [\text{BUN in mg/dL}]/2.8 \quad (5.2)$$

Here, BUN stands for blood urea nitrogen.

Although this formula is commonly used, a stricter approach to calculate plasma osmolality takes into account other osmotically active substances in plasma such as potassium, calcium, and proteins by adding 9 mOsm/kg to yield Equation 5.3:

$$\text{Plasma osmolality} = 1.86\,[\text{Sodium in mmol/L}] \\ + [\text{Glucose in mg/dL}]/18 + [\text{BUN in mg/dL}]/2.8 + 9 \\ (5.3)$$

Plasma osmolality increases with dehydration and decreases with over hydration. Plasma osmolality regulates secretion of antidiuretic hormone (ADH). Another important laboratory parameter is the osmolar gap, defined in Equation 5.4:

$$\text{Osmolar gap} = \text{Observed osmolality} - \text{Calculated osmolality} \quad (5.4)$$

If the measured osmolality is higher than the calculated osmolality then this is referred to as the osmolar gap and can be due to the presence of abnormal osmotically active substances such as overdose with ethanol, methanol, and ethylene glycol, or if fractional water content of plasma is reduced, due to hyperlipidemia or paraproteinemia. Although normal urine osmolality of random urine is relatively low, fluid restriction can raise urine osmolality to 850 mOsm/kg or higher (although within the normal range of urine osmolality). However, greater than normal urine osmolality may be seen when:

- There is reduced renal perfusion (e.g. dehydration, shock, renal artery stenosis).
- Excessive water retention without renal hypoperfusion (e.g. SIADH).
- Osmotically active substances in urine (e.g. glycosuria).

5.3 HORMONES INVOLVED IN WATER AND ELECTROLYTE BALANCE

Antidiuretic hormone (ADH) and aldosterone play important roles in the water and electrolyte balance of the human body. ADH along with oxytocin are produced in the supraoptic and paraventricular nuclei of the hypothalamus. These hormones are stored in the posterior pituitary and released in response to appropriate stimuli. ADH secretion is regulated by plasma osmolality. If plasma osmolality increases, it stimulates secretion of ADH, which acts at the collecting duct of the nephron where it causes reabsorption of only water and produces concentrated urine. In this process water is conserved in the body, and as a result, plasma osmolality should be reduced. A low serum osmolality, on the other hand, reduces secretion of ADH and more water is excreted as urine (diluted urine) and plasma osmolality is corrected. However, ADH at high concentrations causes vasoconstriction, thus raising blood pressure. Increased water retention due to ADH can result in the following conditions:

- Concentrated urine
- Increased plasma volume
- Reduced plasma osmolality.

Therefore, it is logical to assume that ADH secretion is stimulated by low plasma volume and increased plasma osmolality. In humans, urine produced during sleep is more concentrated than urine produced during waking hours. Usually urine in the morning (first void) is most concentrated. This may be partly due to less or no fluid intake during sleeping hours, but plasma ADH concentration is also higher during the night than during the day. It has been postulated that rapid eye movement (REM) sleep or dreaming sleep induces ADH secretion.

5.4 RENIN–ANGIOTENSIN–ALDOSTERONE SYSTEM

With low circulating blood volume, the juxtaglomerular apparatus of the kidney secretes renin, a peptide hormone, into the blood stream. Renin converts angiotensinogen released by the liver into angiotensin I, which is then converted into angiotensin II in the lungs by angiotensin-converting enzyme (ACE). Angiotensin II is a vasoconstrictor and also stimulates release of aldosterone from the adrenal cortex. This is defined as the "Renin–Angiotensin–Aldosterone" system. Aldosterone is a mineralocorticoid secreted from the zona glomerulosa of the adrenal cortex. It acts on the distal tubules and collecting ducts of the nephron and causes:

- Retention of water
- Retention of sodium
- Loss of potassium and hydrogen ions.

Retention of water and sodium results in increased plasma volume and blood pressure. An increase in plasma potassium is a strong stimulus for aldosterone synthesis and release. Atrial natriuretic peptide (ANP) and brain natriuretic peptide (BNP) are secreted by the right atrium and ventricles, respectively. The main stimulus for secretion of these peptides is volume overload.

5.5 DIABETES INSIPIDUS

Diabetes insipidus is an uncommon condition that occurs when the kidneys are unable to concentrate urine properly. As a result, diluted urine is produced, affecting plasma osmolality. The cause of diabetes insipidus is lack of secretion of ADH (cranial diabetes insipidus, also known as central diabetes insipidus) or is due to the inability of ADH to work at the collecting duct of the kidney (nephrogenic diabetes insipidus). Cranial diabetes insipidus is due to hypothalamic damage or pituitary damage. The major causes of such damage include the following conditions:

- Head injury
- Stroke
- Tumor
- Infections affecting the central nervous system
- Sarcoidosis
- Surgery involving the hypothalamus or pituitary.

Diabetes insipidus due to viral infection is rarely reported, but one report illustrates diabetes insipidus due to type A (sub-type: H1N1, swine flu) influenza virus infection in a 22-year-old man who produced up to 9 liters of urine per day [2]. Neuroendocrine complication following meningitis in neonates may also cause diabetes insipidus [3]. Pituitary abscess is a rare life-threatening condition that may also cause central diabetes insipidus. Autoimmune diabetes insipidus is an inflammatory non-infectious form of diabetes insipidus that is rare and is presented with antibodies to ADH secreting cells.

CASE REPORT

A 48-year-old woman with diffuse large cell lymphoma and severe hepatic involvement presented with herpes zoster infection on the right eye and was treated with acyclovir orally. When she was undergoing chemotherapy, on the ninth day she developed a fever, weakness, hypotension, pancytopenia, renal failure, and a highly elevated C-reactive protein. A diagnosis of Gram-negative sepsis was made and she was treated with intravenous antibiotic along with acyclovir, catecholamine, and hydrocortisone. Three days later she developed hypotonic polyuria (12 liters of urine per day) and a diagnosis of diabetes insipidus was made based on low urine osmolality of 153 mmol/kg and undetectable vasopressin (ADH) levels. However, a brain MRI showed no pituitary abnormality, but encephalitis was present as evidenced by hyperintensities in the area of the left lateral ventricle of the cerebrum. Analysis of cerebrospinal fluid showed herpes zoster infection. The authors concluded that central diabetes insipidus was due to herpes encephalitis in this patient. The patient responded to desmopressin (synthetic analog of vasopressin, also known as ADH) therapy [4].

Nephrogenic diabetes insipidus is due to the inability of the kidney to concentrate urine in the presence of ADH. The major causes of nephrogenic diabetes include:

- Chronic renal failure
- Polycystic kidney disease
- Hypercalcemia, hypokalemia
- Drugs such as amphotericin B, demeclocycline, lithium.

In both types of diabetic insipidus, patients usually present with diluted urine with low osmolality, but plasma osmolality should be higher than normal. These patients also experience excessive thirst and drink lots of fluid to compensate for the high urine output. Even if a patient is not allowed to drink fluid, urine still remains diluted with a possibility of dehydration. In contrast, in a normal healthy individual fluid deprivation results in concentrated urine. This observation is the basis of the water deprivation test to establish the presence of diabetes insipidus in a patient. In order to differentiate cranial diabetes insipidus from nephrogenic diabetes insipidus, intranasal vasopressin is administered. If urine osmolality increases then the diagnosis is cranial diabetes insipidus, but if urine is still dilute with no change in urine osmolality, then the diagnosis is nephrogenic diabetes insipidus. The congenital form of nephrogenic diabetes is a rare disease and most commonly inherited in an X-linked manner with mutations of the arginine vasopressin receptor type 2 (AVPR2) [5].

5.6 THE SYNDROME OF INAPPROPRIATE ANTIDIURETIC HORMONE SECRETION (SIADH)

The syndrome of inappropriate antidiuretic hormone secretion (SIADH, also known as Schwartz–Bartter syndrome) is due to excessive and inappropriate release of antidiuretic hormone (ADH). Usually reduction of plasma osmolality causes reduction of ADH secretion, but in SIADH reduced plasma osmolality does not inhibit ADH release from the pituitary gland, causing water overload. The main clinical features of SIADH include:

- Hyponatremia (plasma sodium <131 mmol/L)
- Decreased plasma osmolality (<275 mOsm/kg)
- Urine osmolality >100 mOsm/kg) and high urinary sodium (>20 mmol/L)
- No edema.

Various causes of SIADH are listed in Table 5.1.

Table 5.1 Causes of SIADH*

Type of Disease	Specific Disease/Comments
Pulmonary diseases	Pneumonia, pneumothorax, acute respiratory failure, bronchial asthma, atelectasis, tuberculosis.
Neurological	Meningitis, encephalitis, stroke, brain tumor infection.
Malignancies	Lung cancer especially small cell carcinoma, head and neck cancer, pancreatic cancer.
Hereditary	Two genetic variants, one affecting renal vasopressin receptor and another affecting osmolality sensing in hypothalamus have been reported.
Hormone therapy	Use of desmopressin or oxytocin can cause SIADH.
Drugs	Cyclophosphamide, carbamazepine, valproic acid, amitriptyline, SSRI, monoamine oxidase inhibitors and certain chemotherapeutic agents may also cause SIADH.

SIADH: Syndrome of Inappropriate Antidiuretic Hormone Secretion.

5.7 HYPONATREMIA, SICK CELL SYNDROME, AND HYPERNATREMIA

Hyponatremia can be either absolute hyponatremia or dilutional hyponatremia, although in a clinical setting, dilutional hyponatremia is encountered more commonly than absolute hyponatremia. In absolute hyponatremia, total sodium content of the body is low. The patient is hypovolemic, which results in activation of the renin–angiotensin system, causing secondary hyperaldosteronism and also increased levels of ADH. In dilutional hyponatremia total body sodium is not low, rather, total body sodium may be increased. The patient is volume overloaded with resultant dilution of sodium levels. Examples of such conditions include congestive heart failure, renal failure, nephrotic syndrome, and cirrhosis of the liver. Although hyponatremia is defined as any sodium value less than reference range (135 mEq/L), usually clinical features such as confusion, restlessness leading to drowsiness, myoclonic jerks, convulsions, and coma are observed at much lower sodium levels. Hyponatremia is common among hospitalized patients, and affects up to 30% of all patients [6]. However, a sodium level below 120 mEq/L is associated with poor prognosis and even a fatal outcome [7]. Major types of hyponatremia include:

- Absolute hyponatremia (patient is hypovolemic) related to loss of sodium through the gastrointestinal tract or loss through the kidneys due to kidney diseases (pyelonephritis, polycystic disease, interstitial disease) or through the kidneys due to glycosuria or therapy with diuretics or less retention of sodium by the kidney due to adrenocortical insufficiency.

■ Dilutional hyponatremia (patient hypervolemic). This condition is related to SIADH or conditions like congestive heart failure, renal failure, nephrotic syndrome, and cirrhosis of the liver.

■ Pseudohyponatremia as seen in patients with hyperlipidemia and hypergammaglobulinemia (also known as factitious hyponatremia).

Sick cell syndrome is defined as hyponatremia seen in individuals with acute or chronic illness where cell membranes leak, allowing solutes normally inside the cell to escape into extracellular fluid. Therefore, leaking of osmotically active solutes causes water to move from intracellular fluid to extracellular fluid, causing dilution of plasma sodium and consequently hyponatremia. Sick cell hyponatremia also produces a positive osmolar gap. Sick patients also produce high levels of ADH, which causes water retention, causing hyponatremia.

CASE REPORT

A 36-year-old man was hospitalized with 3 days history of malaise, drowsiness, and jaundice. He had a history of agoraphobia and alcohol abuse. On admission there was no meningismus, focal neurological signs, or liver failure. However, later the patient became unconscious and developed hypotension and grand mal seizure and was transferred to the ICU. His serum sodium level was 101 mEq/L and potassium was 3.6 mmol/L, but all liver function tests were abnormally high. His serum osmolality was 259 mOsm/kg, but calculated osmolality was 214 mOsm/kg with an osmolar gap of +35 mOsm/kg. His serum albumin was 2.8 mg/dL. The patient deteriorated despite aggressive therapy and later died. The patient suffered from critical illness with multi-organ failure. Standard causes of hyponatremia were ruled out, and he showed a markedly positive osmolar gap with severe hyponatremia due to sick cell syndrome [8].

Hypernatremia is due to elevated serum sodium levels (above 150 mEq/L). Symptoms of hypernatremia are usually neurological due to intraneuronal loss of water to extracellular fluid. Patients exhibit features of lethargy, drowsiness, and eventually become comatose. Hypernatremia may be hypovolumic or hypervolumic. The most common cause of hypovolemic hypernatremia is dehydration, which may be due to decreased water intake or excessive water loss through the skin (heavy sweating), kidney, or gastrointestinal tract (diarrhea). Patients usually present with concentrated urine (osmolality over 800 mOsm/kg) and low urinary sodium (<20 mmol/L). Hypervolemic hypernatremia may be observed in hospitalized patients receiving sodium bicarbonate or hypertonic saline. Hyperaldosteronism, Cushing's syndrome, and Conn's disease may also cause hypervolemic hypernatremia.

5.8 HYPOKALEMIA AND HYPERKALEMIA

Hypokalemia is defined as a serum potassium concentration <3.5 mEq/L, which may be caused by loss of potassium or redistribution of extracellular potassium into the intracellular compartment. Hypokalemia may occur due to the following:

- Loss of potassium from the gastrointestinal tract due to vomiting, diarrhea, and active secretion of potassium from villous adenoma of rectum.
- Loss of potassium from the kidneys due to diuretic therapy, and glucocorticoid and mineralocorticoid excess. Increased levels of lysozyme (seen in monocytic leukemia) may also cause renal loss of potassium. Bartter's, Liddle and Gitelman syndromes are rare inherited disorders due to mutations in the ion transport proteins of the renal tubules that may cause hypokalemia.
- Intracellular shifts due to drug therapy with beta-2 agonists (salbutamol), which drives potassium into the cell, or due to alkalosis (hydrogen ions move out of the cell in exchange with potassium), or insulin therapy or familial periodic paralysis and hypothermia.

Clinically, patients with hypokalemia present with muscle weakness, areflexia, paralytic ileus, and cardiac arrhythmias. Electrocardiogram findings include prolonged PR interval, flat T, and tall U.

CASE REPORT

A 69-year-old white man with a history of high-grade prostate carcinoma and widely metastatic adenocarcinoma presented to the hospital with metabolic alkalosis (arterial blood pH of 7.61, pO2 of 45, and pCO2 of 48), hypokalemia (potassium 2.1 mEq/L), and hypertension secondary to ectopic ACTH (adrenocorticotropic hormone) and CRH (corticotropin-releasing hormone) secretion. His serum cortisol was also markedly elevated (135 μg/dL) along with ACTH (1,387 pg/dL) and CRH (69 pg/dL). As expected, his urinary cortisol was also elevated (16,267 μg/24 h). An abdominal CT scan and MRI study showed multiple small liver lesions and multiple thoracic and lumbar intensities consistent with diffuse metastatic disease. The severe metabolic alkalosis secondary to glucocorticoid-induced excessive mineralocorticoid activity and hypokalemia were treated with potassium supplements, spironolactone, and ketoconazole. This patient had Cushing's syndrome, most likely as a result of ectopic ACTH and CRH secretion from metastatic adenocarcinoma of the prostate gland [9].

Most of potassium of the body resides intracellularly. Hyperkalemia presents as elevated serum or plasma potassium levels; a common cause is hemolysis of blood, where potassium leaks from red blood cells into serum, thus artificially increasing potassium levels.

Causes of hyperkalemia include:

- Lysis of cells: in vivo hemolysis, rhabdomyolysis, and tumor lysis.
- Intracellular shift. In acidosis, intracellular potassium is exchanged with extracellular hydrogen ions, causing hyperkalemia. Thus hyperkalemia typically accompanies metabolic acidosis. An exception is renal tubular acidosis (RTA) types I and II where acidosis without hyperkalemia is observed. Acute digitalis toxicity (therapy with digoxin or digitoxin) may cause hyperkalemia (please note digitalis toxicity is precipitated in the hypokalemic state).
- Renal failure.
- Pseudohyperkalemia. Although pseudohyperkalemia or artificial hyperkalemia is most commonly seen secondary to red cell hemolysis, it is also seen in patients with thrombocytosis and rarely in patients with familial pseudohypokalemia. Patients with highly elevated white blood cell counts, such as patients with chronic lymphocytic leukemia (CLL), may also show pseudohyperkalemia. Diagnosis of pseudohyperkalemia can be made from observation of higher serum potassium than plasma potassium (serum potassium exceeds plasma potassium by 0.4 mEq/L provided both specimens are collected carefully and analyzed within 1 h), or measuring potassium in whole blood (using a blood gas machine) where whole blood potassium is within normal range.

Clinical features of hyperkalemia include muscle weakness, cardiac arrhythmias, and cardiac arrest. EKG findings include flattened P, prolonged PR interval, wide QRS complex, and tall T waves. Drugs that may cause hyperkalemia are listed in Table 5.2.

Table 5.2 Drugs that may Cause Hyperkalemia

Potassium supplement and salt substitute

Beta-blockers

Digoxin and digitoxin (acute intoxication)

Potassium sparing diuretics (spironolactone and related drugs)

NSAIDs (non-steroidal antiinflammatory drugs)

ACE inhibitors

Angiotensin II-blockers

Trimethoprim/sulfamethoxazole combination (Bactrim)

Immunosuppressants (cyclosporine and tacrolimus)

Heparin

CASE REPORT

A 51-year-old male patient with CLL demonstrated high plasma potassium of 6.8 mEq/L, but no abnormality was observed in his electrocardiogram. He showed normal creatinine (1.1 mg/dL), low hemoglobin (7.3 g/dL), and high white blood cell count (273.9 k/microliter). He was treated in the emergency room with a presumed diagnosis of hyperkalemia with calcium gluconate, sodium bicarbonate, albuterol aerosol, glucose, insulin, and Kayexalate. His potassium remained high for the next two days (in the range of low 6 s), but his whole blood potassium was normal (2.7 mEq/L). Based on these observations, diagnosis of pseudohyperkalemia was established. Interestingly, his plasma potassium was increased to 9.0 mEq/L, but his whole blood potassium was still 3.6 mEq/L [10].

5.9 INTRODUCTION TO ACID–BASE BALANCE

In general, an acid is defined as a compound that can donate hydrogen ions, and a base is a compound that can accept hydrogen ions. In order to determine if a solution is acidic or basic, the pH scale is used, which is the abbreviation of the power of hydrogen ions; pH is equal to the negative log of hydrogen ion concentration in solution. Neutral pH is 7.0. If a solution is acidic, pH is below 7.0, and basic if above 7.0. Therefore, a physiological pH of 7.4 is slightly basic. Concentration of hydrogen ions that are present in both the extracellular and intracellular compartments of the human body are tightly controlled. Although the normal human diet is almost at a neutral pH and contains very low amounts of acid, the human body produces about 50–100 mEq of acid in a day, principally from the cellular metabolism of proteins, carbohydrates, and fats; this generates sulfuric acid, phosphoric acid, and other acids. Although excess base is excreted in feces, excess acid generated in the body must be neutralized or excreted in order to tightly control near normal pH of the blood (arterial blood 7.35–7.45 and venous blood 7.32–7.48). Carbonic acid (H_2CO_3) is generated in the human body due to dissolution of carbon dioxide in water present in the blood (Equation 5.5):

$$CO_2 + H_2O = H_2CO_3 = H^+ + HCO_3^- \qquad (5.5)$$

The hydrogen ion concentration of human blood can be calculated from the Henderson–Hasselbalch equation (Equation 5.6):

$$pH = pKa + \log[\text{salt}]/[\text{acid}] \qquad (5.6)$$

Here, salt is the concentration of bicarbonate [HCO_3^-] and the concentration of acid is the concentration of carbonic acid, which can be calculated from the measured partial pressure of carbon dioxide. The value of pKa is 6.1, which is the dissociation constant of carbonic acid at physiological temperature. The concentration of carbonic acid can be calculated by multiplying the partial pressure of carbon dioxide (pCO_2) by 0.03. Therefore the Henderson–Hasselbalch equation can be expressed as Equation 5.7:

$$pH = 6.1 + \log\frac{[HCO_3^-]}{0.03 \times pCO_2} \tag{5.7}$$

The body has three mechanisms to maintain acid–base homeostasis:

- A physiological buffer present in the body that consists of a bicarbonate–carbonic acid buffer system, phosphate in the bone, and intracellular proteins.
- Respiratory compensation, where the lungs can excrete more carbon dioxide or less depending on the acid–base status of the body.
- The kidneys can also correct acid–base balance of the human body if other mechanisms are ineffective.

Respiratory compensation to correct acid–base balance is the first compensatory mechanism. It is effective immediately, but it may take a longer time for initiation of the renal compensatory mechanism. At the collecting duct, sodium is retained in exchange for either potassium or hydrogen ions, and if excess acid is present, more hydrogen ions should be excreted by the kidney to balance acid–base homeostasis. In the presence of excess acid (acidosis), kidneys excrete hydrogen ions and retain bicarbonate, while during alkalosis, kidneys excrete bicarbonate and retain hydrogen ions. However, when there is excess acid, hydrogen ions may also move into the cells in exchange for potassium moving out of the cell. As a result, metabolic acidosis usually causes hyperkalemia. Concurrently, the bicarbonate concentration is reduced because hydrogen ions react with bicarbonate ions to produce carbonic acid. The kidneys need to reabsorb more of the filtered bicarbonate, which takes place at the proximal tubule.

5.10 DIAGNOSTIC APPROACH TO ACID–BASE DISTURBANCE

Major acid–base disturbances can be divided into four categories: metabolic acidosis, respiratory acidosis, metabolic alkalosis, and respiratory alkalosis. In general, metabolic acidosis or alkalosis is related to abnormalities in regulation of bicarbonate and other buffers in blood, while abnormal removal of carbon dioxide may cause respiratory acidosis or alkalosis. Both states may also co-exist. However, it is important to know normal values of certain parameters measured in blood for diagnosis of acid–base disturbances:

- Normal pH of arterial blood is 7.35–7.45.
- Normal pCO_2 is 35–45 mmHg.

- Normal bicarbonate level is 23−25 mmol/L.
- Normal chloride level is 95−105 mmol/L.

The first question is whether the pH value is higher or lower than normal. If the pH is lower than normal, then it is acidosis, and if the pH is higher than normal, the diagnosis of alkalosis can be made. If the diagnosis is acidosis, then the next question to ask is whether the acidosis is metabolic or respiratory in nature. Similarly, if the pH is above normal, the question is whether the alkalosis is metabolic or respiratory in nature. In general, if the direction of change from normal pH is the same direction for change of pCO_2 and bicarbonate, then the disturbance is metabolic in nature, and if the direction of change from normal pH is in the opposite direction of change for pCO_2 and bicarbonate, then the disturbance is respiratory. Therefore four different scenarios are possible:

- Metabolic acidosis, where the value of pH is decreased along with decreases in the values of pCO_2 and bicarbonate (both values below normal range).
- Respiratory acidosis, where the value of pH is decreased but values of both pCO_2 and bicarbonate are increased from normal values.
- Metabolic alkalosis, where the value of pH is increased along with values of both pCO_2 and bicarbonate (both values above reference range).
- Respiratory alkalosis, where the value of pH is increased, but values of both pCO_2 and bicarbonate are decreased.

5.10.1 Metabolic acidosis

Metabolic acidosis may occur with an increased anion gap (high) or normal anion gap. Anion gap is defined as the difference between measured cations (sodium and potassium) and anions (chloride and bicarbonate) in serum. Sometimes concentration of potassium is omitted because it is low compared to sodium ion concentration in serum (Equation 5.8):

$$\text{Anion gap} = [\text{sodium}] - ([\text{chloride}] + [\text{bicarbonate}]) \qquad (5.8)$$

The normal value is 8−12 mmol/L (mEq/L).

In metabolic acidosis bicarbonate should decrease, resulting in increased anion gap metabolic acidosis. If the chloride level increases, then even with a decline in bicarbonate, the anion gap may remain normal. This is normal anion gap metabolic acidosis. Thus, normal anion gap metabolic acidosis is also referred to as hyperchloremic metabolic acidosis. Causes of normal anion gap metabolic acidosis include loss of bicarbonate buffer from the gastrointestinal tract (chronic diarrhea, pancreatic fistula, and sigmoidostomy), or renal loss of bicarbonate due to kidney disorders such as renal tubular acidosis and renal failure. Causes of increased anion gap metabolic acidosis can be remembered by the mnemonic MUDPILES (M for methanol, U for uremia, D for diabetic ketoacidosis, P for paraldehyde, I for isopropanol, L for lactic acidosis, E for ethylene glycol and S for salicylate). In addition, alcohol abuse and other toxins such as formaldehyde, toluene, and certain drug overdoses may also cause metabolic acidosis with an increased anion gap.

In general, if any other metabolic disturbance co-exists with increased anion gap metabolic acidosis, this can be diagnosed from the corrected bicarbonate level (Equation 5.9):

$$\text{Corrected bicarbonate} = \text{measured value of bicarbonate} + (\text{anion gap} - 12)$$

$$(5.9)$$

If corrected bicarbonate is less than 24 mmol/L, then there exists additional metabolic acidosis, and if corrected bicarbonate is greater than 24 mmol/L, then there exists additional metabolic alkalosis.

Winter's formula is used to assess whether there exists adequate respiratory compensation with metabolic disturbance (Equation 5.10):

$$\text{Winter's formula: Expected } pCO_2 = [1.5 \times \text{Bicarbonate}] + (8 \pm 2) \quad (5.10)$$

If pCO_2 is as expected by Winter's formula, then there is adequate respiratory compensation, but if pCO_2 is less than expected, then additional respiratory alkalosis may be present. However, if pCO_2 is more than expected, then there is additional respiratory acidosis.

5.10.2 Metabolic alkalosis

Metabolic alkalosis is related to the loss of hydrogen ions or is due to the gain of bicarbonate or alkali:

- Loss of acid, from the gastrointestinal tract (GIT) (e.g. vomiting, diarrhea).
- Loss of acid from kidneys (e.g. glucocorticoid or mineralocorticoid excess, diuretics).
- Gain of alkali (e.g. "milk-alkali syndrome," also called Burnett's syndrome, caused by excess intake of milk and alkali leading to hypercalcemia).

In general, the body attempts to compensate metabolic acidosis by using respiratory compensation mechanisms where enhanced carbon dioxide elimination can be achieved by hyperventilation (Kussmaul respiration), but this process may lead to respiratory alkalosis. In the case of metabolic alkalosis, depression of the respiratory mechanism causes retention of carbon dioxide to compensate for metabolic alkalosis. However, respiratory response to metabolic alkalosis may be erratic. In addition, during metabolic alkalosis, kidneys try to compensate increased pH by decreasing excretion of hydrogen ion and sodium ions. When an adequate compensation mechanism is absent, mixed acidosis may occur.

5.10.3 Respiratory acidosis

Respiratory acidosis is due to carbon dioxide retention due to type II respiratory failure. Causes include:

- CNS disorders which damage or suppress the respiratory center (e.g. stroke, tumor, drugs, alcohol).
- Neuropathy or myopathy affecting muscles of ventilation (e.g. Guillain–Barré syndrome, myasthenia gravis).
- Reduced movement of chest wall (e.g. flail chest, severe obesity (Pickwickian syndrome)).
- Airway obstruction (e.g. severe acute asthma).

5.10.4 Respiratory alkalosis

Major causes of respiratory alkalosis include:

- CNS stimulation (e.g. drugs such as aspirin, ketamine).
- Hysteria.
- Bronchial asthma (early stage).

If the acid–base disturbance is related to respiratory disturbance, then it is important to establish whether such disturbance is acute or chronic. In acute respiratory disturbance, for any 10 mmHg pCO_2 change (assuming a normal value of 40 mmHg), the change in pH is 0.08 units. In chronic respiratory disturbance, for any 10 mmHg pCO_2 change, the change in pH is 0.03 units.

5.11 SHORT CASES: ACID–BASE DISTURBANCES

CASE 1

A patient overdosed on aspirin in an attempted suicide and was brought to the ER. Her arterial blood pH was 7.57, pCO_2 was 20 mmHg, and bicarbonate was 22 mmol/L. Because the pH was above the normal range, the patient presented with alkalosis. In addition, both pCO_2 and bicarbonate were also decreased, but these two values were opposite in direction of pH (which was increased). Therefore, the patient had respiratory alkalosis. In addition to establishing the diagnosis of respiratory alkalosis, it was also important to establish if this was an acute or chronic respiratory disturbance. The decrease of pCO_2 was 20 (normal value is 40 mmHg). Multiplying 20 by 0.08 (in acute respiratory disturbance, for any 10 mmHg pCO_2 change the change in pH is 0.08 units) yields a value of 0.16. The increase of pH was 0.18 (assuming a normal pH value of 7.4). This was comparable to 0.16, and the patient showed acute respiratory alkalosis as expected with acute aspirin overdose.

CASE 2

A patient with myasthenia gravis admitted to the hospital showed arterial blood gas pH of 7.13, pCO_2 of 80 mmHg, and bicarbonate of 26 mmol/L. Because blood pH was below the reference range, the patient suffered from acidosis. Moreover, both pCO_2 and bicarbonate were increased, but pH was decreased (change in opposite direction), indicating that the patient had respiratory acidosis. Moreover, pCO_2 was 80, and assuming (for purposes of calculation) 40 was normal, the change was 40, which when multiplied by 0.08 was equal to 0.32. The patient's pH was 7.13, which was lower by 0.27 from a normal value of 7.4. Therefore, the patient had acute respiratory disturbance (respiratory acidosis).

CASE 3

An adult pregnant female with persistent vomiting was brought to the ER and her arterial blood pH was 7.62, pCO_2 was 47 mmHg, and bicarbonate was 38 mmol/L. Because pH was increased from the normal value, the patient presented with alkalosis. In addition, because both pCO_2 and bicarbonate were increased along with pH (all changes in the same direction), the patient had metabolic alkalosis. Using Winter's formula, expected pCO_2 should be $1.5 \times$ bicarbonate) $+ 8 \pm 2$ or 65 ± 2 (i.e. between 63 and 67). If pCO_2 was as expected by Winter's formula, then adequate respiratory compensation was present, but this patient showed a pCO_2 of 47, indicating that in addition to metabolic alkalosis, additional respiratory alkalosis was also present.

CASE 4

An adult male with insulin-dependent diabetes mellitus (IDDM) was admitted with altered mental status and had the following values: pH 7.22, pCO_2 25 mmHg, bicarbonate 10 mmol/L, sodium 130 mmol/L, and chloride 80 mmol/L. Because pH was lower than normal he had acidosis. In addition, all three parameters (pH, pCO_2, and bicarbonate) were decreased (changed in the same direction), establishing the diagnosis as metabolic acidosis. The anion gap of the patient was 40 (elevated). Therefore, the patient presented with metabolic acidosis with increased anion gap. The corrected bicarbonate of the patient was 38 (using Equation 5.11):

Because the corrected bicarbonate was higher than 24, the patient had additional metabolic alkalosis (corrected bicarbonate < 24 mmol/L, additional metabolic acidosis present; corrected bicarbonate > 24 mmol/L, additional metabolic alkalosis present). Moreover, using Winter's formula, the expected pCO_2 was $[1.5 \times \text{Bicarbonate}] + (8 \pm 2)$; the expected pCO_2 should be between 21 and 25. Because measured pCO_2 was 25, adequate respiratory compensation was present in the patient. In summary, this patient had increased anion gap metabolic acidosis with additional metabolic alkalosis but adequate respiratory compensation.

$$\text{Corrected bicarbonate} = \text{measured value of bicarbonate} + (\text{anion gap} - 12)$$

$$(5.11)$$

KEY POINTS

- Plasma osmolality = $2 \times$ [Sodium in mmol/L] + [Glucose mg/dL]/18 + [BUN mg/dL]/2.8 (BUN: blood urea nitrogen; Osmolar gap = Observed osmolality − Calculated osmolality).

- Higher osmolar gap can be due to the presence of abnormal osmotically active substances such as ethanol, methanol, and ethylene glycol (overdosed patients), or, if fractional water content of plasma is reduced, can be due to hyperlipidemia or paraproteinemia.
- Diabetes insipidus is due to lack of secretion of ADH (cranial diabetes insipidus, also known as central diabetes insipidus) or due to the inability of ADH to work at the collecting duct of the kidney (nephrogenic diabetes insipidus).
- The main clinical features of SIADH (syndrome of inappropriate antidiuretic hormone secretion) include: hyponatremia (plasma sodium <131 mmol/L), decreased plasma osmolality (<275 mOsm/kg), urine osmolality >100 mOsm/kg, and high urinary sodium (>20 mmol/L) with no edema.
- Major categories of hyponatremia include: absolute hyponatremia (patient is hypovolemic) due to loss of sodium through gastrointestinal tract and kidneys, and dilutional hyponatremia (patient hypervolemic) related to SIADH, volume overload state, and pseudohyponatremia.
- Hypokalemia may occur due to loss of potassium from the gastrointestinal tract and intracellular shift.
- Causes of hyperkalemia include: lysis of cells, intracellular shift, renal failure, or pseudohyperkalemia.
- Metabolic acidosis: The value of pH is decreased along with decreases in the values of pCO_2 and bicarbonate (both values below normal range). May be normal anion gap or increased anion gap where anion gap = [sodium] − ([chloride] + [bicarbonate]) (normal value is 8−12 mmol/L (mEq/L)). Causes of normal anion gap metabolic acidosis include loss of bicarbonate buffer from the gastrointestinal tract (chronic diarrhea, pancreatic fistula, and sigmoidostomy), or renal loss of bicarbonate due to kidney disorders such as renal tubular acidosis and renal failure. Causes of increased anion gap metabolic acidosis can be remembered by the mnemonic MUDPILES (M for methanol, U for uremia, D for diabetic ketoacidosis, P for paraldehyde, I for isopropanol, L for lactic acidosis, E for ethylene glycol, and S for salicylate).
- Metabolic alkalosis: The pH value is increased along with values of both pCO_2 and bicarbonate (both values above reference range). Metabolic alkalosis is related to loss of hydrogen ion or is due to gain of bicarbonate or alkali for any of the following reasons: loss of acid from gastrointestinal tract issues (vomiting, diarrhea), loss of acid from kidneys (glucocorticoid or mineralocorticoid excess, diuretics), or gain of alkali (e.g. "milk-alkali syndrome," also called Burnett's syndrome, caused by excess intake of milk and alkali leading to hypercalcemia).
- Winter's formula is used to assess whether there exists adequate respiratory compensation with metabolic disturbance. Winter's formula: expected $pCO_2 = [1.5 \times$ Bicarbonate$] + (8 \pm 2)$. If pCO_2 is as expected by Winter's formula, then there is adequate respiratory compensation, but if pCO_2 is less than expected, then additional respiratory alkalosis may be present. However, if pCO_2 is more than expected, then there is additional respiratory acidosis.
- Respiratory acidosis: The value of pH is decreased but values of both pCO_2 and bicarbonate are increased from normal values. Respiratory acidosis is due to carbon

dioxide retention due to type II respiratory failure. Causes include: CNS disorders that damage or suppress the respiratory center (e.g. stroke, tumor, drugs, alcohol, neuropathy), or myopathy affecting muscles of ventilation (e.g. Guillain—Barré syndrome, myasthenia gravis), reduced movement of chest wall (e.g. flail chest, severe obesity (Pickwickian syndrome)), or airway obstruction (e.g. severe acute asthma).

- Respiratory alkalosis: The value of pH is increased, but values of both pCO_2 and bicarbonate are decreased. Major causes of respiratory alkalosis include: CNS stimulation due to drugs such as aspirin, ketamine, hysteria, or bronchial asthma (early stage). If the acid—base disturbance is related to respiratory disturbance, then it is important to establish whether such disturbance is acute or chronic. In acute respiratory disturbance, for any 10 mmHg pCO_2 change (assuming a normal value of 40 mmHg), the change in pH is 0.08 units. In chronic respiratory disturbance, for any 10 mmHg pCO_2 change, the change in pH is 0.03 units.

REFERENCES

[1] Turban S, Thompson CB, Parekh RS, Appel LJ. Effects of sodium intake and diet on racial difference in urinary potassium excretion: results from the Dietary Approaches to Stop Hypertension (DASH) sodium trial. Am J Kidney Dis 2013;61:88—95.

[2] Kobayashi T, Miwa T, Odawara M. A case of central diabetes insipidus following probable type A/H1N1 influenza infection. Endocr J 2011;58:913—8.

[3] Cohen C, Rice EN, Thomas DE, Carpenter TO. Diabetes insipidus as a hallmark neuroendocrine complication of neonatal meningitis. Curr Opin Pediatr 1998;10:449—52.

[4] Sceinpflug K, Schalk E, Reschke K, Franke A, et al. Diabetes insipidus due to herpes encephalitis in a patient with diffuse large cell lymphoma: a case report. Exp Clin Endocrinol Diabetes 2006;114:31—4.

[5] Devuyst O. Physiology and diagnosis of nephrogenic diabetes insipidus. Ann Endocrinol (Paris) 2012;73:128—9.

[6] Pillai B, Unnikrishnan AG, Pavithran P. Syndrome of inappropriate antidiuretic hormone secretion: revisiting a classical endocrine disorder. Indian J Endocrinol 2011;15(Suppl. 3):S208—15.

[7] Gill GV, Osypiw JC, Shearer ES, English PJ, et al. Critical illness with hyponatremia and impaired cell membrane integrity-the sick cell syndrome revisited. Clin Biochem 2005;38:1045—8.

[8] Richman T, Garmany R, Doherty T, Benson D, et al. Hypokalemia, metabolic alkalosis, and hypertension: Cushing's syndrome in a patient with metastatic prostate. Am J Kidney Dis 2001;37:838—46.

[9] Perazella MA. Drug induced hyperkalemia: old culprits and new offenders. Am J Med 2000;109:307—14.

[10] Rifkin S. Pseudohyperkalemia in a patient with chronic lymphocytic leukemia. Int J Nephrol 2011; [Article ID 759749].

CHAPTER 6

Lipid Metabolism and Disorders

6.1 LIPIDS AND LIPOPROTEINS

Along with proteins, carbohydrates, and nucleic acids, various lipids are also vital building blocks of life. However, in contrast to some proteins, carbohydrates, and nucleic acids, all lipids are insoluble in water. This is essential because lipids are integral structural parts of cell membranes in animals and human. Because lipids are insoluble in water, when transported in blood these molecules must combine with water-soluble proteins to form lipoproteins. Carbohydrates and lipids (especially fatty acids) are major sources of energy. Steroids are also lipids, and many steroids also act as hormones. Major lipids are listed below:

- Triglycerides: Formed when a glycerol molecule that has three hydroxy groups is esterified with three fatty acid molecules. There are two main sources of triglycerides, exogenous and endogenous. Exogenous triglycerides refer to dietary triglycerides, which are the main lipids in diet.
- Fatty acids: These molecules are an integral part of the triglyceride molecule, but a small amount may exist in the circulation. Metabolism of fatty acids is a major energy source of the body.
- Phospholipids: Integral building blocks of cell membrane where two hydroxyl groups are esterified with a fatty acid but the third hydroxyl group is esterified with a phosphorus-containing ester.
- Cholesterol: An integral component of cell membranes and also acts as a precursor of steroid hormones. In contrast to the structures of triglycerides and phospholipids, cholesterol has a four-membered ring structure with a side chain containing a hydroxyl group. In circulation, most cholesterol molecules exist as cholesterol ester, where the hydroxyl group is esterified with a fatty acid.

In addition to these lipids that are also found in circulation, there are some specific lipids known as sphingolipids found in cell membranes, especially in the central nervous system and gray matter of the brain. Sphingolipids are formed

CONTENTS

A. Dasgupta and A. Wahed: Clinical Chemistry, Immunology and Laboratory Quality Control
DOI: http://dx.doi.org/10.1016/B978-0-12-407821-5.00006-1

when amino alcohol sphingosine is esterified with fatty acids. When sphingosine is bound to one fatty acid molecule containing 18 or more carbons, it is called ceramide. When a ceramide binds with a phosphocholine it forms sphingomyelin, which is also found in cell membranes. Sphingolipids are complex molecules that play an important role in communication between cells. These molecules may accumulate in certain lipid disorders.

When lipoproteins are devoid of lipids, they are called apolipoproteins. Apolipoproteins can be classified under several groups:

- Apolipoprotein A (Apo A): Consist of Apo AI and Apo AII.
- Apolipoprotein B (Apo B): Most abundant is large Apo B known as Apo B-100, while the less abundant is a smaller particle known as Apo B-48.
- Apolipoprotein C (Apo C): Three forms are found, Apo CI, Apo CII and Apo CIII.
- Apolipoprotein D (Apo D).
- Apolipoprotein E (Apo E).

Characteristics of various apolipoproteins are summarized in Table 6.1.

6.2 CLASSES OF LIPOPROTEINS

Lipoproteins are classified based on their density following ultracentrifugation of serum, preferably overnight. In general, lipids are lighter than water. Therefore, as the protein content of the lipoprotein increases, the particle becomes denser. The gold standard for separation and analysis of plasma lipoprotein fractions is

Table 6.1 Characteristics of Various Apolipoproteins

Apolipoprotein	Characteristics
AI	Activates LCAT; found in HDL and chylomicron.
Apo AII	Found only in HDL.
Apo AIV	Activates LCAT and found in HDL.
Apo B-100	LDL receptors recognize Apo B-100 and remove cholesterol from circulation; found in LDL, IDL and VLDL.
Apo B-48	Smaller particle than Apo B-100; mostly found in chylomicron but may also be associated with LDL.
Apo CI	Activated LCAT; found in chylomicron, VLDL and HDL.
Apo CII	Cofactor for lipoprotein lipase; found in chylomicrons, VLDL and HDL.
Apo CIII	Inhibits activation of lipoprotein lipase (opposite action of Apo II); found in chylomicrons, VLDL and HDL.
Apo E	Facilitates uptake of chylomicrons remnant and IDL; found in chylomicrons, VLDL and HDL.

LCAT: lecithin cholesterol acyltransferase. HDL: High density lipoprotein; LDL: low density lipoprotein; IDL: intermediate density lipoprotein; VLDL: very low density lipoprotein.

"ultracentrifugation"[1]. Then, after isolation and quantification of individual fractions, plasma cholesterol, triglycerides, and apolipoprotein can be measured. Usually low protein-containing lipoproteins such as VLDL (very low density lipoprotein) stay at the top of the specimen while other denser lipoproteins are found in various other fractions. In reality, a lipid ultracentrifugation test is offered in relatively few reference laboratories. The following major lipoproteins are found in plasma:

- Chylomicrons: Lightest fraction containing approximately 2% lipoprotein and mostly lipids, especially triglycerides. This fraction is absent in fasting specimens (preferred specimen for lipid analysis) unless the patient is suffering from a lipid disorder.
- Very low density lipoprotein (VLDL): Denser than chylomicron but lighter than LDL, this fraction contains 4−10% proteins and the rest lipids, most notably triglycerides (45−60%).
- Intermediate density lipoprotein (IDL): This fraction is lighter than LDL and contains approximately 15% proteins and the rest lipids. This fraction is usually a transient fraction and is absent in fasting specimens, except for certain lipid disorders.
- Low density lipoprotein (LDL): This fraction is denser than VLDL but lighter then HDL and contains approximately 25% proteins and the rest lipids, most commonly esterified cholesterol (approximately 50%).
- High density lipoprotein (HDL): Most dense fraction containing approximately 50% protein and 50% lipids.

Characteristics of various lipoprotein fractions are summarized in Table 6.2.

Table 6.2 Characteristics of Various Lipoproteins

Lipoproteins	Lipid: Protein Ratio	Major Lipids
Chylomicron (lowest density)	99:1	Triglycerides (86%), coming from diet.
VLDL (higher than chylomicron)	90:10	Triglycerides (55%), endogenously synthesized.
IDL (higher density than VLDL)	85:15	Contain less triglycerides and more cholesterol ester than VLDL, a transitory particle between VLDL and HDL.
LDL (higher density than IDL)	80:20	Cholesterol-rich lipoprotein and higher levels are associated with higher risk for cardiovascular diseases.
HDL	50:50	Highest content of phospholipid among all lipoprotein particles. HDL removes cellular cholesterol from peripheral cells to liver for excretion by reverse cholesterol transport pathway.

6.3 LIPID METABOLISM

There are two sources of lipids in the human body: exogenous lipids from diet and endogenous lipids that are synthesized mostly in the liver. Dietary triglycerides are broken down into fatty acids, glycerol, and monoglycerides in the small intestine. In the intestinal epithelial cell, triglycerides are re-synthesized and are then incorporated into chylomicrons and finally enter into the systematic circulation. Chylomicrons contain very small amounts of lipoproteins. In the circulation, lipoprotein lipase breaks down the triglyceride component of chylomicron into glycerol and free fatty acids. Lipoprotein lipase is found in the capillary endothelium of adipose tissue, and skeletal and cardiac muscle. Apo CII, which is present in chylomicron, plays an important role in activating lipoprotein lipase. The resultant particle is a chylomicron remnant that is quickly removed by hepatic lysosomes. Free fatty acids generated during catabolism of chylomicrons are taken up by cells for oxidation to produce energy or can be utilized for re-synthesis of triglycerides for storage. In this process chylomicron particles are converted into chylomicron remnants. Therefore, chylomicron is absent in a fasting specimen except in the case of a specific lipid disorder. If chylomicrons are present, they are found to float at the top of serum of plasma as a creamy layer.

Chylomicron is the major transport form of exogenous triglycerides. Liver is the site of endogenous triglyceride synthesis, but endogenously produced triglycerides are not incorporated into chylomicrons. Instead they are incorporated into VLDL, which is the major transport form of endogenous triglycerides. Triglycerides present in VLDL are also hydrolyzed by lipoprotein lipase and as a result IDL is formed, which is a transitory particle because IDL is eventually converted into LDL. VLDL, IDL, and LDL all have the apoprotein B-100. LDL is removed from the circulation by the liver and other tissues. Uptake of LDL is receptor-dependent and Apo B-100 interacts with the LDL receptor present in the liver.

Cholesterol is present in the diet and is also synthesized in the liver. The rate-limiting step is catalyzed by 3-hydroxy-3-methylglutaryl-CoA reductase (HMG-CoA reductase). Cholesterol is an integral part of cell membranes and is a precursor for steroid hormones and bile acids. Fatty acids are derived from triglycerides. Cholesterol, after synthesis, is released into circulation as lipoprotein and approximately 70% of cholesterol is esterified because esterified cholesterol can be more readily transported by lipoproteins. Essential fatty acids cannot be synthesized by the human body and must be obtained via diet. Free fatty acids are transported by albumin in the plasma. There are two essential fatty acids; linoleic acid and α-linolenic acid.

6.4 LOW DENSITY LIPOPROTEIN METABOLISM

Of all the lipoproteins, LDL has the highest amount of cholesterol. LDL is taken up by tissue with LDL receptors. Apo B-100 interacts with the LDL receptors present mostly in the liver. Lysosomal degradation of LDL releases free cholesterol. Cholesterol released from LDL then inhibits HMG-CoA reductase (3-hydroxy-3-methylglutaryl-CoA reductase), thus preventing endogenous synthesis of cholesterol by the liver. If the Apo B-100 protein is defective, uptake of LDL by an LDL receptor is impaired. Patients with familial hypercholesterolemia have a defect in the gene that codes for the LDL receptor. As a result, the LDL receptor may be absent or deficient, causing elevated plasma cholesterol levels. These patients are very susceptible to coronary atherosclerosis at a very early age.

6.5 HIGH DENSITY LIPOPROTEIN METABOLISM

HDL is principally produced in the liver. HDL particles can be classified under the sub-classes, HDL_2 and HDL_3. Nascent HDL acquires free cholesterol from tissues, chylomicrons, and VLDL. The major role of HDL is to remove cholesterol from peripheral cells and then return it to the liver for excretion, a pathway called reverse cholesterol transport. Cholesterol transfer from cell membranes to HDL is stimulated by ATP-binding cassette protein A1 (ABCA1). The free cholesterol is converted to cholesteryl esters by the enzyme lecithin cholesterol acyltransferase (LCAT). This enzyme is present in nascent HDL. Apo A-1, also present in HDL, activates this enzyme. The cholesteryl esters are then transferred to chylomicron remnants and IDL. Cholesteryl ester transport proteins (CETP) are involved in this transfer. Chylomicron remnants and IDL are removed from the circulation by the liver.

6.6 LIPID PROFILE AND RISK OF CARDIOVASCULAR DISEASE

The relationship between plasma cholesterol and the risk of atherosclerosis was extensively investigated in the Framingham Heart Study. This study was initiated in 1948 in Framingham, Massachusetts; 5,209 men and women were enrolled in a study of risk factors for heart diseases. The study was under the direction of the National Heart Institute, now known as the National Heart, Lung and Blood Institute. Many important guidelines regarding risk of cardiovascular diseases emerged from the Framingham Heart Study. These led to publication of many scientific papers in leading medical journals.

Cardiovascular diseases (including myocardial infarction) are leading causes of morbidity and mortality throughout the world. Numerous studies have demonstrated the link between elevated cholesterol levels and the risk of cardiovascular diseases. In the large international INTERHEART study, lipid disorders and smoking were shown to be the two most important risk factors for cardiovascular diseases. Other important risk factors are hypertension, obesity, and diabetes [2]. Risk factors for cardiovascular diseases are listed in the following sections.

6.6.1 Un-Modifiable Risk Factors

- Male sex.
- Advanced age (Male > 45 years, Female > 55 years).
- Postmenopausal.
- Family history: Myocardial infarction or sudden death below 55 years of age in father or other male first-degree relative or below 65 years of age in mother or other first-degree female relative.
- Genetic factors: African Americans, Mexican Americans, Native Indians, and people from the Indian subcontinent all have a higher risk of heart diseases than Caucasians.

6.6.2 Modifiable Risk Factors

- Abnormal lipid profile (can also be genetic).
- Hypertension.
- Diabetes.
- Smoking.
- Obesity (more than 20% of ideal body weight).
- Physical inactivity.
- Excessive use of alcohol; moderate drinking, however, protects against cardiovascular disease and stroke.
- Poor diet (no fruits and vegetables and high in carbohydrate).
- Excessive stress.

According to the World Health Organization (WHO), the majority of cardiovascular diseases can be prevented by risk factor modification and a change in lifestyle. Unfortunately, approximately 70% of Americans are overweight and fewer than 15% of children and adults exercise sufficiently. Among American adults, 11−13% have diabetes and 34% have hypertension, indicating the depth of the problem and risk of cardiovascular diseases in Americans [3]. Initial results of the Framingham study established the link between high total cholesterol and low HDL cholesterol and the risk for cardiovascular disease, but elevated triglyceride was thought to play little role in elevating risks of cardiovascular diseases [4]. However, later reports observed the link between elevated triglycerides and a risk of heart disease. Currently

Table 6.3 Lipid Profile and Risk of Cardiovascular Diseases

Analyte	Value	Comment
Total cholesterol	<200 mg/dL	Desirable
	200–239 mg/dL	Borderline high
	>240	High
Low density lipoprotein cholesterol	<100 mg/dL	Optimal
(LDL cholesterol)	100–129 mg/dL	Near optimal
	130–159	Borderline high
	>160 mg/dL	Highly elevated
High density lipoprotein cholesterol	<40 mg/mL	Low
(HDL cholesterol)	≥60 mg/dL*	High (desirable)
Triglycerides	<150 mg/dL	Desirable
	150–199	Borderline high
	200–499	High
	>500 mg/dL	Very high

*High HDL cholesterol is not a risk factor for cardiovascular disease because only low HDL cholesterol (<40 mg/dL) is a risk factor for both men and women.

the basis of treatment for lipid disorders is the third report of the expert panel of the National Cholesterol Education program, and currently the desirable total cholesterol level of less than 200 mg/dL has been universally accepted. Desirable and elevated lipid parameters related to risk of cardiovascular diseases are listed in Table 6.3. Unfortunately, in the U.S. approximately 16.3% of the population suffers from cholesterol levels of 240 mg/dL or higher. This population has a cardiovascular risk factor twice as high as people with an optimal cholesterol level of 200 mg/dL or lower [5].

6.6.3 High LDL and Risk for Cardiovascular Disease

Although a desirable total cholesterol level of less than 200 mg/dL is universally accepted, guidelines for desirable LDL levels have changed significantly over time. Older guidelines of desirable LDL cholesterol levels of less than 130 mg/dL have been lowered to less than 100 mg/dL in the National Cholesterol Education Program expert recommendation for Adult Treatment Panel III (ATP III). Even if LDL cholesterol levels are near optimal (100–129 mg/dL), some atherogenesis occurs, and at levels above 130 mg/dL the process is accelerated. Therefore, in a high-risk patient, drug therapy may be initiated with an LDL cholesterol level over 100 mg/dL. However, some investigators reported that desirable LDL cholesterol in high-risk patients is around 70 mg/dL because LDL cholesterol level is the most important in predicting the risk of cardiovascular diseases; especially oxidized LDL which infiltrates the intima where it stimulates inflammation, endothelial dysfunction, and, finally, atherosclerosis. Atherosclerosis has been observed in individuals with relatively low LDL cholesterol levels (90–130 mg/dL) [6]. Therefore, it has been recommended that very high-risk patients target an LDL cholesterol level below 70 mg/dL as a valid therapeutic option [7]. Multiple statin trials and meta-analyses support a treatment target of LDL cholesterol less than 70 mg/dL in very high-risk patients [8]. Because LDL level is tightly associated with cardiovascular disease, the primary goal of lipid-lowering therapy using statins is targeting lower LDL levels.

6.6.4 High Triglycerides and Risk for Cardiovascular Disease

The extent to which high triglycerides directly promote cardiovascular disease has been debated over three decades. The most recent guidelines consider serum triglyceride concentrations of 150 mg/dL or less as optimal, but approximately 31% of the adult U.S. population has triglyceride levels over 150 mg/dL [9]. There are speculations that some triglyceride-rich lipoproteins may be atherogenic, especially remnant lipoproteins such as remnant VLDL and IDL. In addition, when triglyceride levels are above 200 mg/dL, the increased concentration of triglyceride-rich remnant lipoproteins may further increase the risk of cardiovascular disease. The major causes of elevated triglycerides are as follows:

- Overweight and obesity.
- Excessive alcohol use.
- Very high carbohydrate diet.
- Diseases such as Type 2 diabetes and nephrotic syndrome.
- Certain drug therapies.
- Genetic factors.

Hypertriglyceridemia is also a risk factor for acute pancreatitis.

6.6.5 HDL Cholesterol and Cardiovascular Disease

Many epidemiological studies have shown a correlation between low HDL cholesterol and higher risk of cardiovascular disease; high HDL cholesterol is associated with lower risk. HDL can remove cholesterol from atherosclerotic plaque by reverse cholesterol transport. In addition, the antioxidant and anti-inflammatory properties of HDL also protect against atherogenesis. Although the level is set at <40 mg/dL for low HDL cholesterol, women typically have higher HDL cholesterol than men. However, if HDL cholesterol is below 50 mg/dL in a woman, it may be a marginal risk factor requiring lifestyle changes in order to elevate the HDL cholesterol level. Sometimes low HDL cholesterol is encountered in individuals with high triglycerides. Causes of low HDL cholesterol include:

- High serum triglycerides.
- Obesity and physical inactivity.
- Cigarette smoking.
- Type 2 diabetes.
- Certain drug therapy such as therapy with beta-blockers.
- Genetic factors.

CASE REPORT

A 48-year-old Caucasian female suffered from myocardial infarction in the past. Her total cholesterol was 102 mg/dL, triglycerides were 120 mg/dL, and measured LDL cholesterol was 86 mg/dL. Despite her favorable total cholesterol, her HDL cholesterol was only 2 mg/dL and she also had extremely low Apo AI (6 mg/dL) and Apo AII (5 mg/dL). However, her Apo B level was 94 mg/dL. She had most of the symptoms associated with Tangier disease, including early corneal opacities, yellow-streaked tonsils, hepatomegaly, and variable degrees of peripheral neuropathy, but she did not show any splenomegaly. Interestingly, plasma levels of HDL cholesterol, Apo AI, and Apo AII, were within normal levels or high in her five relatives. The authors analyzed in detail the composition of her lipoprotein particles and found abnormalities. Most of her triglycerides were associated with the LDL fraction rather than VLDL fraction. In addition, the authors found abnormalities in the composition of her HDL particles [10].

6.6.6 Non-HDL Cholesterol, Lp(a), and Risk of Cardiovascular Disease

More recently, the role of non-HDL cholesterol in risk stratification for coronary artery diseases has been investigated extensively (Equation 6.1):

$$\text{Non-HDL cholesterol} = \text{VLDL cholesterol} + \text{LDL cholesterol} \qquad (6.1)$$

Therefore, non-HDL cholesterol is equal to total cholesterol minus HDL cholesterol (this is the way it is measured in a laboratory), and non-HDL cholesterol includes all lipoproteins that contain Apo B. The amount of non-HDL cholesterol is important in patients with elevated triglyceride levels. In persons with triglyceride levels between 200−499 mg/dL, most cholesterol found in the VLDL faction is associated with smaller (remnant) VLDL particles, making such particles atherogenic. Some studies have indicated better correlation between non-HDL cholesterol and cardiovascular mortality than LDL cholesterol in patients with high triglycerides [11]. In addition, non-HDL cholesterol is also highly correlated with Apo B, the major atherogenic lipoprotein. In patients not requiring therapy, the target for non-HDL cholesterol is <130 mg/dL (Apo B target <90 mg/dL). However, during statin therapy it may be necessary to reduce non-HDL cholesterol to <100 mg/dL to get optimal benefits [12].

Total cholesterol-to-HDL cholesterol ratio is also used for calculating risk factors for cardiovascular disease. Usually if the ratio is above 5, it is considered high. Similarly, the Apo B/Apo A1 ratio has also been used for calculating the risk of cardiovascular disease. Goswami et al. reported that the mean total cholesterol-to-HDL cholesterol ratio was 5.15 in 100 patients who suffered from myocardial infarction, but in 100 controls the ratio was 3.45, which indicated that high total cholesterol-to-HDL cholesterol ratio increased risk of myocardial infarction. The Apo B-to-Apo AI ratio was also higher in patients (0.96) than in controls (0.71) [13].

Lipoprotein(a), also known as Lp(a), is structurally related to LDL because both particles contain Apo B. Lp(a) is a modification of LDL with the addition of the "lipoprotein antigen," also synthesized by the liver. This antigen is attached with Apo B through a disulfide bond. The lipoprotein antigen is highly variable in molecular weight (300,000 to 800,000 Daltons) because of duplication of a sequence in the coding region of the gene that produces a repeat amino acid sequence. The serum level of Lp(a) is controlled by genetic makeup and is an independent risk factor for cardiovascular disease. The normal range of Lp(a) is up to 30 mg/dL. Elevated plasma Lp(a) levels have been reported in patients with nephrotic syndrome [14]. In one study, the authors reported that the mean Lp(a) level in patients with nephrotic syndrome was 69 mg/dL, but in controls the mean value was 18 mg/dL [14].

6.7 VARIOUS TYPES OF HYPERLIPIDEMIA

Hyperlipidemia is also called hyperlipoproteinemia and can be primary or secondary in origin. Various primary hyperlipidemias include:

- Familial hypercholesterolemia: This disease is transmitted as an autosomal dominant disorder. Mutations affect LDL receptor synthesis or its proper function as well as mutation of the Apo B-100 gene, which results in decreased binding of LDL with Apo B-100. Total cholesterol and LDL cholesterol are highly elevated in these individuals, making them susceptible to myocardial infarction at a young age.
- Polygenic hypercholesterolemia: In these individuals both genetics and environmental factors play important roles in producing high cholesterol levels.
- Familial hypertriglyceridemia: This disease is also transmitted as an autosomal dominant disorder characterized by an increased production of VLDL by the liver.
- Familial hyperchylomicronemia: This disease is transmitted as autosomal recessive and is caused by deficiency of the enzyme lipoprotein lipase or Apo CII. Triglyceride levels are high in these individuals.

- Familial dysbetalipoproteinemia: In these individuals there is an increased level of IDL and chylomicron remnants. Both cholesterol and triglycerides are subsequently increased. The apoprotein E exhibits polymorphism, showing three isoforms: Apo E2, Apo E3, and Apo E4. The common phenotype is E3/E3. Individuals with familial dysbetalipoproteinemia tend to have an E2/E2 phenotype. This phenotype results in impaired hepatic uptake of chylomicron remnants and IDL by the liver.
- Familial combined hyperlipidemia: In these individuals either cholesterol or triglyceride or both are elevated. It is possibly transmitted as autosomal dominant.

Secondary hyperlipidemia is common, and causes include diabetes mellitus, hypothyroidism, nephrotic syndrome, cholestasis, and alcohol abuse.

Lipid analysis can also be performed using electrophoresis where chylomicron is observed at the point of application, and the second band above the point of application is VLDL (pre-beta band) followed by the LDL band; the band furthest from the point of application is HDL. Lipid disorders are also classified according to the Fredrickson classification, which is an older classification. In this classification there are five types of hyperlipidemia:

- Type I: In these individuals elevated chylomicrons are found due to lipoprotein lipase or Apo CII deficiency, causing elevated levels of triglycerides. This is actually familial hyperchylomicronemia.
- Type IIa: In these individuals elevated LDL cholesterol and total cholesterol are observed due to familial hypercholesterolemia, polygenic hypercholesterolemia, familial combined hyperlipidemia, as well as nephritic syndrome and hypothyroidism.
- Type IIb: In these individuals elevated LDL and VLDL are seen as observed in individuals with familial combined hyperlipidemia. Both cholesterol and triglyceride levels may be elevated.
- Type III: These individuals have elevated IDL, and this disorder is actually dysbetalipoproteinemia due to the Apo E/Apo E2 profile. Both cholesterol and triglyceride levels may be elevated. In type III lipid disorder, the VLDL/triglyceride ratio is usually above 0.3 while the normal value is around 0.2.
- Type IV: These individuals have elevated VLDL as seen in familial hypertriglyceridemia or familial combined hyperlipidemia. As a result, triglyceride levels are elevated. Type IV disorder also may be due to secondary causes such as diabetes and nephrotic syndrome.
- Type V: These individuals have elevated VLDL and chylomicrons causing elevated triglycerides.

Type IIa, IIb, and also type III, are associated with significantly increased risk for cardiovascular diseases.

CASE REPORT

A 2-month-old girl was admitted to the hospital due to acute bronchiolitis. Her respiratory distress subsided quickly with supportive care. On admission, her serum cholesterol was 432 mg/dL. After her recovery a detailed analysis of her lipid profile was performed along with a lipid profile of her parents due to suspicion of familial hypercholesterolemia. During that follow-up analysis, her total cholesterol was 507 mg/dL, HDL cholesterol was 66 mg/dL, LDL cholesterol was 423 mg/dL, and triglyceride was 89 mg/dL. However, her father's total cholesterol was 235 mg/dL and her mother's cholesterol was 190 mg/dL. Her father had an HDL level of 60 mg/dL, a triglyceride level of 100 mg/dL and an LDL cholesterol level of 150 mg/dL. Her mother showed HDL cholesterol of 63 mg/dL, a triglyceride level of 124 mg/dL and an LDL cholesterol level

of 102 mg/dL. Her grandfather did not show an abnormal lipid profile. Observing no lipid abnormality in her parents, the girl was diagnosed with pseudo-homozygous type II hyperlipoproteinemia. In this disease, a normal lipid profile is observed in parents and lipid parameters in offspring can be improved following diet and oral cholestyramine therapy. One month after therapy her total cholesterol was reduced to 270 mg/dL and her LDL cholesterol was reduced to 170 mg/dL. At age 30 months, oral cholestyramine therapy was discontinued and diet restrictions were lifted because the cholesterol level was reduced below 200 mg/dL. Total cholesterol remained between 170 and 260 mg/dL during the next six years of follow up, and no cutaneous or tendon xanthoma developed [15].

6.8 VARIOUS TYPES OF HYPOLIPIDEMIA

Hypolipidemias (also called hypolipoproteinemia) are also classified as primary and secondary. Secondary hypolipidemias can be seen in severe liver diseases, malabsorption of protein, and energy and malnutrition states. However, primary hypolipidemias are rare. Primary hypolipidemias include:

- Tangier disease.
- Abetalipoproteinemia.
- Familial hypobetalipoproteinemia.
- Chylomicron retention disease.

Tangier disease is due to loss of function of ABCA1 protein as a result of mutation of the *ABCA1* gene, causing very low levels of HDL in serum. This disease is inherited in an autosomal recessive pattern. Normally ABCA1 protein helps in the uptake of cholesterol by HDL. Therefore, cholesteryl esters cannot be removed from peripheral cells by a reverse cholesterol transport mechanism. The tonsils appear hyperplastic and orange in color. Individuals with Tangier disease have a higher risk for cardiovascular disease. Low HDL levels in the absence of orange tonsils may indicate use of anabolic steroids. Anabolic steroids not only decrease HDL cholesterol levels but also increase LDL cholesterol levels.

Abetalipoproteinemia is a rare inherited disease with approximately 100 cases reported worldwide. In this disease there is a total absence of Apo B-100, thus concentrations of triglycerides and cholesterol carrying lipoprotein concentrations (chylomicrons, VLDL, IDL, and LDL) are highly reduced. This causes malabsorption of dietary fats, cholesterol, and fat-soluble vitamins such as A, D, E, and K. The signs and symptoms of this disease appear within the first few months of life, including failure to grow and gain weight, steatorrhea, and red cell acanthocytosis. Eventually people develop poor muscle balance and ataxia. Mutation of the *MTTP* gene causes abetalipoproteinemia.

Familial hypobetalipoproteinemia is inherited as a recessive form as a result of two mutations of the *MTTP* gene. This disease is more common than abetalipoproteinemia, and affects one in 1,000 individuals. Due to the partial absence of Apo B-100 lipoproteins, reduced levels of chylomicron, VLDL, IDL, and LDL are observed, causing malabsorption of fats and fat-soluble vitamins. The progress of this disease is severe if manifested in early childhood.

Chylomicron retention disease is also an inherited disease (autosomal recessive pattern), and affects absorption of dietary fats and cholesterol and fat-soluble vitamins. Chylomicrons are essential for the transport of dietary fats. In this disease, which is very rare (approximately 40 cases worldwide), mutation of the *SAR1B* gene causes impaired release of chylomicrons in the blood.

6.9 NEWER LIPID PARAMETERS AND OTHER FACTORS RELATED TO RISK FOR CARDIOVASCULAR DISEASE

In addition to traditional lipid parameters such as cholesterol, triglycerides, HDL cholesterol, and Lp(a), there are also other lipid markers and non-lipid markers that can be used for assessing risk of cardiovascular disease in individuals. These markers include:

- Lipoprotein-associated phospholipase A_2 (Lp-PLA2, lipid parameter).
- LDL particle size (lipid parameter).
- C-reactive protein (non-lipid parameter).
- Homocysteine (non-lipid parameter).
- Myeloperoxidase.

Lipoprotein-associated phospholipase A_2 (Lp-PLA2) is a monomeric enzyme that catalyzes oxidized phospholipid found in LDL into lysophosphatidylcholine and oxidized fatty acid, both of which are atherogenic. Therefore, Lp-PLA2 is an inflammatory biomarker like C-reactive protein. Lp-PLA2 is mostly associated with LDL (particularly small, dense LDL), but may be also associated with HDL. Lp-PLA2 levels are elevated in patients with elevated cholesterol, especially LDL-C. However, measurement of this parameter is not recommended for routine screening of patients for assessing risk of cardiovascular disease.

Two different phenotypes of LDL particles have been described: pattern B with mostly small, dense LDL particles (peak diameter <25.5 nm), and pattern A with a higher proportion of large, more buoyant LDL particles (peak diameter >25.5 nm). Small LDL particles contain more Apo B, and tend to coexist with elevated triglyceride HDL cholesterol and Apo AI concentration (atherogenic dyslipidemia), and are heritable. Women tend to have less small LDL particles than men. Small LDL particle size is associated with several other cardiovascular risk factors, including metabolic syndrome, type 2 diabetes mellitus, and postprandial hypertriglyceridemia [16].

C-reactive protein is an inflammation marker and is also a predictor for risk of cardiovascular disease. C-reactive protein is found in low levels (1 mg/L) in normal individuals and may increase 100-fold in response to acute phase. Levels usually return to normal in 8−10 days. Traditional assays are not sensitive enough to measure C-reactive protein for risk assessment; highly sensitive C-reactive protein assay (analytical measurement range of 0.1−100 mg/L) is used for this purpose. A C-reactive protein level of <1 mg/L is associated with low risk, 1−3 mg/L with moderate risk, and >3 mg/L with high risk for cardiovascular disease.

Homocysteine is a thiol-containing amino acid intermediate formed during methionine metabolism. McGully in 1992 reported the presence of atherosclerosis in children and young adults with inborn errors of homocysteine metabolism such as cystathionine−beta-synthase deficiency. The disorders are associated with markedly elevated plasma homocysteine levels (>100 μmol/L). McGully's work raised the possibility that mild to moderate elevation in homocysteine concentrations could contribute to atherosclerotic vascular disease. Such increases in homocysteine levels can occur with aging, menopause, hypothyroidism, low plasma level of vitamin cofactors (B_6, B_{12}, and folate) and chronic renal failure [17]. Genetic variation in enzymes involved in the metabolism of homocysteine may contribute to the difference in homocysteine levels in different individuals. One such polymorphism in methylene tetrahydrofolate reductase may lead to mild or moderate elevation in homocysteine levels; about 15% of Caucasians may have that genetic polymorphism [18]. A homocysteine level over 15 μmol/L is associated with an increased risk for cardiovascular disease and each 5 μmol/L increase is equivalent to an increase in 20 mg/dL in cholesterol concentration. Before treatment of elevated homocysteine levels is considered, vitamin B_{12} status should be evaluated to ensure that there is no such deficiency. Treatment with 1 mg/day of folic acid is effective in reducing mild to moderately elevated homocysteine levels. Supplementing folic acid therapy with B_{12} and B_6 is also practiced. However, patients with renal failure may need a much higher dose of folic acid (up to 20 mg/day) for effective reduction of their elevated homocysteine levels.

Myeloperoxidase, an enzyme released by activated neutrophils, has pro-oxidant and pro-inflammatory properties. It is stored in azurophilic granules

Table 6.4 Laboratory Parameters for Assessing Cardiovascular Disease Risk Factors

Parameter	Comments
Lipid Parameters	
Total cholesterol	Values over 240 mg/dL indicate high risk.
LDL cholesterol	Optimal value is <100 mg/dL, values >160 mg/dL indicate high risk.
HDL cholesterol	Value < 40mg/dL indicates risk, while value >60 mg/dL is good as it reduces the risk for cardiovascular disease (negative risk).
Cholesterol/HDL-C ratio	High ratios, especially value of 5 or more, is associated with increased risk.
High non-HDL-C	Optimal is <130 mg/dL but in high-risk patient optimal level is <100 mg/dL. Higher levels are associated with increased risk.
Lp(a)	Value over 30 mg/dL indicates increased risk.
Non-Lipid Parameters	
C-reactive protein	Desirable level is <1 mg/L, while > 3 mg/L indicates high risk.
Homocysteine	Value >15 μmol/L indicates high risk.

of polymorphonuclear neutrophils and macrophages. Because myeloperoxidase is involved in oxidative stress and the inflammatory process, it is a biomarker for inflammation in ischemic heart disease and acute coronary syndromes. Various laboratory parameters used for assessing risk factors for cardiovascular diseases are listed in Table 6.4.

6.10 LABORATORY MEASUREMENTS OF VARIOUS LIPIDS

Lipid profiles that consist of total cholesterol, triglycerides, LDL, and HDL are measured routinely. Blood specimens should be collected after an overnight fast of 10−12 hours. This ensures that chylomicrons are cleared from plasma. In serum, the majority of cholesterol exists as cholesterol ester. Therefore, in the first step cholesterol ester is hydrolyzed by cholesterol ester hydrolase enzyme. Then cholesterol is oxidized by cholesterol oxidase, generating cholest-4-en-3-one and hydrogen peroxide. Hydrogen peroxide generated is proportional to serum cholesterol concentration and is measured by its reaction with a suitable compound, for example, 4-aminoantipyrene (reaction catalyzed by peroxidase) to form a colored dye. HDL is usually measured as HDL cholesterol after precipitating out other lipoprotein fractions using polyanions such as dextran sulfate-magnesium chloride, phospho-tungstate-magnesium chloride or heparin sulfate-manganese chloride.

For serum triglyceride measurement, lipase enzyme is used, which converts triglyceride into glycerol and free fatty acid. Then glycerol is oxidized by glycerokinase into glycerophosphate. Glycerophosphate is then measured by either its reaction with nicotinamide adenine dinucleotide (NAD, no absorption at 340 nm) to form NADH (absorbs at 340 nm) or its oxidation by glycerophosphate oxidase enzyme, generating dihydroxyacetone and hydrogen peroxide.

Plasma LDL values are typically calculated with the Friedewald formula (Equation 6.2):

$$LDL\ cholesterol = Total\ cholesterol - HDL\ cholesterol - Triglyceride/5$$

$$(6.2)$$

All measurements are in mg/dL. This formula is invalid if triglyceride values are above 400 mg/dL. In such situations direct measurement of LDL is indicated. In addition, this is only applicable for calculating LDL cholesterol in an overnight fasting specimen. For certain patients, calculated LDL cholesterol may not reflect true LDL cholesterol levels. If there is a discrepancy between measured LDL cholesterol level and calculated LDL cholesterol level, it indicates that there is a modification of lipoprotein metabolism.

CASE REPORT

A 64-year-old female patient with total cholesterol of 331 mg/dL, triglycerides of 307 mg/dL, and HDL cholesterol of 47 mg/dL, was seen by the authors in their lipid clinic due to a history of hypertension for 20 years. She also suffered from hyperlipidemia and diabetes for 11 years. Her calculated LDL cholesterol was 222.6 mg/dL. However, direct measurement of her LDL cholesterol using an enzyme assay showed a value of 97 mg/dL. On detailed analysis of her lipoprotein composition it was revealed that only 30% of her serum cholesterol was associated with LDL particles and that she also had higher Apo AI (148 mg/dl) than Apo B (91 mg/dL). Interestingly, her LDL particles were resistant to oxidation. Therefore, atherogenic dyslipidemia in this patient could be compensated by her altered lipoprotein metabolism and the enhanced antioxidant properties of her lipoproteins [19].

Elevated chylomicrons cause the plasma to appear as milky, and when plasma is allowed to stand, a creamy layer is visible at the top. Elevated triglycerides cause the entire plasma to appear turbid. The various lipoproteins have distinct electrophoretic patterns as seen with serum protein electrophoresis. Chylomicron has minimum protein and is found at the origin. The HDL fraction is seen in the alpha-1 region. LDL migrates in the beta region and VLDL is present in the pre-beta region. Apolipoproteins are measured by using appropriate immunoassays.

6.11 DRUGS FOR TREATING LIPID DISORDERS

Several drugs are available for treating lipid disorders, including 3-hydroxy-3-methylglutaryl-CoA reductase (HMG-CoA reductase) inhibitors (statins), nicotinic acid, fibrates, and bile acid sequestering agents. The most commonly used drugs for treating lipid disorders are statins. Common statins such as pravastatin, lovastatin, and simvastatin can significantly lower total cholesterol, mostly LDL cholesterol, which is the primary goal of lipid lowering therapy. In addition, statin therapy also increases HDL cholesterol and lowers triglyceride levels. Cholesterol absorption inhibitors such as ezetimibe modestly lower LDL cholesterol and can be used in combination with statins. Fibrates are primarily used in treating patients with hypertriglyceridemia, but these agents may also increase HDL cholesterol levels. Nicotinic acid (niacin) raises HDL cholesterol very effectively with modest reduction of LDL cholesterol. Nicotinic acid may also lower Lp(a) levels [20]. Tolerability is a major issue with nicotinic acid. Bile acid sequestrants such as cholestyramine are capable of reducing LDL cholesterol by mild to moderate amounts. Use is limited by major side effects.

KEY POINTS

- Apolipoprotein A (Apo A): Consists of Apo AI and Apo AII.
- Apolipoprotein B (Apo B): Most abundant is large Apo B known as Apo B-100, while the less abundant is a smaller particle known as Apo B-48. Apo B-48 is more atherogenic than Apo B-100.
- Apolipoprotein C (Apo C): Exists in three forms (Apo CI, Apo CII, and Apo CIII).
- Apolipoprotein D (Apo D).
- Apolipoprotein E (Apo E).
- Major lipoproteins found in plasma are chylomicrons, VLDL, LDL, IDL, and HDL.
- Chylomicrons: Lightest fraction containing approximately 2% lipoprotein and mostly lipids, especially triglycerides. This fraction is absent in fasting specimens (preferred specimen for lipid analysis) unless the patient is suffering from a lipid disorder.
- Very low density lipoprotein (VLDL): Denser than chylomicron but lighter than LDL, this fraction contains 4–10% proteins and the rest lipids, most notably triglycerides (45–60%).
- Intermediate density lipoprotein (IDL): This fraction is lighter than LDL and contains approximately 15% proteins and the rest lipids. This fraction is usually a transient fraction and is absent in fasting specimens except for certain lipid disorders.
- Low density lipoprotein (LDL): This fraction is denser than VLDL but lighter than HDL and contains approximately 25% proteins and the rest lipids, most commonly esterified cholesterol (approximately 50%).

- High density lipoprotein (HDL): Most dense fraction containing approximately 50% protein and 50% lipids.
- In the circulation, lipoprotein lipase found in the capillary endothelium of adipose tissue, skeletal, and cardiac muscle breaks down the triglyceride component of chylomicrons into glycerol and free fatty acids. Apo CII present in chylomicrons activates lipoprotein lipase. Chylomicrons, if present, are found to float at the top of serum of plasma as a creamy layer.
- Chylomicron is the major transport form of exogenous triglycerides. Liver is the site of endogenous triglyceride synthesis, but endogenously produced triglycerides are not incorporated into chylomicron. Instead they are incorporated into VLDL, which is the major transport form of endogenous triglycerides. Triglycerides present in VLDL are also hydrolyzed by lipoprotein lipase and as a result IDL is formed, which is a transitory particle because IDL is eventually converted into LDL. VLDL, IDL, and LDL all have the apoprotein B-100. LDL is removed from the circulation by the liver and other tissues. Uptake of LDL is receptor-dependent, and Apo B-100 interacts with the LDL receptor present in the liver.
- Cholesterol is present in the diet and is also synthesized in the liver. The rate-limiting step is catalyzed by 3-hydroxy-3-methylglutaryl-CoA reductase (HMG-CoA reductase). Cholesterol is an integral part of cell membranes and is a precursor for steroid hormones and bile acids.
- Of all the lipoproteins, LDL has the highest amount of cholesterol. LDL is taken up by tissue with LDL receptors. Apo B-100 interacts with the LDL receptors present mostly in the liver. Lysosomal degradation of LDL releases free cholesterol. Cholesterol released from LDL then inhibits HMG-CoA reductase, thus preventing endogenous synthesis of cholesterol by the liver. If the Apo B-100 protein is defective, uptake of LDL by LDL receptor is impaired. In addition, patients with familial hypercholesterolemia have a defect in the gene that codes for the LDL receptor. Plasma LDL values are typically calculated using the Friedewald formula: LDL cholesterol = Total cholesterol − HDL cholesterol − Triglyceride/5, where all measurements are in mg/dL. This formula is invalid if triglyceride values are above 400 mg/dL. In such situations direct measurement of LDL is indicated.
- The major role of HDL is to remove cholesterol from peripheral cells and then return it to the liver for excretion, a pathway called reverse cholesterol transport. Cholesterol transfer from cell membranes to HDL is stimulated by ATP-binding cassette protein A1 (ABCA1). The free cholesterol is converted to cholesteryl esters by the enzyme lecithin cholesterol acyltransferase (LCAT). This enzyme is present in nascent HDL. Apo A1 also present in HDL activates this enzyme. The cholesteryl esters are then transferred to chylomicron remnants and IDL. Cholesteryl ester transport proteins (CETP) are involved in this transfer.
- Currently desirable levels of total cholesterol, LDL cholesterol, and triglycerides are < 200 mg/dL, <100 mg/dL (<70 mg/dL in high-risk patients), and <150 mg/dL, respectively. Hypertriglyceridemia is also a risk factor for acute pancreatitis.

- Many epidemiological studies have shown correlation between low HDL cholesterol and a higher risk of cardiovascular disease. The desired level of HDL is set at >40 mg/dL in males and >50 mg/dL in females.
- Non-HDL cholesterol = Total cholesterol − HDL cholesterol (this is the way it is measured in a laboratory). Some studies have indicated better correlation between non-HDL cholesterol and cardiovascular mortality than LDL cholesterol in patients with high triglycerides. In patients not requiring therapy, the target for non-HDL cholesterol is <130 mg/dL. However, during statin therapy, it may be necessary to reduce non-HDL cholesterol to <100 mg/dL to get optimal benefit.
- Total cholesterol-to-HDL cholesterol ratio is also used for calculating risk factor for cardiovascular disease. Usually if the ratio is above 5, it is considered high.
- Hyperlipidemia (also called hyperlipoproteinemia) can be primary or secondary in origin. Various primary hyperlipidemias include:
  - Familial hypercholesterolemia: This disease is transmitted as an autosomal dominant disorder. Mutations affect LDL receptor synthesis or its proper function as well as mutation of the Apo B-100 gene, which results in decreased binding of LDL with Apo B-100. Total cholesterol and LDL cholesterol are highly elevated in these individuals, making them susceptible to myocardial infarction at a young age.
  - Polygenic hypercholesterolemia: In these individuals both genetics and environmental factors play important roles in producing high cholesterol levels.
  - Familial hypertriglyceridemia: This disease is also transmitted as an autosomal dominant disorder where there is an increased production of VLDL by the liver.
  - Familial hyperchylomicronemia: This disease is transmitted as autosomal recessive where there is a deficiency of the enzyme lipoprotein lipase or Apo CII. Triglyceride levels are high in these individuals.
  - Familial dysbetalipoproteinemia: In these individuals increased levels of IDL and chylomicron remnants are observed in circulation. Both cholesterol and triglycerides are also subsequently increased. Apoprotein E exhibits polymorphism showing three isoforms: Apo E2, Apo E3, and Apo E4. The common phenotype is E3/E3. Individuals with familial dysbetalipoproteinemia tend to have the E2/E2 phenotype. This phenotype results in impaired hepatic uptake of chylomicron remnants and IDL by the liver.
  - Familial combined hyperlipidemia: In these individuals either cholesterol or triglyceride or both are elevated. It is possibly transmitted as autosomal dominant.
  - Secondary hyperlipidemia is common and its causes include diabetes mellitus, hypothyroidism, nephrotic syndrome, cholestasis, and alcohol.
- Lipid analysis can also be performed using electrophoresis where chylomicron is observed at the point of application. The second band above the point of application is VLDL (pre-beta band) followed by the LDL band, and the band furthest from the point of application is HDL.
- Lipid disorders are also classified according to the Fredrickson classification, which is an older classification. In this classification there are five types of hyperlipidemia:

- Type I: In these individuals elevated chylomicrons are found due to lipoprotein lipase or Apo CII deficiency, causing elevated levels of triglycerides. This is actually familial hyperchylomicronemia.
- Type IIa: In these individuals elevated LDL cholesterol and total cholesterol are observed due to familial hypercholesterolemia, polygenic hypercholesterolemia, familial combined hyperlipidemia, as well as nephritic syndrome and hypothyroidism.
- Type IIb: In these individuals elevated LDL and VLDL are seen as observed in individuals with familial combined hyperlipidemia. Both cholesterol and triglyceride levels may be elevated.
- Type III: These individuals have elevated IDL, and this disorder is actually dysbetalipoproteinemia due to its Apo E/Apo E2 profile. Both cholesterol and triglyceride levels may be elevated. In type III lipid disorder, the VLDL/triglyceride ratio is usually close to 0.3 while the normal value is around 2.0.
- Type IV: These individuals have elevated VLDL as seen in familial hypertriglyceridemia or familial combined hyperlipidemia. As a result, triglyceride level is elevated. Type IV disorder also may be due to secondary causes such as diabetes and nephrotic syndrome.
- Type V: These individuals have elevated VLDL and chylomicrons that cause elevated triglycerides.
- Type IIa, IIb, and also type III are associated with significantly increased risk for cardiovascular diseases.

- In Tangier disease there is a loss of function of ABCA1 protein due to mutation of the *ABCA1* gene, causing very low levels of HDL in serum. This disease is inherited in an autosomal recessive pattern. The tonsils appear hyperplastic and orange in color. Individuals with Tangier disease have a higher risk for cardiovascular disease.
- Abetalipoproteinemia is a rare inherited disease with approximately 100 cases reported worldwide. In this disease there is a total absence of Apo B-100, thus concentrations of triglycerides and cholesterol carrying lipoprotein concentrations (chylomicrons, VLDL, IDL, and LDL) are highly reduced.
- In addition to traditional lipid parameters such as cholesterol, triglycerides, HDL cholesterol, and Lp(a), there are also other lipid markers and non-lipid markers that can be used for assessing risk of cardiovascular disease in individuals. These markers include: lipoprotein-associated phospholipase A_2 (Lp-PLA2, lipid parameter), LDL particle size (lipid parameter), C-reactive protein (non-lipid parameter), homocysteine (non-lipid parameter), and myeloperoxidase.

REFERENCES

[1] Sawle A, Higgins MK, Olivant MP, Higgins JA. A rapid single step centrifugation method for determination of HDL, LDL and VLDL cholesterol and TG and identification of predominant LDL subclass. J Lipid Res 2002;43:335–43.

[2] Yusuf S, Hawken S, Ounpuu S, Dans T, et al. Effect of potentially modifiable risk factors associated with myocardial infarction in 52 countries (the INTERHEART study): case control study. Lancet 2004;364:937−52.

[3] Kones R. Primary prevention of coronary heart disease: integration of new data, evolving views, revised goals, and role of rosuvastatin in management: A comprehensive survey. Drug Des Devel Ther 2011;5:325−80.

[4] Gordon T, Kannel WB, Castelli WP, Dawber TR. Lipoproteins, cardiovascular disease and death. The Framingham study. Arch Intern Med 1981;141:1128−31.

[5] Vijayakrishnan R, Kalyatanda G, Srinivasan I, Abraham GM. Compliance with the adult treatment panel III guidelines for hyperlipidemia in a resident run ambulatory clinic: a retrospective study. J Clin Lipidol 2013;7:43−7.

[6] Law MR, Wald NJ. Risk factor thresholds: their existence under scrutiny. BMJ 2003;327:518.

[7] Grundy SM, Cleeman JI, Merz CN, Brewer HB, et al. Implications of recent clinical trials for National Cholesterol Education Program Adult Treatment Panel III guidelines. J Am Coll Cardiol 2004;44:720−32.

[8] Martin SS, Blumenthal RS, Miller M. LDL cholesterol: lower the better. Med Clin North Am 2012;96:13−26.

[9] Miller M, Stone NJ, Ballantyne C, Bittner V, et al. Triglycerides and cardiovascular disease: a scientific statement from American Heart Association. Circulation 2011;123:2292−333.

[10] Cheung MC, Mendez AJ, Wolf AC, Knopp RH. Characteristics of apolipoprotein A-1 and A-II containing lipoproteins in a new case of high density lipoprotein deficiency resembling Tangier disease and their effects on intracellular cholesterol efflux. J Clin Invest 1993;91:522−9.

[11] Liu J, Sempos CT, Donahue RP, Dorn J, et al. Non high density lipoprotein and very low density lipoprotein cholesterol and their risk predictive values in coronary heart disease. Am J Cardiol 2006;98:1363−8.

[12] Ballantyne CM, Raichlen JS, Cain VA. Statin therapy alters the relationship between apolipoprotein B and low density lipoprotein cholesterol and non-high density lipoprotein cholesterol targets in high risk patients: the MERCURY II (measuring effective reductions in cholesterol using rosuvastatin) trial. J Am Coll Cardiol 2008;52:626−32.

[13] Goswami B, Rajappa M, Malika V, Kumar S, et al. Apo B/Apo AI ratio, a better discriminator for coronary artery disease risk than other conventional lipid ratios in Indian patients with acute myocardial infarction. Acta Cardiol 2008;63:749−55.

[14] Warner C, Rader D, Bartens W, Kramer J, et al. Elevated plasma lipoprotein (a) in patients with nephrotic syndrome. Ann Intern Med 1993;119:263−9.

[15] Tamaki W, Fujieda M, Madda M, Hasokawa T, et al. A case of pseudohomozygous type II hyperlipoproteinemia in early infancy. Pediatr Int 2011;110−113.

[16] Roheim PS, Asztalos BF. Clinical significance of lipoprotein size and risk for coronary atherosclerosis. Clin Chem 1995;41:147−52.

[17] McGully KS. Homocystinuria, arteriosclerosis, methylmalonic aciduria and methyltransferase deficiency: a key case revisited. Nutr Rev 1992;50:7−12.

[18] Jacques PF, Bostom AG, Williams RR, Ellison RC, et al. Relation between folate status a common mutation in methylene tetrahydrofolate reductase and plasma homocysteine concentration. Circulation 1996;93:7−9.

[19] Lee JH, Park JE, Lee SH, Kim JR, et al. Elevated HDL2-paraoxonase and reduced CETP activity are associated with a dramatically lower ratio of LDL cholesterol/total cholesterol in a hypercholesterolemic and hypertriglyceridemic patient. Int J Mol Med 2010;25:945−51.

[20] Julius U, Fischer S. Nicotinic acid as a lipid modifying drug-a review. Atheroscler Suppl 2013;14:7−13.

CHAPTER 7

Carbohydrate Metabolism, Diabetes, and Hypoglycemia

7.1 CARBOHYDRATES: AN INTRODUCTION

Carbohydrates, including sugar and starch, are important in human physiology because glucose provides more than half of the total energy requirements of the human body. Glucose is a breakdown product of dietary carbohydrates. In addition, glucose is also produced by glycogenolysis and gluconeogenesis. From an organic chemistry point of view, carbohydrates contain only carbon, hydrogen, and oxygen, but the ratio of hydrogen to oxygen must be 2:1, the same proportion present in water. Therefore, the component of DNA and RNA, deoxyribose and ribose, is also a sugar. In biochemistry, carbohydrates are often referred to as saccharides (derived from the Greek word meaning sugar) and can be sub-classified into four categories:

- Monosaccharides: Simplest carbohydrates that cannot be further hydrolyzed. The common monosaccharides are glucose, fructose, and galactose; all contain six carbons. Ribose, a monosaccharide containing five carbons, is an integral part of RNA and several cofactors such as ATP, NAD, etc.
- Disaccharides: When two monosaccharide molecules are joined together, a disaccharide is formed that can be hydrolyzed to monosaccharides. The most abundant disaccharide in blood is sucrose, which can be hydrolyzed to glucose and fructose. Lactose, the disaccharide present in milk, is composed of galactose and glucose. Maltose is another common disaccharide that breaks down into two glucose molecules.
- Oligosaccharides: These are less complex molecules than polysaccharides and usually contain ten or less monosaccharide molecules in one oligosaccharide molecule. Oligosaccharides are often found in glycoprotein molecules.
- Polysaccharides: Complex carbohydrates containing many monosaccharides (200–2500). Polysaccharides can serve as energy storage components (starch, glycogen, etc.) and/or can act as

107

A. Dasgupta and A. Wahed: Clinical Chemistry, Immunology and Laboratory Quality Control
DOI: http://dx.doi.org/10.1016/B978-0-12-407821-5.00007-3

structural components (e.g. cellulose, an integral component of plant cell membranes).

7.2 REGULATION OF BLOOD GLUCOSE CONCENTRATION

Blood glucose concentration is tightly controlled, and increased blood glucose concentration is encountered in patients suffering from diabetes. After eating a meal, starch and glycogen present in food are partially digested by salivary amylase and then further digested by pancreatic amylase and disaccharidase present in intestinal mucosa; they break down into monosaccharides (glucose, fructose, and galactose). These monosaccharides are absorbed into circulation by an active carrier-mediated transfer process. After absorption into the portal vein, these monosaccharides are transported into the liver, and, depending on physiological needs, glucose can be metabolized completely into carbon dioxide and water to provide immediate energy, or stored in the liver as glycogen. Major biochemical processes involved in the regulation of blood glucose levels include:

- Glycolysis: Metabolism of glucose through a series of biochemical reactions into lactate or pyruvate is called glycolysis. In this process two molecules of ATP (adenosine triphosphate) are formed, thus providing energy. However, fructose and galactose can also enter the glycolysis process after phosphorylation. Oxidation of glucose into carbon dioxide and water can also take place through hexose monophosphate shunt pathways.
- Glycogenesis: Conversion of glucose into glycogen for storage in liver. Glycogen can also be found in skeletal muscle.
- Glycogenolysis: A process by which glycogen breaks down into glucose when needed for energy during a fasting period.
- Gluconeogenesis: During an extended fasting period this biochemical pathway is used for production of glucose from non-carbohydrate sources such as lactate, glycerol, pyruvate, glucogenic amino acids, and odd chain fatty acids.

The major hormones involved in blood glucose regulation are insulin and glucagon. Insulin is a 51-amino acid polypeptide secreted by the beta cells of the islets of Langerhans in the pancreas. The molecule proinsulin is cleaved to form insulin and C-peptide. Whenever insulin is secreted, the C-peptide molecule is also found in blood. The ratio of C-peptide to insulin in a normal individual is 5:1. In type 1 and type 2 diabetes, there is an inappropriate increase in the release of proinsulin. Proinsulin has 10 to 15% of the biological activity of insulin.

Insulin induces:

- Cellular uptake of glucose.
- Glycogen and protein synthesis.
- Fatty acid and triglyceride synthesis.

Insulin inhibits:

- Glycogenolysis.
- Gluconeogenesis.
- Proteolysis and lipolysis.

Cell membranes are not permeable to glucose. Specialized glucose transporter (GLUT) proteins transport glucose through the cell membranes. Insulin-mediated cellular uptake of glucose is regulated through GLUT-4 proteins. The insulin receptor consists of alpha and beta subunits. The insulin molecule binds with the alpha subunit. The beta subunits traverse the cell membrane and conformational changes take place in the beta subunit when insulin binds with the alpha subunit. This results in various intracellular responses, including translocation of vesicles carrying GLUT-4 proteins to the cell membrane.

Glucagon is a 29-amino acid polypeptide secreted by the alpha cells of the islets of Langerhans in the pancreas. It has the opposite action to that of insulin. Thus, it increases:

- Glycogenolysis.
- Gluconeogenesis.
- Lipolysis.

In addition to insulin and glucagon, other hormones are also involved in the regulation of blood glucose concentration. Somatostatin, a polypeptide found in several organs but mostly in the hypothalamus and in delta cells of pancreatic islets, can regulate secretion of insulin and glucagon, thus modulating actions of these two important hormones. In general, insulin decreases blood glucose but glucagon increases blood glucose, thus counteracting the effect of insulin. Other hormones that can increase blood glucose include epinephrine, cortisol, and growth hormone.

7.3 DIABETES MELLITUS: BASIC CONCEPTS

Diabetes mellitus is a syndrome characterized by hyperglycemia due to relative insulin deficiency or insulin resistance. It is important to note that diabetes insipidus, which is also characterized by polyuria, is different from diabetes mellitus because diabetes insipidus is not related to insulin secretion or insulin resistance but is an uncommon condition that occurs when the kidney is unable to concentrate urine properly. As a result, diluted urine is produced, affecting plasma osmolality. The cause of diabetes insipidus is lack of secretion of ADH (cranial diabetes insipidus, also known as central diabetes insipidus) or the inability of ADH to work at the collecting duct of the kidney (nephrogenic diabetes insipidus). Please see Chapter 5 for more detail.

Diabetes mellitus can be primary or secondary in nature. Primary diabetes mellitus can be monogenic or polygenic. Monogenic diabetes mellitus covers a heterogenous group of diabetes caused by a single gene mutation and characterized by impaired insulin secretion by beta cells of the pancreas. Maturity onset diabetes of the young (MODY), mitochondrial diabetes, and neonatal diabetes are examples of monogenic diabetes mellitus. The diagnosis of monogenic diabetes mellitus, and differentiating this type of diabetes from type 1 and type 2 diabetes mellitus, is essential. Monogenic diabetes accounts for 2–5% of all diabetes and is less common than type 1 and type 2 diabetes mellitus, which are the most common forms of diabetes mellitus encountered in clinical practice [1].

Polygenic diabetes mellitus can be either type 1 or type 2. Type 1 diabetes mellitus (formerly called insulin-dependent diabetes) is characterized by an absolute deficiency of insulin due to islet cell destruction and usually presents in younger people with acute onset. Type 2 diabetes mellitus (formerly called non-insulin-dependent diabetes mellitus) is characterized by insulin resistance and beta cell dysfunction in the face of insulin resistance and hyperglycemia. There is also secondary diabetes, which may be drug-related or due to various diseases.

7.4 MONOGENIC DIABETES MELLITUS

All cases of maturity onset diabetes of the young (MODY) and most cases of neonatal diabetes are due to defects in insulin secretion. MODY is the most common form of monogenic diabetes. The clinical pattern of MODY is characterized by young age onset of diabetes (10–45 years, but most likely before age 25) and a marked family history of diabetes in every generation due to autosomal dominant inheritance, absence of obesity and insulin resistance, negative autoantibody against pancreatic beta cells, and mild hyperglycemia. Usually patients respond to sulfonylurea therapy if needed. Many genetic mutations have been reported in patients with MODY, but most commonly encountered mutations are due to mutations in genes encoding the enzyme glucokinase (GCK) and mutations of genes encoding nuclear transcription factors of the hepatocyte nuclear factors (HNF). At present, sequencing of common genes causing MODY, such as GCK, HNF1A (transcription factor-1), and HNF1B (transcription factor-2), is available in reference laboratories for confirming diagnosis of MODY. However, new gene mutations related to MODY are regularly described in the medical literature [2]. MODY can be further sub-classified as MODY 1, 2, and 3, with MODY 3 the most commonly encountered form; MODY 3 is caused by mutation of HNF1A. It is important to note that patients with MODY require lower doses of sulfonylurea than other groups of patients.

Neonatal diabetes mellitus, a rare disease, may occur up to the age of six months due to mutation of different genes involved in organogenesis, formation of beta cells, and insulin synthesis. Depending on the genetic mutation, neonatal diabetes can be transient or permanent. Neonatal diabetes diagnosed before six months of age is frequently due to the mutation of genes that encode Kir6.2 (ATP-sensitive inward rectifier potassium channel) or the sulfonylurea receptor 1 subunit of the ATP-sensitive potassium channel; these patients respond to high doses of sulfonylurea therapy rather than insulin therapy [3]. Mitochondrial diabetes mellitus is due to the mutation of mitochondrial DNA and the disease can be manifested as early as 8 years of age; mean onset is 35 years. Because diabetes develops due to failure of insulin secretion, most patients will eventually require insulin therapy [4].

7.5 TYPE 1 DIABETES MELLITUS

Type 1 diabetes, formerly known as insulin-dependent diabetes or juvenile onset diabetes, is encountered in 5–10% of all patients with diabetes mellitus and is characterized by polyuria, polydipsia, and rapid weight loss. Type 1 diabetes is due to autoimmune destruction of pancreatic beta cells by T lymphocytes. Markers of immune destruction of beta cells in these patients include islet cell autoantibodies as well as autoantibodies to insulin, glutamic acid decarboxylase (GAD), and tyrosine phosphatases. Usually 85–90% of patients with type 1 diabetes mellitus have one or more autoantibodies. Individuals with antibodies are classified as type 1A, and other individuals with type 1 diabetes, but without any evidence of autoimmunity or any known cause of islet cell destruction, are classified as type 1B. This form of type 1 diabetes mellitus is less common and is often referred to as idiopathic diabetes.

Both genetic susceptibility and environmental factors play important roles in the pathogenesis. Genetic susceptibility is polygenic, with the greatest contribution coming from the HLA region; however, no single gene responsible for type 1 diabetes has been characterized. More than 90% of patients with type 1 diabetes mellitus carry HLA-DR3-DQ2, HLA-DR4-DQ8, or both. Environmental factors that have been implicated include dietary constituents, Coxsackie viruses, and vaccinations. The rate of destruction of beta cells in patients with type 1 diabetes mellitus is variable, with more rapid destruction usually observed in infants and children compared to adults. Children and adolescents may first present with ketoacidosis as the manifestation of the disease. Patients are usually dependent on insulin as very little or no insulin is produced.

7.6 TYPE 2 DIABETES MELLITUS

Type 2 diabetes mellitus is the most common form of diabetes mellitus and accounts for over 90% of all cases. It was formerly referred to as non-insulin-dependent diabetes mellitus. Type 2 diabetes mellitus is adult onset, is characterized by insulin resistance, and may also be accompanied by beta cell dysfunction causing insulin deficiency. Many patients with type 2 diabetes mellitus are obese because obesity itself can cause insulin resistance. However, these patients may not need insulin initially after diagnosis or even throughout their life. Timely diagnosis of type 2 diabetes mellitus is important because early intervention can prevent many complications of type 2 diabetes mellitus (including neuropathy, nephropathy, and retinopathy), but such diagnosis is often difficult because hyperglycemia develops gradually and at an early stage a patient may not notice any classical symptoms of diabetes. Ketoacidosis seldom occurs in type 2 diabetes, and, when seen, it is usually associated with a stress factor such as infection.

Insulin resistance in type 2 diabetes mellitus includes down-regulation of the insulin receptor, abnormalities in the signaling pathway, and impairment of fusion of GLUT-4 (glucose transporter type 4)-containing vesicles with the cell membrane. Initially with insulin resistance, hyperinsulinemia is observed, which attempts to compensate for the insulin resistance. With time, beta cell dysfunction may be encountered (both quantitative and qualitative), thus causing hyperglycemia. Type 2 diabetes mellitus is a polygenic disorder. Various features of type 1 diabetes mellitus, type 2 diabetes mellitus, and MODY are summarized in Table 7.1.

7.7 METABOLIC SYNDROME OR SYNDROME X

Metabolic syndrome (syndrome X) was first described in 1988 by Gerald Reaven. He proposed the existence of a new syndrome, or syndrome X, characterized by insulin resistance, hyperinsulinemia, hyperglycemia, dyslipidemia, and arterial hypertension. Following this description of the syndrome, it became a major theme of research and public health concern. Individuals with this syndrome have higher risk of coronary artery disease, stroke, and type 2 diabetes. The American Heart Association/National Heart, Lung and

Table 7.1 Major Features of Type 1/2 Diabetes Mellitus and MODY

Clinical Feature	Type 1 DM	Type 2 DM	MODY
Typical age of diagnosis	<25 years	>25 years	>25 years
Body weight	Usually not obese	Overweight to obese	No obesity
Autoantibodies	Present (90%)	Absent	Absent
Insulin dependence	Yes	No	No
Family history	Infrequent	Frequent	Yes, in multiple generations
Diabetic ketoacidosis	High risk	Low risk	Low risk

Abbreviations: Type 1 DM, Type 1 diabetes mellitus; Type 2 DM, Type 2 diabetes mellitus; MODY, Maturity onset diabetes of young.

Blood Institute (AHA/NHLBI) criteria for metabolic syndrome include three or more of the following risk factors:

- Central obesity (waist circumference: 40 inches or more in men, 35 inches or more in women).
- Insulin resistance: Fasting glucose over 100 mg/dL.
- Elevated triglycerides (>150 mg/dL).
- Reduced high density lipoprotein cholesterol (HDL cholesterol: <40 mg/dL for men, <50 mg/dL for women).
- Elevated blood pressure (≥130 mm of Hg for systolic, or ≥85 mm of Hg for diastolic) or drug treatment for hypertension.

Other risk factors for metabolic syndrome include genetic makeup, advanced age, lack of exercise, and hormonal changes. Weight control, daily exercise, and healthy food habits are the primary goals of therapy. Drug therapy may be initiated depending on the clinical judgment of the physician [5]. Smokers with metabolic syndrome are advised to quit smoking.

7.8 COMPLICATIONS OF DIABETES

Diabetic complications can be divided into two broad categories: acute and chronic complications. Acute complications include diabetic ketoacidosis (DKA), hyperosmolar non-ketosis, and lactic acidosis. Chronic complications could be either macrovascular (stroke, myocardial infarction, gangrene, etc.) or microvascular (such as diabetic retinopathy, diabetic eye diseases, and diabetic neuropathy).

Diabetic ketoacidosis may be the presenting feature of type 1 diabetes mellitus, or it may occur in a diabetic individual managed on insulin who does not take insulin or whose insulin requirement has been increased due to infection, myocardial infarction, or other causes. Diabetic ketoacidosis is a medical emergency, and, if not treated on time, can be fatal. In typical diabetic ketoacidosis, absolute insulin deficiency, along with increased secretion of glucagon and other counter regulatory hormones, results in decreased uptake of glucose into cells with increased glycogenolysis and gluconeogenesis. This produces more sugars, which are released in the circulation. As a result, marked hyperglycemia, glycosuria, and osmotic diuresis may result, causing water and salt loss through the kidneys. Reduction in plasma volume may cause renal hypoperfusion and eventually acute renal failure. Absence of insulin also leads to release of fatty acids, mostly from adipose tissue (lipolysis), causing generation of excess fatty acids that result in the formation of ketone bodies, including acetoacetic acid, beta hydroxybutyric acid, and acetone. Acetoacetic acid and beta-hydroxybutyric acid contribute to the acidosis. The body attempts to neutralize such excess acid by bicarbonate compensatory mechanisms. As bicarbonate is depleted, the body attempts other mechanisms of compensation, such as hyperventilation, and, in some patients, extreme forms of hyperventilation (Kussmaul respiration). Acetone is volatile and responsible for the typical ketone odor present in patients with diabetic ketoacidosis. Typically, in patients with diabetic ketoacidosis, blood glucose is above 250 mg/dL, arterial blood pH < 7.3, and bicarbonate is between 15 and 18 mmol/L (but may be lower than 10 mmol/L in severe cases). In addition, ketone bodies are present in urine. Diabetic ketoacidosis is more commonly encountered in patients with type 1 diabetes, although under certain circumstances (e.g. trauma, surgery, infection) severe stress diabetic ketoacidosis may also be observed in patients with type 2 diabetes. Other than diabetes, ketoacidosis may be observed in patients with alcoholism (alcoholic ketoacidosis), starvation, and also may be drug-induced (e.g. salicylate poisoning).

Hyperosmolar non-ketosis is typically seen in patients with type 2 diabetes mellitus. Insulin deficiency is not absolute and as a result ketosis is not significant. There is also minimal acidosis. Characteristic clinical features include significant hyperglycemia with high plasma osmolality and dehydration. Lactic acidosis is an uncommon situation with diabetics. It may be seen in patients on biguanide therapy (e.g. on phenformin) with liver or renal impairment.

Macrovascular complications of diabetes mellitus are related to atherosclerosis, and diabetes is a major risk factor for cardiovascular diseases. There are multiple reasons for this. Diabetic individuals have abnormal lipid metabolism with increased low density lipoprotein cholesterol (LDL), and decreased high density lipoprotein cholesterol (HDL). Triglyceride levels are typically increased in patients with diabetes mellitus. In addition, glycation of lipoproteins may lead to altered functions of these lipoproteins.

Microvascular complications are related to the following mechanisms:

■ Non-enzymatic glycosylation of proteins.
■ Activation of protein kinase C.
■ Disturbance in the polyol pathway.

The degree of non-enzymatic glycosylation is related to the blood glucose level and measurement of glycosylated hemoglobin, and, less frequently, the glycosylated fructosamine level in blood is measured on a regular basis in patients with diabetes. Hyperglycemia-induced activation of protein kinase C results in the production of pro-angiogenic molecules that can cause neovascularization and also the formation of pro-fibrogenic molecules that lead to the deposition of extracellular matrix and basement membrane material. An increase in intracellular glucose leads to increased production of sorbitol by the enzyme aldose reductase. Sorbitol is a polyol, which is converted to fructose. Increased accumulation of sorbitol and fructose can cause cellular injury.

CASE REPORT

A 15-year-old female with known type 1 diabetes mellitus was admitted to the hospital with complaints of abdominal pain and fatigue for the past 24 h. Her glycemic control in the morning showed hypoglycemia (glucose: 61 mg/dL) and she omitted her morning insulin dose due to low glucose level and poor appetite. On admission, she was alert but showed marked Kussmaul breathing and the smell of ketones on her breath. She also showed severe hyperglycemia (glucose: 414 mg/dL) and blood gas analysis revealed severe metabolic acidosis (pH 6.99); her bicarbonate level was 5.0 mmol/L. She also showed an elevated anion gap of 29.8 mmol/L and increased base excess. However, blood urea, electrolytes, and liver enzymes were within normal limits. At the sixth hour of treatment with intravenous fluid and insulin, the patient became delirious. A brain image study did not reveal any edema or abnormal intracranial pathology. At the 18th hour of treatment, the patient developed a high fever and further laboratory investigation indicated that the patient had vulvovaginitis. Treatment with fluconazole was initiated. At the 24th hour of therapy, her acidosis was resolved completely, but she was still unconscious with little response to verbal stimuli. Finally, at the 36th hour, the patient was able to respond to commands and sit up. She was discharged a few days later and recovered completely. The authors established the diagnosis as severe diabetic ketoacidosis associated with infection [6].

CASE REPORT

A 26-year-old-African-American man presented to the emergency department with a three-week history of increased urination, thirst, fatigue, mild nausea, and weight loss. On admission, his random glucose level was 615 mg/dL, hemoglobin A1c was 15.8%, C-peptide was 0.4 ng/mL, and he had an elevated anion gap of 25 mmol/L. His venous blood gas showed a pH of 7.19, bicarbonate of 13.9 mmol/L, and pCO_2 of 38 mmHg. A urine dipstick analysis showed a small amount of ketones. A diagnosis of type 1 diabetes mellitus was established and the patient was given insulin. On discharge he was advised to stop taking ephedra and to start eating a balanced diet. He returned to the clinic after eight weeks and stated that he was not suffering from diabetes. His hemoglobin A1c was 6.2% and his blood glucose was 120 mg/dL, and he also said that he had discontinued the insulin. He was advised to continue monitoring his blood glucose. He returned to the clinic three months later with casual glucose between 90 and 140 mg/dL and continued to do well. At that time antibodies to islet cell and anti-glutamic acid decarboxylase antibody tests were ordered and both were negative. Lack of antibodies and his blood glucose levels raised suspicions regarding the initial diagnosis of type 1 diabetes mellitus. Finally the patient was managed with metformin, an oral hypoglycemic agent, and insulin was discontinued. The most likely diagnosis for this patient was ketosis prone diabetes, a disease most commonly encountered in African-Americans, but also observed in Hispanics, Asians, and sometimes in the Caucasian population [7].

7.9 SECONDARY CAUSES OF DIABETES MELLITUS

Gestational diabetes mellitus is typically seen during the second or third trimester of pregnancy, most likely due to increased levels of hormones such as estrogen, progesterone, cortisol, etc., which counteract the action of insulin. Although gestational diabetes may resolve after delivery, these women often have a higher risk of developing type 2 diabetes mellitus later. Pancreatic disease, endocrine diseases, and various drugs can also cause diabetes; these are all considered secondary diabetes. Various drugs can also cause diabetes. The causes of secondary diabetes are summarized in Table 7.2.

7.10 DIAGNOSTIC CRITERIA FOR DIABETES

The classic clinical presentation of diabetes mellitus patients includes:

- Polyuria
- Polydipsia
- Weight loss.

Table 7.2 Various Causes of Secondary Diabetes

Causes	Specific Examples
Pancreatic diseases	Cystic fibrosis, chronic pancreatitis, hemochromatosis
Endocrine diseases	Cushing's syndrome, acromegaly, pheochromocytoma
Drugs	Thiazide diuretics, glucocorticoids, thyroid hormones

The American Diabetes Association (ADA) recommends screening for diabetes mellitus for any individual over 45 years of age. Fasting blood glucose and glycosylated hemoglobin A1C are the best criteria for diagnosis of diabetes mellitus. Guidelines of Expert Committee on Diagnosis and Classification of Diabetes Mellitus indicate that the normal fasting glucose level should be 70–99 mg/dL (3.9–5.5 mmol/L). Individuals with fasting glucose levels between 100 (5.6 mmol/L) and 125 mg/dL (6.9 mmol/L) are classified as having impaired fasting glucose. Impaired glucose tolerance and impaired fasting glucose are both considered prediabetic conditions because individuals are at higher risk of developing diabetes. Impaired glucose tolerance is present in an individual when, in a glucose tolerance test, the two-hour glucose value is in the range of 140 mg/dL (7.8 mmol/L) to 199 mg/dL (11.0 mmol/L). These conditions are regarded as prediabetic because individuals are at higher risk of developing diabetes later. HbA1c values between 5.7 and 6.4% can also be considered prediabetic in adults.

The criteria for diagnosis of diabetes mellitus are a fasting plasma glucose level of 126 mg/dL (7.0 mmol/L) or higher on more than one occasion with no calorie intake in the last eight hours, or in a patient with classic symptoms of hyperglycemia, a random plasma glucose level of 200 mg/dL (11.1 mmol/L) or higher. In addition, a glucose tolerance test after an oral dose of 75 gm of glucose, with a two-hour plasma glucose level of 200 mg/dL (11.1 mmol/L) or higher, indicates diabetes mellitus. However, oral glucose tolerance tests to establish diagnosis of diabetes mellitus are only recommended in pregnant women to establish the diagnosis of gestational diabetes. An International Expert Committee recommended use of a hemoglobin A1C (glycated hemoglobin) test for diagnosis of diabetes mellitus with a cut-off value of 6.5% [8]. Criteria for diagnosis of diabetes mellitus are the same for adults and children.

For diagnosis of gestational diabetes, typically a glucose tolerance test is performed using 75 g of oral anhydrous glucose during weeks 24–28 of gestation in pregnant women. The glucose tolerance test is typically performed in the morning after at least eight hours of overnight fasting. Various laboratory-based criteria for diagnosis of diabetes mellitus and gestational diabetes are summarized in Table 7.3. If a woman has gestational diabetes mellitus, then the risk of developing diabetes later is higher for both mother and child.

7.11 HYPOGLYCEMIA

Hypoglycemia is arbitrarily defined as blood glucose below 50 mg/dL (2.8 mmol/L), but some authors favor a cut-off value of 60 mg/dL.

Table 7.3 Laboratory-Based Criteria for Diagnosis of Diabetes Mellitus and Gestational Diabetes Mellitus

Laboratory Test	Value	Comment
Diabetes Mellitus		
Fasting blood glucose*	70–99 mg/dL	Normal value
	100–125 mg/dL	Impaired fasting glucose
	>126 mg/dL	Determined in at least two occasions indicative of diabetes
Random blood glucose	>200 mg/dL	Indicative of diabetes in a patient with suspected diabetes mellitus
Glucose tolerance test (2 h)#	<140 mg/dL	Normal
	140–199 mg/dL	Impaired glucose tolerance
	>200 mg/dL	Indicative of diabetes
Hemoglobin A1 C^	5.7–6.4%	Increased risk of diabetes
	>6.5%	Indicative of diabetes
Gestational Diabetes		
Glucose tolerance test	>92 mg/dL (fasting)	Any of the criteria (fasting glucose or 1 h or 2 h glucose in glucose tolerance test is equal or exceeding the limit
	>180 mg/dL (1 h)	
	>153 mg/dL (2 h)	

*Fasting glucose means no calorie intake for at least 8 h.
#Glucose tolerance test is typically performed using 75 g of glucose given orally; the test is preferably performed in the morning in ambulatory patients after overnight fasting.
^International Federation of Clinical Chemistry (IFCC) recommends expressing the hemoglobin A1c value in mmol of HbA1c/mol of hemoglobin unit.

Epinephrine is mostly responsible for symptoms of hypoglycemia, including trembling, sweating, lightheadedness, hunger, and possibly epigastric discomfort. The brain is dependent on glucose for energy, but under prolonged fasting conditions, ketones may be used for energy. In neonates, blood glucose is usually lower than in adults, and a blood glucose level of 30 mg/dL may be encountered in a neonate without any symptoms of hypoglycemia. Hypoglycemia can be related to prolonged fasting, but other disease conditions may precipitate such a condition. Postprandial or reactive hypoglycemia may be related to insulin therapy, inborn errors of metabolism, or other causes. Various causes of fasting and reactive hypoglycemia are summarized in Table 7.4. Hypoglycemia can occur in both type 1 and type 2 diabetes, but is more common in type 1 diabetic patients receiving insulin. Some type 1 diabetic patients may experience hypoglycemia even once or twice a week. Insulinomas are tumors of the insulin-secreting beta cells of the islets of

Table 7.4 Common Causes of Hypoglycemia

Fasting Hypoglycemia
Endocrine diseases, for example hypoadrenalism, hypopituitarism, etc.
Hepatic failure
Renal failure
Neoplasms
Inborn errors of metabolism, for example, glycogen storage disease type 1
Insulinoma
Reactive Hypoglycemia
Postprandial due to gastric surgery or idiopathic
Inborn errors of metabolism, for example, galactosemia, fructose 1,6-diphosphate deficiency
Drug-induced but most likely due to insulin therapy or oral hypoglycemic agents such as sulfonylureas
Alcohol abuse

Langerhans. Insulin levels are high along with high levels of C-peptide. Individuals with insulinomas usually have characteristic features of hypoglycemia and high insulin, as well as high C-peptide levels. Imaging techniques are used to detect the tumor. In a patient with symptoms of hypoglycemia, measurement of blood glucose should confirm low blood glucose and administration of glucose should alleviate the symptoms. These three characteristics together are known as Whipple's triad.

7.12 LABORATORY METHODS

In a clinical laboratory, glucose concentration is usually measured in serum or plasma. However, glucose concentration can be measured in whole blood. It is important to note that fasting whole blood glucose concentration is approximately 10−12% lower than corresponding plasma or serum glucose concentration. Glycolysis reduces the blood sugar level in an uncentrifuged specimen by 5 to 10 mg/dL per hour, and timely centrifugation of the specimen is required. However, the best practice is to collect blood in tubes containing sodium fluoride and potassium oxalate (fluoride/oxalate tube; gray top) because sodium fluoride inhibits glycolysis and the glucose level is stable for up to 3 days at room temperature. A less commonly used preservative for glucose collection tubes is sodium iodoacetate.

Measurement of glucose can be done using either the hexokinase method or the glucose oxidase method. The hexokinase method is considered the reference method.

The hexokinase method is based on the following reactions (Equations 7.1 and 7.2):

$$\text{Glucose} + \text{ATP} \xrightarrow{\text{Hexokinase}} \text{Glucose 6-phosphate} + \text{ADP} \tag{7.1}$$

$$\text{Glucose 6-phosphate} \xrightarrow[\text{NAD} \;\diagup\diagdown\; \text{NADH}]{\text{Glucose 6-phosphate dehydrogenase}} \text{6-Phosphogluconate} \tag{7.2}$$

In the second reaction, NAD (which has no absorption at 340 nm) is converted into NADH (which absorbs at 340 nm) and the absorbance is proportional to the glucose concentration.

In the glucose oxidase method, the following reaction takes place (Equation 7.3):

$$\text{Glucose} \xrightarrow{\text{Glucose oxidase}} \text{Gluconic acid} + \text{Hydrogen peroxide} \tag{7.3}$$

In this enzymatic reaction, hydrogen peroxide is generated and its concentration is measured to determine glucose concentration. The concentration of hydrogen peroxide can be measured by the addition of peroxidase enzyme and an oxygen receptor such as *o*-dianisidine, which, when oxidized, forms a colored complex that can be measured spectrophotometrically.

In the glucose dehydrogenase method, glucose is oxidized to gluconolactone, and NAD is converted into NADH (Equation 7.4):

$$\text{Glucose} + \text{NAD} \xrightarrow{\text{Glucose dehydrogenase}} \text{Gluconolactone} + \text{NADH} \tag{7.4}$$

Mutarotase is added to shorten the time needed to reach the end point. The amount of NADH formed (signal at 340 nm) is proportional to the glucose concentration. Although NAD is a cofactor required for the catalytic reaction for glucose oxidase, other forms of glucose oxidase can use either pyrroloquinoline quinone (PQQ) or flavin dinucleotide (FAD) as a cofactor. These methods are used in point of care glucose meters.

7.13 GLUCOSE METERS

Since the 1980s, portable glucose meters (glucometers) have been available for blood glucose monitoring both in point of care testing sites and in home monitoring of glucose, especially for patients receiving insulin or patients with type 2 diabetes who have difficulty maintaining good glucose control. In order to perform a measurement, a sample of blood (usually a fingerstick) is placed on the test pad and then this test strip is inserted into the meter (or the test strip may already be inserted in the meter and there is a point for application of a drop of blood). After a short period, a digital reading is taken. Some more recent meters also have memory where a value can be stored for a period of time. Glucose meters also utilize glucose oxidase, hexokinase, or glucose dehydrogenase with pyrroloquinoline quinone cofactor (PQQ) or glucose oxidase combined with nicotinamide adenine dinucleotide (NAD) for glucose measurement; a final reading can be made by reflectance photometry or electrochemical measurement methods.

Accuracy of glucose meters is of major concern. The following criteria must be met by a glucose home monitoring method:

- Current Food and Drug Administration (FDA) criteria for acceptability of a glucometer are that 95% of all values should fall within ±15 mg/dL of the glucose value obtained by a clinical laboratory-based reference method at a glucose concentration <75 mg/dL. For glucose values over 75 mg/dL, 95% of the individual values must fall within ±20% of the glucose value determined by a reference method.
- Glucose readings obtained by a glucose meter should not be used for diagnosis of diabetes.
- ADA criteria suggest that the glucose value measured by a glucose meter should be within ±5% of the value obtained by a laboratory-based glucose assay.

Major limitations of glucose meters: inaccuracy of measurement compared to a reference glucose method, as well as various interferences. During peritoneal dialysis, infuse may contain icodextrin, which is converted into maltose by the human body. Maltose falsely increases the glucose value if the glucose meter is based on the glucose oxidase method. In 2009, the deaths of 13 patients who were on peritoneal dialysis were reported to the FDA; the deaths were due to falsely elevated glucose readings using glucose meters. Falsely elevated glucose values can result in insulin overdose where a clinician may think that a patient was severely hyperglycemic when in reality the glucose value may be within acceptable limits. Insulin overdose can cause death. Even failure to dry hands after hand washing prior to the prick for glucose monitoring can falsely decrease glucose readings due to hemodilution [9]. Major interferences include:

- An elevated concentration of ascorbic acid (vitamin C) can falsely elevate a reading in glucose dehydrogenase-based methods, but small changes in value in both directions (positive/negative) may occur with glucose meters using the glucose oxidase method.
- Maltose, xylose, or galactose can falsely increase the glucose value in the glucose meter using the glucose dehydrogenase−PQQ method (but not the glucose dehydrogenase−NAD-based method).

- Hematocrit affects the reading, and, regardless of methods, anemia can falsely elevate the glucose reading.
- Hypoxia or increased altitude can falsely elevate readings in glucose meters using the glucose oxidase method.
- Diabetic ketoacidosis can falsely decrease glucose readings regardless of the method.

CASE REPORT

A 76-year-old man had three unexpectedly high readings (over 400 mg/dL) on a point of care glucose meter using the glucose dehydrogenase–PQQ method (Roche). Glucose kinase measured using a laboratory-based analyzer (Abbott Architect, hexokinase method) at about the same time showed blood glucose between 129–202 mg/dL. Therefore, interference in the point of care glucose meter was suspected, but none of the medications the patient was taking accounted for the interference. The patient was taking Nepro with Carb Steady and an Abbott Nutrition Product containing maltitol and Fibersol (produced by a combination of acidification and heating of maltodextrin). Further investigation showed that these compounds were responsible for interference with the glucose measurement using the glucose meter [10].

Glucose monitoring in urine is usually performed with a urine dipstick that has a pad for the detection of glucose. Many such test strips use the glucose oxidase method. However, glucose tests in urine lack specificity because false positive results can be encountered if hydrogen peroxide or a strong oxidizing agent is present, and false negative results can occur if reducing substances such as ascorbic acid, ketones, or salicylate are present in urine. Urine dipsticks usually provide qualitative results, but quantitative methods for measuring glucose in urine are also available. Tests for detecting ketone bodies in urine usually detect acetoacetate but not beta-hydroxybutyric acid, the major ketone in urine. However, enzyme assays are available for estimation of beta-hydroxybutyric acid.

Glucose and galactose are major reducing monosaccharides present in urine. The most common cause of galactose in urine is galactosemia, a rare inborn error of metabolism. Usually Clinitest is used to detect the presence of reducing sugars in urine. This test utilizes the ability of a reducing sugar to reduce copper sulfate to cuprous oxide in the presence of sodium hydroxide; the characteristic orange color of cuprous oxide (copper sulfate is blue) qualitatively indicates the presence of a reducing sugar in urine. If the Clinitest is positive and the urine dipstick indicates negative glucose, then most likely galactose is present and an enzymatic test should be performed to establish the diagnosis of galactosemia.

Glycated hemoglobin (hemoglobin A1c) in whole blood can be determined by various methods, including high-performance liquid chromatography, ion exchange microcolumn, affinity chromatography, capillary electrophoresis, and immunoassays. However, liquid chromatography combined with mass spectrometry is usually considered the reference method (although most clinical laboratories do not use this method). Patients do not need to fast, and specimens can be collected in EDTA tubes or oxalate and fluoride tubes. However, the glycosylated hemoglobin value may be falsely lowered due to autoimmune hemolysis where life spans of erythrocytes are shorter than normal. Ribavirin and other drugs that may cause anemia can also cause a false decline in the glycosylated hemoglobin value [11]. In contrast, iron deficiency anemia leads to falsely increased glycosylated hemoglobin levels, and even normal individuals with iron deficiency anemia can have abnormally elevated glycosylated hemoglobin levels [12].

In selected diabetic patients, fructosamine can be measured. Fructosamine is a generic name for plasma protein ketoamines (non-enzymatic attachment of glucose to amino groups of proteins). In principle, glycated albumin is the major fructosamine present in serum, but fructosamine provides long-term blood glucose only for the previous 2−3 weeks, where glycosylated hemoglobin can indicate glucose control over the past 2−3 months. The normal level of fructosamine in adults is 160−240 µmol/L [13]. However, the normal reference range of 200−285 µmol/L has also been reported.

KEY POINTS

- Common monosaccharides are glucose, fructose, and galactose. Lactose, the disaccharide present in milk, is composed of galactose and glucose. Maltose is composed of two glucose molecules.
- Insulin is a 51-amino acid polypeptide secreted by the beta cells of the islets of Langerhans in the pancreas. The molecule proinsulin is cleaved to form insulin and C-peptide. Insulin induces cellular uptake of glucose and glycogen, and protein, fatty acid, and triglyceride synthesis. Insulin inhibits glycogenolysis, gluconeogenesis, proteolysis, and lipolysis.
- Diabetes mellitus is a syndrome characterized by hyperglycemia due to relative insulin deficiency or insulin resistance. Diabetes mellitus can be primary or secondary in nature. Primary diabetes mellitus can be monogenic or polygenic. Monogenic diabetes mellitus covers a heterogenous group of diabetes caused by a single gene mutation and characterized by impaired insulin secretion by beta cells of the pancreas. Maturity onset diabetes of the young (MODY), mitochondrial diabetes, and neonatal diabetes are examples of monogenic diabetes mellitus. Polygenic diabetes mellitus can be either type 1 or type 2.
- Type 1 diabetes mellitus (formerly called insulin-dependent diabetes) is characterized by an absolute deficiency of insulin due to islet cell destruction, and usually presents in younger people with acute onset.

- Type 1 diabetes is due to autoimmune destruction of pancreatic beta cells by T lymphocytes. Markers of immune destruction of beta cells in these patients include islet cell autoantibodies as well as autoantibodies to insulin, glutamic acid decarboxylase (GAD), and tyrosine phosphatases. Usually 85–90% of patients with type 1 diabetes mellitus have one or more autoantibodies.
- Type 2 diabetes mellitus (formerly called non-insulin-dependent diabetes mellitus) is characterized by insulin resistance and beta cell dysfunction in the face of insulin resistance and hyperglycemia.
- Type 2 diabetes mellitus is adult onset and is characterized by insulin resistance and also may be accompanied by beta cell dysfunction causing insulin deficiency. Many patients with type 2 diabetes mellitus are obese because obesity itself can cause insulin resistance.
- Diabetic complications can be divided into two broad categories: acute and chronic complications. Acute complications include diabetic ketoacidosis (DKA), hyperosmolar non-ketosis, and lactic acidosis. Chronic complications can be either macrovascular (stroke, myocardial infarction, gangrene, etc.) or microvascular (diabetic retinopathy, diabetic eye diseases, and diabetic neuropathy).
- Microvascular complications are related to non-enzymatic glycosylation of protein, activation of protein kinase C, and disturbance in the polyol pathway. The degree of non-enzymatic glycosylation is related to blood glucose level; measurement of glycosylated hemoglobin (and, less frequently, glycosylated fructosamine) level in blood is measured on a regular basis in patients with diabetes.
- The American Diabetes Association (ADA) recommends screening for diabetes mellitus for any individual over 45 years of age. Fasting blood glucose and glycosylated hemoglobin A1C are the best criteria for diagnosis of diabetes mellitus. Guidelines of Expert Committee on Diagnosis and Classification of Diabetes Mellitus indicate that normal fasting glucose levels should be 70–99 mg/dL (3.9–5.5 mmol/L). Individuals with fasting glucose levels between 100 (5.6 mmol/L) and 125 mg/dL (6.9 mmol/L) are classified as having impaired fasting glucose. The criteria for diagnosis of diabetes mellitus are a fasting plasma glucose level of 126 mg/dL (7.0 mmol/L) or higher on more than one occasion with no calorie intake in the previous eight hours; in a patient with classic symptoms of hyperglycemia, a random plasma glucose level of 200 mg/dL (11.1 mmol/L) or higher is an indication of diabetes mellitus. In addition, in a glucose tolerance test after an oral dose of 75 g of glucose, a two-hour plasma glucose of 200 mg/dL (11.1 mmol/L) or higher indicates diabetes mellitus. However, oral glucose tolerance tests to establish diagnosis of diabetes mellitus are only recommended in a pregnant woman to establish the diagnosis of gestational diabetes. An International Expert Committee recommended use of a hemoglobin A1C (glycated hemoglobin) test for diagnosis of diabetes mellitus with a cut-off value of 6.5%. Criteria for diagnosis of diabetes mellitus are the same in both adults and children.

- For diagnosis of gestational diabetes, typically a glucose tolerance test is performed using 75 g of oral anhydrous glucose during week 24–28 of gestation in pregnant women.
- In a clinical laboratory, glucose concentration is usually measured in serum or plasma. However, occasionally glucose concentration may be measured in whole blood. Fasting whole blood glucose concentration is approximately 10–12% lower than corresponding plasma or serum glucose concentration. Glycolysis reduces blood sugar levels in an uncentrifuged specimen by 5 to 10 mg/dL per hour. The best practice is to collect blood in tubes containing sodium fluoride and potassium oxalate (fluoride/oxalate tube; gray top) because sodium fluoride inhibits glycolysis and the glucose level is stable up to 3 days at room temperature.
- Measurement of glucose can be done using either the hexokinase method or glucose oxidase method. The hexokinase method is considered as the reference method.
- Glucose meters also utilize glucose oxidase, hexokinase, glucose dehydrogenase with pyrroloquinoline quinone (PQQ) or glucose oxidase combined with nicotinamide adenine dinucleotide (NAD) for glucose measurement. A glucose reading obtained by a glucose meter should not be used for diagnosis of diabetes.
- Major interferences while using glucose meters can occur from vitamin C, acetaminophen, hematocrit, hypoxia, maltose, xylose, galactose, and diabetic ketoacidosis.
- Glucose monitoring in urine is usually performed using urine dipsticks, which have a pad for detection of glucose. Many such test strips use the glucose oxidase method. However, a glucose test in urine lacks specificity because a false positive result may be encountered if hydrogen peroxide or a strong oxidizing agent is present and a false negative result may occur if reducing substances such as ascorbic acid, ketones, or salicylate are present in urine.
- Glucose and galactose are major reducing monosaccharides present in urine. The most common cause of galactose in urine is galactosemia, a rare inborn error of metabolism. Usually Clinitest is used to detect the presence of reducing sugars in urine. If Clinitest is positive and the urine dipstick indicates negative glucose, then most likely galactose is present.
- Tests for detecting ketone bodies in urine usually detect acetoacetate but not beta-hydroxybutyric acid, the major ketone in urine. However, enzyme assays are available for estimation of beta-hydroxybutyric acid.

REFERENCES

[1] Fajans SS, Graeme IB, Polonksy KS. Molecular mechanisms and clinical pathophysiology of maturity-onset diabetes in the young. N Eng J Med 2001;345:971–80.

[2] Johansson S, Irgens H, Chudssama KK, Molnes J, et al. Exon sequencing and genetic testing for MODY. PLoS One 2012;7:e38050.

[3] Murphy R, Ellard S, Hattersley AT. Clinical implications of a molecular genetics classification of monogenic beta-cell diabetes. Nat Clin Pract Endocrinol Metab 2008;4:200–13.

[4] Henzen C. Monogenic diabetes mellitus due to defect in insulin secretion. Swiss Med Wkly 2012;142:w13690.

[5] Balkau B, Valensi P, Eschwege E, Slama G. A review of metabolic syndrome. Diabetes Metab 2007;33:405—13.

[6] Cebeci A, Guven A. Delirium in diabetic ketoacidosis: a case report. J Clin Res Pediatr Endocrinol 2012;4:39—41.

[7] Palmer C, Jessup A. Ketoacidosis in patients with type 2 diabetes. Nurs Pract 2012;37:13—7.

[8] International Expert Committee. International expert committee report on the role of the A1C assay in the diagnosis of diabetes mellitus. Diabetes Care 2009;32:1327—34.

[9] Hellman R. Glucose meter inaccuracy and the impact on the care of patients. Diabetes Metab Res Rev 2012;28:207—9.

[10] Kelly BN, Haverstick DM, Bruns D. Interference in a glucose dehydrogenase based glucose meter. Clin Chem 2010;56:1038—40.

[11] Trask L, Abbott D, Lee HK. Low hemoglobin A1c: good diabetic control? Clin Chem 2012;58:648—9.

[12] Shanthi B, Revathy C, Manjula Devi AJ, Subhashree. Effect of iron deficiency on glycation of haemoglobin. J Clin Diagn Res 2013;7:15—7.

[13] Chen HS, Wu TE, Lin HD, Jap TS, et al. Hemoglobin A (1c) and fructosamine for assessing glycemic control in diabetic patients with CKD stage 3 and 4. Am J Kidney Dis 2010;55:867—74.

CHAPTER 8

Cardiac Markers

8.1 MYOCARDIAL INFARCTION

Cardiovascular disease accounts for approximately 37% of all deaths in the United States, making it the number one cause of mortality. Acute coronary syndrome is a broad term that covers unstable angina, non-ST-segment elevated myocardial infarction, and ST-segment elevated myocardial infarction. Although every year several million individuals report to the emergency department with symptoms suggestive of acute coronary syndrome, an estimated 1.4 million individuals are admitted to the hospital with acute coronary syndrome in U.S. hospitals alone. Nearly 70% of these patients suffer from unstable anginal or non-ST-segment elevated myocardial infarction. In patients with non-ST-segment elevated myocardial infarction, assessment of clinical symptoms and measurement of cardiac biomarkers are critical for proper diagnosis [1]. In addition, myocardial infarction (MI) is responsible for approximately 500,000 deaths per year in the U.S. For the diagnosis of MI, two of the following three criteria are required:

- Typical symptoms.
- Characteristic rise-and-fall pattern of a cardiac marker (e.g. serum troponin I or troponin T).
- A typical electrocardiogram (ECG) pattern involving the development of Q waves.

However, newer guidelines indicate that elevated troponin in the context of ischemia can be considered as an indication of myocardial infarction. In this chapter, emphasis is on cardiac markers and the laboratory aspect of diagnosis of myocardial infarction. The earlier a myocardial infarction is diagnosed, the better the outcome. Because several hours are needed for a cardiac biomarker to achieve relatively high concentration in blood, early diagnosis of myocardial infarction is still a clinical diagnosis.

CONTENTS

A. Dasgupta and A. Wahed: Clinical Chemistry, Immunology and Laboratory Quality Control
DOI: http://dx.doi.org/10.1016/B978-0-12-407821-5.00008-5

8.2 OVERVIEW OF CARDIAC MARKERS

In a broad sense, cardiac markers are endogenous substances released in the circulation when the heart is damaged or stressed. Acute coronary syndrome is caused by a plaque formed due to atherosclerosis, which causes thrombus formation in the damaged coronary artery and results in a sudden decrease in the amount of blood and oxygen reaching the heart. Angina is due to reduced blood supply in the heart, and when such blood flow is interrupted for 30−60 min, it can cause necrosis of heart muscle, resulting in myocardial infarction. Cardiac biomarkers are released in the circulation due to damage or death of cardiac myocytes, and measuring these biomarkers in serum or plasma is useful in the diagnosis of myocardial infarction. There are four established biomarkers for myocardial necrosis:

- Myoglobin
- Creatine kinase isoenzymes
- Cardiac troponin I
- Cardiac troponin T

Creatine kinase is an enzyme that is often called cardiac enzyme. However, it is important to note that troponin I and troponin T are proteins but not enzymes. Out of all these biomarkers, troponin I is the most specific for myocardial necrosis. Characteristics of various established cardiac markers are listed in Table 8.1. There are other markers for myocardial damage, but they are less often used and are currently not considered established biomarkers (as are troponin I or troponin T). These markers include:

- Glycogen phosphorylase BB
- Ischemic-modified albumin

Table 8.1 Various Established Cardiac Markers

Cardiac Marker	Increases	Peak	Return to Baseline	Comments
Myoglobin	1−4 h	4−12 h	24−36 h	Earliest marker, but non-specific with negative predictive value.
CK-MB	4−9 h	24 h	48−72 h	Gold standard before troponin was introduced. Mostly found in cytosol but may increase in non-MI situation.
Troponin I/T	4−9 h	12−24 h	7−14 days	Most specific marker. Found in small amounts in cytosol, but mostly in sarcomere of cardiac myocytes (both early and late marker). Troponin T is less specific than troponin I because troponin T is also found in muscle.

- Pregnancy-associated plasma protein A
- Heart-type fatty acid binding protein.

There are also biomarkers to assess myocardial stress that may aid in the diagnosis and prognosis of acute coronary syndrome. There are many biomarkers in this category that have been described in the literature, but currently B-type natriuretic peptide (BNP) or its precursor N-terminal pro-B-type natriuretic peptide (NT-proBNP) are two of the main ones. Biomarkers that indicate stress of myocardium include:

- BNP and NT-proBNP
- C-reactive protein
- Myeloperoxidase.

Lipid parameters are useful in determining risk factors for cardiovascular diseases. Please see Chapter 6 for an in depth discussion on this topic.

Cardiac markers are used in the diagnosis and risk stratification of patients with chest pain and suspected acute coronary syndrome, and cardiac troponin (especially troponin I) is particularly useful in diagnosis of myocardial infection due to its superior specificity compared to other cardiac biomarkers. Individuals with symptoms of myocardial infarction and elevated troponin I or troponin T, but without electrocardiogram (EKG) changes, are now classified as non-ST-segment elevation MI (NSTEMI). Only one elevated troponin level above the established cutoff is required to establish the diagnosis of acute myocardial infarction according to the American College of Cardiology Foundation/American Heart Association guidelines for NSTEMI [2]. Cardiac markers are not necessary for the diagnosis of patients who present with ischemic chest pain and diagnostic EKGs with ST-segment elevation. These patients may be candidates for thrombolytic therapy or primary angioplasty. Treatment should not be delayed to wait for cardiac marker results, especially since the sensitivity is low in the first 6 hours after onset of initial symptoms. Timing of release of various cardiac markers in circulation after myocardial infarction is shown in Figure 8.1.

8.3 MYOGLOBIN

Myoglobin is a heme protein found in both skeletal and cardiac muscle. Myoglobin is typically released in the circulation as early as 1 h after myocardial infarction, with a gradual increase that reaches a peak at 4−12 hours and returns to normal within 24−36 hours. Rapid release of myoglobin probably reflects its low molecular weight (17 kDa) and cytoplasmic location. Myoglobin is an early marker of acute myocardial

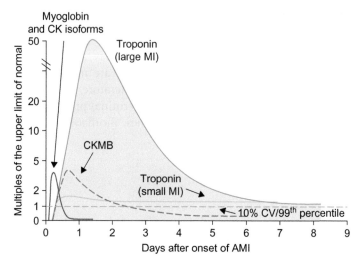

FIGURE 8.1

Timing of release of various cardiac biomarkers after myocardial injury. This figure is reproduced in color in the color plate section.
© *American Heart Association. Reprinted with permission [3].*

infarction and exhibits a high negative predictive value. Myoglobin has poor clinical specificity (60–90%) due to the presence of large quantities of myoglobin in skeletal muscle. Myoglobin, therefore, is potentially useful for ruling out myocardial infarction but not for confirming the diagnosis of acute myocardial infarction. However, some studies suggest adding the myoglobin test to the troponin I test in order to improve diagnostic value [4].

Myoglobin, being a small protein, is excreted in urine, and a high level of serum myoglobin is encountered in patients with acute renal failure (uremic syndrome). Acute renal failure is also a complication of rhabdomyolysis, and very high quantities of myoglobin in serum in the range of 1,253 to 15,450 (median value 3,335 ng/mL) have been observed in patients with rhabdomyolysis-induced acute renal failure. However, the peak value of blood myoglobin can be a good predictor of rhabdomyolysis-induced acute renal failure [5]. Therefore, elevated myoglobin in the absence of normal levels of other cardiac biomarkers is not indicative of myocardial infarction. Woo et al. reported that in a group of 42 patients, 22 patients were later diagnosed with myocardial infarction, but in 11 patients who did not have myocardial infarction, myoglobin concentrations were falsely elevated and these patients with false positive myoglobin levels had various degrees of muscular trauma or renal disorder [6].

8.4 CREATINE KINASE ISOENZYME: CK-MB

Prior to the introduction of cardiac troponin T or troponin I, the biochemical marker of choice for the diagnosis of acute myocardial infarction was creatine kinase isoenzyme. Creatine kinase is a dimeric enzyme that catalyzes the transfer of high-energy phosphate groups, and is found in many tissues that consume large amounts of energy, including myocardial and skeletal muscle. Creatine kinase has two subunits, including M type (for muscle) and B type (for brain). The CK-MM isoenzyme is dominant in adult skeletal muscle (98% of total muscle CK) while CK-BB is found mostly in the central nervous system. In the myocardium, CK-MB is present in a relatively high concentration (15−20% of the total myocardial CK), while about 85% or more is CK-MM. Therefore, CK-MM is the most abundant isoenzyme of creatine kinase.

The molecular weight of CK-MB is roughly 86,000, and during myocardial injury CK-MB first rises, then reaches a peak level and returns to normal, usually according to the following pattern[7]:

- CK-MB concentration gradually rises from 4 to 9 h after the onset of chest pain.
- Peak concentration is reached in approximately 24 h.
- Concentration of CK-MB begins to decline, reaching baseline level in 48 to 72 h.

The criterion most commonly used for the diagnosis of acute myocardial infarction using CK-MB is two serial elevations above the diagnostic cutoff level, or a single result more than twice the upper limit of normal. Although CK-MB is more concentrated in the myocardium, it is also present in skeletal muscle; false positive elevations occur in a number of clinical settings, including trauma, heavy exertion, and myopathy. Elevation of the total CK level is not cardiac-specific, and may be observed in patients with skeletal muscle injury and other disorders.

Because CK-MB remains elevated for a longer period of time following myocardial infarction, it is useful to detect re-infarction using serial CK-MB measurements. Following myocardial or skeletal muscle injury, both the total CK and CK-MB levels gradually increase. In order to differentiate from cardiac and skeletal muscle as the source of elevation of CK-MB, calculation of the relative index (RI) can be used (Equation 8.1):

$$\text{CK-MB RI} = \frac{\text{CK-MB(ng/mL)}}{\text{Total CK(U/L)}} \times 100 \qquad (8.1)$$

It is important to note that in calculating the index, the CK-MB concentration is expressed in ng/mL, while total CK level is expressed in U/L; in a strict mathematical sense, the ratio can only be calculated when both numerator and denominator are expressed in the same units. However, when this index was originally proposed, CK-MB was measured as activity (U/L) just like total CK, but due to interferences in the CK-MB assay, more specific assays for CK-MB were developed using two different antibodies. Such specific assays measure CK-MB mass. Nevertheless, this approach is useful and an index of less than 3% is consistent with a skeletal muscle source, while an index greater than 5% is indicative of a cardiac source suggestive of acute myocardial infarction. Ratios between 3 and 5 represent a gray zone. An exception may occur for patients with chronic myopathic disorders in which the skeletal muscle CK-MB content may be increased. Chronic myopathic disorder usually produces a persistent elevation in the CK-MB level as opposed to the typical rise-and-fall temporal pattern seen in patients with acute myocardial infarction. The CK-MB level may also be elevated in patients with non-ischemic cardiac injury (including myocarditis), as well as in non-cardiac diseases such as seizure, pulmonary embolism, skeletal muscle trauma, etc. Marathon and long distance runners usually have elevated CK-MB. In addition, alcohol and various drug abuse (e.g. opiates) may increase CK-MB level. Various common causes of elevated CK-MB other than myocardial infarction are summarized in Table 8.2. A false negative CK-MB is rarely encountered in clinical situations and is most likely due to the time when the specimen was collected (too soon after infarction) or when the infarction was small, causing only an intra-normal bump [8].

The CK-MB isoenzyme exists as two isoforms: CK-MB1 and CK-MB2. Laboratory determination of CK-MB actually represents the simple sum of the isoforms CK-MB1 and CK-MB2. However, CK-MB2 is the tissue form and is initially released from the myocardium after MI. It is converted peripherally in serum to the CK-MB1 isoform rapidly after onset of symptoms. Normally, the tissue CK-MB1 isoform predominates; thus, the CK-MB2/CK-MB1 ratio is typically less than 1. A result is positive if the CK-MB2 is elevated and the ratio is greater than 1.7. CK-MB2 can be detected in serum within 2−4 hours after onset and peaks at 6−9 hours, making it an early marker for acute myocardial infarction.

Macro CK is a high-molecular-weight complex of one of the CK isoenzymes and immunoglobulin, but most commonly CK-BB with IgG. Two atypical macro CKs with molecular masses over 200 kDa have been described (macro CK types 1 and 2), but their clinical significance is unknown. Lee et al. reported that the prevalence of macro CK type 1 is only 0.43% while the prevalence of macro CK type 2 is only 1.2% and is usually associated with autoimmune disease or malignancy [9]. Macro CK can persist in serum for a

Table 8.2 Various Common Causes of Elevated CK-MB Other Than Myocardial Infarction

Disease/Cause	Comments
Myocarditis	During active inflammation, CK-MB may be significantly elevated (but not always).
Myositis	Elevated CK-MB; may have cardiac involvement.
Cardiac surgery	May increase both total CK and CK-MB concentration.
Muscular dystrophy	Elevated CK-MB may be found.
Muscle trauma	Both total CK and CK-MB elevated.
Pulmonary embolism	Elevated CK-MB usually observed.
Rhabdomyolysis	Increases both total CK and CK-MB.
Hypothermia	May cause elevation of CK-MB level due to possible myocardial damage.
Hypothyroidism	Elevated level due to reduced clearance.
Seizure	Elevated CK-MB may be related to cardiac involvement.
Long distance runner	Probably due to muscle damage.
Renal failure	Mechanism not clearly characterized.
Alcohol overdose	Toxic effect of excess alcohol on muscle and myocardium.
Opiate overdose	Morphine and other opiate overdose may increase both total CK and CK-MB.

long time and may interfere with CK-MB measurement, especially if a mass assay (CK-MB concentration expressed in U/L) is used. However, electrophoresis of CK-isoenzyme can resolve this issue.

CASE REPORT

A 49-year-old woman was admitted to the emergency department with significant hyperglycemia (glucose: 720 mg/dL), elevated total CK (196 U/L), and elevated CK-MB (84.7 U/L), but her troponin I level (0.005 ng/mL) was within normal range. Her EKG was normal and her total CK value returned to normal within a week of her hospital stay. Her troponin I value was never elevated. Interference was suspected in the CK-MB measurement, and CK enzyme electrophoresis demonstrated that the cause of interference in the CK-MB assay was the presence of macro CK type 2 in the specimen. However, this patient had no malignancy. The 3-year follow up still showed elevated CK-MB (30.9 U/L), although the value was reduced significantly from her initial presentation. CK-isoenzyme electrophoresis again showed the presence of macro CK type 2 [10].

8.5 TROPONIN I AND TROPONIN T

Cardiac troponin I and troponin T are effective in identifying myocardial damage, but in addition to being useful for diagnosis, they also permit the estimation of prognosis and risk stratification of patients with acute myocardial infarction. Cardiac troponin T and troponin I (especially troponin I) have been accepted as the "gold standard" in the evaluation of patients with acute myocardial infarction. Troponin is a regulatory complex of three protein subunits located on the thin filament of the myocardial contractile apparatus, and is composed of three subunits encoded by different genes. The three subunits are designated as follows:

- Troponin C (calcium-binding component; molecular weight of 18 kDa).
- Troponin T (tropomyosin-binding component; molecular weight of 21 kDa).
- Troponin I (inhibitory component; molecular weight of 37 kDa).

The majority of both cardiac troponin T and I are stored in the sarcomere, and a small amount (4−6%) is found in a cytosolic pool. Following myocardial damage, cytosolic troponin is released first, and as further damage occurs, troponin present in the sarcomere is released into circulation. This makes troponin both an early and late marker of acute myocardial infarction. The kinetics of release of troponin T and troponin I after myocardial damage are as follows:

- Levels of troponin T and I start increasing 4 to 9 hours after acute myocardial infarction.
- They peak at 12 to 24 hours.
- They can remain elevated for up to 14 days.

Cardiac troponin T and I have many advantages over CK-MB. First, levels of troponin in normal individuals are very low or non-detectable. Therefore, significant elevation of troponin is indicative of injury to the myocardium. A small elevation of troponin (but a normal CK-MB level) may indicate a microscopic zone of myocardial necrosis (microinfarction). Troponin I is very specific for myocardium because only one isoform of cardiac troponin I has been identified; it is found exclusively in cardiac myocytes. Cardiac troponin I is not expressed in skeletal muscle. Although cardiac troponin T has a different amino acid sequence when compared to other troponins, small amounts of cardiac troponin T have been identified in skeletal muscle. In humans, cardiac troponin T isoform expression has been reported in patients with muscular dystrophy, polymyositis, dermatomyositis, and end-stage renal disease. Troponin C is not a useful marker as it is not cardiac-specific. Elevated levels of both troponin T and I provide independent prognostic information regarding myocardial infarction, and following myocardial injury multiple forms of troponin appear in the blood:

- Complexes of cardiac troponin T, I, and C (the T-I-C or ternary complex).
- Complexes of cardiac troponin I and C (the I-C or binary complex).
- Free troponin I.

Unexplained true elevation of troponin is relatively uncommon and in the emergency room setting pulmonary embolism and perimyocarditis are the most common differential diagnoses for such elevated troponin levels. The release of troponin is of shorter duration in unstable angina. However, truly elevated troponin levels have also been documented in tachyarrhythmias, hypertension, myocarditis, and myocardial contusion. Patients with chronic renal failure (CRF) who are on hemodialysis are at increased risk of cardiovascular disease, which accounts for about 50% of deaths in these patients. Studies have revealed a high prevalence of elevated cardiac troponin levels in patients with chronic renal failure, and especially of cardiac troponin T, but its clinical significance is unclear. It has been suggested that chronically elevated troponin levels represent chronic structural cardiovascular disease and that these patients are at higher cardiac risk. Nevertheless, a single elevated cardiac troponin T level in patients with chronic renal failure is non-diagnostic for acute myocardial infarction in the absence of other findings. Therefore, serial determinations are usually required, with a focus on a rise in the troponin level to confirm the diagnosis. A number of studies have demonstrated that cardiac troponin T can be used for risk stratification of patients with chronic renal failure without ischemia [11]. In addition, elevated cardiac troponins are associated with decreased left ventricular ejection fraction and poor prognosis in patients with CHF; these are related to the severity of heart failure. Dialysis does not affect cardiac troponin T or I levels.

According to the American College of Cardiology/European Society of Cardiology (ACC/ESC) guidelines, any elevated measure of troponin at the 99th percentile upper reference limit in the appropriate clinical setting is defined as an indication of acute myocardial infarction. However, it is also important that imprecision (CV) at the 99th percentile limit be less than 10%. The appropriate clinical indicators where troponin I elevation indicates myocardial damage include[12]:

- Ischemic symptoms.
- Development of pathological Q waves in the EKG.
- EKG changes indicating ischemia (ST segment elevation or depression).
- Coronary artery intervention (e.g. angioplasty).

Unfortunately, only a few commercially available troponin I assays can achieve these criteria. Moreover, standardization of the troponin I assay is

Table 8.3 Common Causes of False Positive Troponin I Results Using Immunoassays

Heterophilic antibody, including human anti-mouse antibody (HAMA)
Rheumatoid factor
Autoantibodies
Monoclonal protein
Macrotroponin
Immunocomplex formation
Interference from high bilirubin, gross hemolysis
Fibrin clot

also a big challenge because no suitable reference material is available and different manufacturers use different standard materials. Therefore, troponin I values obtained by using one method may not match those of the other methods. In fact, one study by the International Federation of Clinical Chemistry (IFCC) reported more than 20-fold differences between different troponin I methods. Therefore, each assay has its own reference limit and interpretation of test results is assay-dependent, making it more confusing [13]. In addition, many interferences have been reported in troponin I assays, including endogenous substances such as hemolysis and high bilirubin. However, more commonly, heterophilic antibodies, rheumatoid factors, and macro-troponin are known to interfere with cardiac troponin I immunoassays (Table 8.3). Due to patent issues only one troponin T assay is available commercially (Roche Diagnostics); it has been repeatedly refined and is reliable.

CASE REPORT

A 35-year-old woman complained about tiredness, shortness of breath, and chest pain, and showed an elevated troponin I of 6.4 ng/mL (reference value <0.03 ng/mL) using an Architect i2000 analyzer (Abbott Laboratories); she was admitted to the coronary care unit. During the following days, her troponin I remained elevated, but her coronary angiogram revealed normal arteries. Although the patient did not have any obvious risk of cardiovascular disease, her uncle died from sudden cardiac death at age 47 and her father had artery bypass graft at age 50. Because no decrease in troponin I was observed over the next four days, interference was suspected and when the specimen was sent to different laboratories for reanalysis using different methods, values of cardiac troponin I were not detected. Further detailed investigation by the authors to characterize the interfering substance revealed that the interfering substance was macrotroponin, which consists of a fragment of troponin complexed to immunoglobulin G [14].

CASE REPORT

A 26-year-old Caucasian woman presented to the emergency department complaining of chest pain but no shortness of breath or diaphoresis. Serial cardiac troponin I levels were 5.3 ng/mL, 5.6 ng/mL, and 5.0 ng/mL, but total creatine kinase and CK-MB levels were normal. However, her arteries were angiographically normal and no heart abnormality was detected using echocardiogram. The patient was discharged with a diagnosis of non-cardiac pain. Symptoms recurred after four weeks and again troponin I was elevated to 5.0 ng/mL but with normal CK-MB. Interference in the troponin I assay was suspected. It was eliminated by using a heterophilic antibody blocking tube, a device that rapidly eliminates false positive test results due to the presence of heterophilic antibody in the serum [15].

8.6 HIGH-SENSITIVE CARDIAC TROPONIN ASSAYS

Although cardiac troponin is a superior marker compared to CK-MB and other established cardiac biomarkers, one limitation of conventional assays is the lack of sensitivity and ability to detect small amounts of cardiac troponin in circulation in the first few hours after acute myocardial infarction. High-sensitive cardiac troponin assays have two common features:

- High-sensitive assays can detect cardiac troponin in the majority (90% or more) of healthy individuals. Some healthy individuals show no detected troponin levels using conventional assays.
- Because troponin levels can be detected in the majority of healthy individuals, a precise definition of normal values and establishing an unambiguous 99th percentile is more accurate using high-sensitive assays than conventional assays.

Recent advances in assay technology have led to the development of high-sensitive cardiac troponin assays. High-sensitive troponin I assays allow quantification of cardiomyocyte necrosis and higher values within 1–3 h after onset of chest pain as an important criteria for differentiating acute myocardial infarction from other causes of myocardial necrosis. Mild elevation of cardiac troponin I indicates a small amount of myocardial injury that could be related to a broad category of acute and chronic disorders. However, applying the 99th percentile criteria using a high-sensitive troponin I assay can also render identification of many patients who may not have an acute myocardial infarction, but other causes of minor myocardial injury. Preliminary data suggests that an absolute change of 30% or more at the 99th percentile within 6 h might be reasonable criteria for diagnosis of acute myocardial infarction using high-sensitive cardiac troponin I assay [16].

8.7 LESS COMMONLY USED CARDIAC MARKERS

Lactate dehydrogenase isoenzymes (LDH) were used widely in the past for diagnosis of myocardial infarction, but more recently, due to availability of troponin immunoassays, lactate dehydrogenase isoenzyme assay has been mostly discontinued in the clinical setting for diagnosis of myocardial infarction. However, it may be used in evaluating certain hepatic disorders. Briefly, LDH exists in five isoenzyme forms:

- LDH-1: Present primarily in cardiac myocytes and erythrocytes.
- LDH-2: Present mostly in white blood cells.
- LDH-3: Present in highest quantity in lung tissue.
- LDH-4: Highest amounts found in pancreas, kidney, and placenta.
- LDH-5: Highest amounts found in liver and skeletal muscle.

Usually LDH isoenzyme levels increase 24−72 hours following myocardial infarction and reach a peak concentration in 3−4 days. The levels remain elevated for 8 to 14 days, making it a late marker for myocardial infarction. Normally, concentration of LDH-1 is lower than LDH-2, but after myocardial infarction, LDH-1 concentration becomes elevated and exceeds the concentration of LDH-2. This phenomenon is called a flipped LDH pattern. However, hemolysis (LDH is present in erythrocytes in a similar concentration) produces this characteristic flip and it is important to ensure that the specimen is not hemolyzed prior to analysis. Moreover, LDH is a nonspecific marker for myocardial infarction, and its concentration can be elevated in hemolytic anemia, stroke, pancreatitis, ischemic cardiomyopathy, and a variety of other diseases.

Glycogen phosphorylase is an essential enzyme in the regulation of glycogen metabolism where this enzyme converts glycogen into glucose 1-phosphate in the first step of glycogenolysis. Three different isoenzymes have been identified in humans: glycogen phosphorylase BB (brain), glycogen phosphorylase LL (liver), and glycogen phosphorylase MM (muscle). Skeletal muscles solely contain glycogen phosphorylase MM while glycogen phosphorylase BB is mainly found in high concentration in both heart muscle and the brain. Glycogen phosphorylase BB is released into circulation 2−4 h after onset of cardiac ischemia and returns to baseline levels 1−2 days after acute myocardial infarction, making it an early marker.

Ischemia-modified albumin is also a relatively new cardiac biomarker capable of detecting myocardial ischemia within minutes. This biomarker continues to increase for 6−12 h following acute myocardial infarction and then returns to the baseline value. Myocardial ischemia results in reduction of the ability of circulating albumin to bind cobalt. The ischemic-modified albumin level in serum is measured by its reduced cobalt-binding capacity. A rapid assay with a 30-minute laboratory turnaround time has been developed and marketed as the first commercially available U.S. Food and Drug Administration (FDA)-approved marker of myocardial ischemia. However, ischemic-modified albumin levels are also elevated in patients with cirrhosis, certain infections, and advanced cancer. These factors reduce the specificity of the assay.

Pregnancy-associated plasma protein A is a metalloproteinase enzyme, and after acute myocardial infarction its value increases between 2 and 30 h. However, its utility as a cardiac biomarker is currently under investigation. An enzyme-linked immunosorbent assay (ELISA) is, however, available for its measurement. Another potential cardiac biomarker is myeloid-related protein 8/14 complex, which is involved in plaque destabilization. Its level is probably increased after acute myocardial injury [17]. Heart-type fatty acid-binding protein is a 15-kDa cytosolic protein present in a very high concentration in myocardial tissue, but present in lower concentrations in skeletal muscle, kidney, and brain. Therefore, heart-type fatty acid-binding protein is a potential new cardiac biomarker. Concentration of heart-type fatty acid-binding protein is increased after acute myocardial infarction, but it is also rapidly cleared from the circulation due to low molecular weight. Free fatty acids are also increased in plasma after myocardial infarction but their role in diagnosis of acute myocardial infarction has not been established.

8.8 B-TYPE NATRIURETIC PEPTIDES (BNP)

B-type natriuretic peptide (BNP) is a hormone secreted primarily by the ventricular myocardium in response to wall stress such as volume expansion and pressure overload. BNP has shown promise as a diagnostic marker of congestive heart failure. In addition, multiple studies have demonstrated that BNP may also be a useful prognostic indicator for myocardial stress and correlated with long-term cardiovascular mortality in patients with acute myocardial infarction. Studies have shown that the BNP level predicted cardiac mortality and other adverse cardiac events across the entire spectrum of acute coronary syndrome. The mortality rate nearly doubled when both cardiac troponin I and BNP levels were elevated. In addition, the BNP level is also a good predictor of left ventricular ejection fraction and heart failure in these patients.

BNP is initially synthesized as pre-proBNP, which contains 134 amino acids but is cleaved into proBNP containing 108 amino acids. On secretion, it splits into biologically active BNP (amino acids 77−108) and the remaining N-terminal proBNP (NT-proBNP: 1−76 amino acids), which is biologically inactive. BNP is a smaller molecule than NT-proBNP, and it is cleared from circulation earlier than NT-proBNP. Therefore, concentration of NT-proBNP in serum or plasma is higher than BNP. In addition, NT-proBNP is more stable in serum or plasma than BNP. Although studies have demonstrated that both BNP and NT-proBNP have similar effectiveness as biomarkers, some authors tend to favor NT-proBNP as a slightly superior biomarker than BNP. Assays are available (both laboratory-based analyzers and point of care) for analysis of both BNP and NT-proBNP, and in general both BNP and NT-proBNP follow a similar pattern after heart failure. However, the quantitative BNP value does not match the quantitative proBNP value. Major points to remember regarding BNP and NT-proBNP include:

- A major application of both BNP and proBNP testing is the evaluation of patients with congestive heart failure. Although a single determination can be helpful in diagnosis, multiple determinations can provide more useful information. If heart failure responds to therapy, concentrations of BNP and NT-proBNP should decline, indicating progress of therapy. If a patient does not respond, values may be increased gradually.
- In general, NT-proBNP is more stable (up to seven days at room temperature and up to four months if stored at $-20°C$) than BNP, which is not stable for a day even if the specimen is stored in a refrigerator. Therefore, BNP analysis must be performed as soon as possible after collecting the specimen.
- The cut-off level of BNP and NT-proBNP depends on age, as values tend to increase with advancing age. In general, heart failure is unlikely if the BNP value is less than 100 pg/mL and heart failure is very likely if the value is over 500 pg/mL. For NT-proBNP, the normal value for a person 50 years or younger is usually 125 ng/mL, but heart failure is unlikely if the NT-proBNP value is <300 pg/mL. However, heart failure is likely if the value is >450 pg/mL (>900 pg/mL in a patient of age 50 and above) [18].
- Patients with end-stage renal disease and dialysis patients usually show higher BNP and NT-proBNP in serum than normal individuals.

8.9 C-REACTIVE PROTEIN

C-reactive protein (CRP), a non-specific marker of inflammation, is considered to be directly involved in coronary plaque atherogenesis. Studies show that an elevated CRP level independently predicts adverse cardiac events at the primary and secondary prevention levels. Data indicate that CRP is a useful prognostic indicator in patients with acute coronary syndrome, as elevated CRP levels are independent predictors of cardiac death, acute myocardial infarction, as well as congestive heart failure. In combination with cardiac troponin I and BNP, CRP may be a useful adjunct, but its non-specific nature limits its use as a diagnostic cardiac marker for acute coronary syndrome in patients presenting at the emergency department. See also Chapter 6 for more detail.

8.10 MYELOPEROXIDASE

Myeloperoxidase (MPO) is a leukocyte enzyme. Initial studies showed significantly increased MPO levels in patients with angiographically documented coronary artery disease. In patients presenting to the emergency department with chest pain, elevated MPO levels independently predicted increased risk for major adverse cardiac events, including myocardial infarction, reinfarction, need for revascularization, or death at 30 days and at 6 months. Among the patients who presented to the emergency department with chest pain but who were ultimately ruled out for myocardial infarction, an elevated MPO level at presentation predicted subsequent major adverse cardiovascular outcomes. MPO may be a useful early marker in the emergency department based on its ability to detect plaque vulnerability that precedes acute coronary syndrome.

KEY POINTS

- There are four established biomarkers for myocardial necrosis: myoglobin, creatine kinase isoenzymes, cardiac troponin I, and cardiac troponin T.
- Myoglobin is a heme protein found in both skeletal and cardiac muscle. Myoglobin is typically released into the circulation as early as 1 h after myocardial infarction, gradually increasing to a peak at 4−12 hours, and returning to normal within 24−36 hours. Rapid release of myoglobin probably reflects its low molecular weight (17 kDa) and cytoplasmic location. Myoglobin is an early marker of acute myocardial infarction and exhibits a high negative predictive value. Myoglobin has poor clinical specificity (60−90%) because it is also found in large quantities in skeletal muscle.
- Myoglobin, being a small protein, is excreted in urine. A high level of myoglobin is encountered in patients with acute renal failure, or uremic syndrome.
- Creatine kinase is an enzyme (often called cardiac enzyme). Creatine kinase has two subunits, including M type (for muscle) and B type (for brain). The CK-MM isoenzyme is dominant in adult skeletal muscle (98% of total muscle CK), while CK-BB is found mostly in the central nervous system. In myocardium, CK-MB is present in a relatively high concentration (15−20% of the total myocardial CK) while about 85% is CK-MM. Therefore, CK-MM is the most abundant isoenzyme of creatine kinase.
- CK-MB concentration gradually rises 4 to 9 h after the onset of chest pain, reaching a peak concentration in approximately 24 h. Concentration of CK-MB begins to decline and reaches a baseline level in 48 to 72 h. Because CK-MB remains elevated for a longer period of time following myocardial infarction, it is useful to detect re-infarction using a serial CK-MB measurement.

- Although CK-MB is more concentrated in the myocardium, it is also present in skeletal muscle and false-positive elevations occur in a number of clinical settings, including trauma, heavy exertion, and myopathy.
- Macro CK is a high-molecular-weight complex of one of the CK isoenzymes and immunoglobulin, but most commonly CK-BB with IgG. Macro CK can be seen in patients with autoimmune diseases or malignancy. Macro CK can persist in serum for a long time and can interfere with CK-MB measurement, especially if a mass assay is used.
- Troponin I and T are proteins but not enzymes. Out of all these biomarkers, troponin I is the most specific for myocardial necrosis.
- Cardiac troponin I and T have a similar capability in identifying myocardial damage, but in addition to being useful for diagnosis, they also permit the estimation of prognosis and risk stratification of patients with acute myocardial infarction. Cardiac troponin T and I (especially troponin I) have been accepted as the "gold standard." Troponin is a regulatory complex of three protein subunits located on the thin filament of the myocardial contractile apparatus, and is composed of three subunits encoded by different genes. The three subunits are designated as follows: troponin C (the calcium-binding component, 18 kDa), troponin T (the tropomyosin-binding component, 21 kDa); and troponin I (the inhibitory component, 37 kDa).
- Levels of both troponin T and I start increasing 4 to 9 hours after acute myocardial infarction, with a peak between 12 and 24 hours. Levels can remain elevated for up to 14 days.
- Cardiac troponin T and I have many advantages over CK-MB as cardiac markers. First, levels of troponin in normal individuals are very low or non-detectable. Therefore, significant elevation of troponin indicates myocardial injury. A small elevation of troponin but normal CK-MB level may indicate a microscopic zone of myocardial necrosis (microinfarction). Troponin I is very specific for myocardium because only one isoform of cardiac troponin I has been identified that is found exclusively in cardiac myocytes. Cardiac troponin I is not expressed in skeletal muscle. Although cardiac troponin T has a different amino acid sequence when compared to other troponins, small amounts of cardiac troponin T have been identified in skeletal muscle. In humans cardiac troponin T isoform expression has been reported in patients with muscular dystrophy, polymyositis, dermatomyositis, and end-stage renal disease.
- According to the American College of Cardiology/European Society of Cardiology (ACC/ESC) guidelines, any elevated measure of troponin at the 99th percentile upper reference limit in the appropriate clinical setting is defined as an indication of acute myocardial infarction.
- Individuals with symptoms of myocardial infarction and elevated troponin I or T, and without electrocardiogram (EKG) changes, are now classified as non-ST-segment elevation MI (NSTEMI). Only one elevated troponin level above the established cutoff is required to establish the diagnosis of acute myocardial

infarction according to the American College of Cardiology Foundation/American Heart Association guidelines for NSTEMI.

- Studies have revealed a high prevalence of elevated cardiac troponin levels (especially troponin T) in patients with chronic renal failure. A single elevated cardiac troponin T level in patients with chronic renal failure is non-diagnostic for acute myocardial infarction in the absence of other findings.
- Future cardiac biomarkers include: glycogen phosphorylase BB (brain type), ischemic-modified albumin, pregnancy-associated plasma protein A, and heart-type fatty acid-binding protein.
- Biomarkers that indicate stress of myocardium include: B-type natriuretic peptide (BNP), N-terminal pro-B-type natriuretic peptide (NT-proBNP), C-reactive protein, and myeloperoxidase.
- A major application of both BNP and proBNP testing is evaluation of patients with congestive heart failure. BNP is initially synthesized as pre-proBNP, which contains 134 amino acids, but is cleaved into proBNP containing 108 amino acids. On secretion, it splits into biologically active BNP (amino acids: 77–108), and the remaining N-terminal proBNP (NT-proBNP: 1–76 amino acids), which is biologically inactive. BNP is a smaller molecule than NT-proBNP, and it is cleared from circulation earlier than NT-proBNP. Therefore, concentration of NT-proBNP in serum or plasma is higher than BNP. In addition, NT-proBNP is more stable in serum or plasma than BNP.

REFERENCES

[1] Movahed MR, John J, Hashemzadeh M, Hashemzadeh M. Mortality trends in non-ST-segment elevation myocardial infarction (NSTEMI) in the United States from 1998 to 2004. Clin Cardiol 2011;34:689–92.

[2] Wright RS, Anderson JL, Adams CD, Bridges CR, et al. ACCF/AHA focused update of the guidelines for the management of patients with unstable angina/non-ST-elevation myocardial infarction (updating 2007 guidelines). J Am Coll Cardiol 2011;57:1920–59.

[3] Adams CD, Antman EM, Bridges CR, et al. ACC/AHA 2007 guidelines for the management of patients with unstable angina/non-ST-elevation myocardial infarction: a report of American College of Cardiology/American Heart Association task force on practice guidelines. Circulation 2007;116:e148–304.

[4] Lackner KJ. Laboratory diagnostics of myocardial infarction-troponins and beyond. Clin Chem Lab Med 2013;51:83–9.

[5] Kasaoka S, Todani M, Kaneko T, Kawamura Y, et al. Peak value of blood myoglobin predicts acute renal failure induced by rhabdomyolysis. J Critical Care 2010;25:601–4.

[6] Woo J, Lacbawan FL, Sunheimer R, LeFever D, et al. Is myoglobin useful in the diagnosis of acute myocardial infarction in the emergency department setting? Am J Clin Pathol 1995;103:725–9.

[7] Lawandrowski K, Chen A, Januzzi J. Cardiac markers for myocardial infarction. Am J Clin Pathol 2002;118(Suppl.):S93–9.

[8] Lee T, Goldman L. Serum enzymes in the diagnosis of acute myocardial infarction. Ann Intern Med 1986;105:221–33.

[9] Lee KN, Casko G, Bernhardt P, Elin R. Relevance of macro creatinine kinase type 1 and type 2 isoenzymes to laboratory and clinical data. Clin Chem 1994;40:1278−83.

[10] Er TK, Ruiz Gines MA, Jong YJ, Tsai LY, et al. Identification of false positive creatinine kinase-MB activity in a patient with nonketonic hyperglycemia. Am J Med 2007;859: e9−e10.

[11] Ferguson JL, Beckett GJ, Stoddart M, Fox KAA. Myocardial infarction redefined: the new ACC/ESC definition, based on cardiac troponin, increases the apparent incidence of infarction. Heart 2002;88:343−7.

[12] Hamm CW, Giannitsis E, Katus HA. Cardiac troponin elevations in patients without acute coronary syndrome. Circulation 2002;106:2871−2.

[13] Panteghini M, Bunk DM, Christenson RH, Katrukha A, et al. Standardization of troponin I measurements: an update. Clin Chem Lab Med 2008;46:1501−6.

[14] Michielsen EC, Bisschops PG, Janssen MJ. False positive troponin result caused by a true macrotroponin. Clin Chem Lab Med 2011;49:923−5.

[15] Salah AK, Gharad SM, Bodiwala K, Booth DC. You can assay it again. Am J Med 2007;120:671−2.

[16] Twernbold R, Reichlin T, Reiter M, Muller C. High-sensitive cardiac troponin: friend or foe? Swiss Med Wkly 2011;141:w13202.

[17] Dekker MS, Mosterd A, van Hof A, Hoes AW. Novel biomarkers in suspected acute coronary syndrome: systematic review and critical appraisal. Heart 2010;96:1001−10.

[18] Weber M, Hamm C. Role of B-type natriuretic peptide (BNP) and NT-proBNP in clinical routine. Heart 2006;92:843−9.

CHAPTER 9

Endocrinology

9.1 INTRODUCTION TO VARIOUS ENDOCRINE GLANDS

Homeostasis is maintained by both the nervous system and endocrine system in the human body. Endocrine activity can be classified as autocrine, paracrine, or classical endocrine. In autocrine activity, chemicals produced by a cell act on the cell itself. In paracrine activity, chemicals produced by a cell act locally. However, in classical endocrine activity, chemicals produced by an endocrine gland act at a distant site after their release into the circulation system, and these chemicals are called hormones. Major endocrine glands include pituitary, thyroid, parathyroid, adrenals, gonads (testis in male, ovary in female), and the pancreas. However, the pineal gland secretes melatonin, which may contribute to regulation of biological rhythm and may induce sleep [1]. Chemical structures of hormones vary widely, and hormones may be polypeptides, glycoproteins, steroids, or amines. Most classical hormones are secreted into the systemic circulation. However, hypothalamic hormones are secreted into the pituitary portal system. Hormones may be bound to certain proteins in blood, and such binding proteins include thyroxine-binding globulin (TBG), sex hormone-binding globulin (SHBG), cortisol-binding globulin (CBG), and insulin-like growth factor (IGF)-binding proteins (IGF-BP). However, a major protein in circulation, albumin, can also bind certain hormones. In addition, prealbumin can also act as a binding protein for a hormone. Hormones usually act by binding to receptors. Receptors for hormones can be either of the following:

- Cell surface or membrane receptors.
- Nuclear receptors.

Cell surface or membrane receptors may be G protein-coupled receptors or dimeric transmembrane receptors. G protein-coupled receptors bind hormones in the extracellular domain that activate the membrane G protein

145

A. Dasgupta and A. Wahed: Clinical Chemistry, Immunology and Laboratory Quality Control
DOI: http://dx.doi.org/10.1016/B978-0-12-407821-5.00009-7

complex. The activated G protein complex is then responsible for generating secondary messengers. Most peptide hormones act via this mechanism. Dimeric transmembrane receptors bind hormones in their extracellular component, and the intracellular component is responsible for phosphorylation of intracellular messengers, which leads to activation of various messengers. Growth hormone (GH) and insulin-like growth factor-1 (IGF-1) act by this mechanism. Steroid and thyroid hormones act via nuclear receptors. These hormones pass through the cell membrane and bind with the receptors in the cytoplasm; the complex is translocated to the nucleus, causing an increased transcription of genes. Secretion of hormones from endocrine glands may be under positive feedback and negative feedback. For example, thyrotropin-releasing hormone (TRH) secreted by the hypothalamus stimulates the release of thyroid-stimulating hormone (TSH) by the anterior pituitary, which in turn causes the thyroid gland to release thyroxine (T4) and tri-iodothyronine (T3). T3 and T4, once released, cause negative feedback on the secretion of TSH and TRH.

Hormone secretion may be continuous or intermittent. Thyroid hormone secretion is continuous. Thus levels may be measured at any time to assess hormonal status. Secretion of follicle-stimulating hormone (FSH), luteinizing hormone (LH), and GH are pulsatile. A single measurement may not reflect hormonal status. Some hormones exhibit biological rhythms. Cortisol exhibits a circadian rhythm, where levels are highest in the morning and lowest during late night. The menstrual cycle is an example of a longer biological rhythm where different levels of a hormone are observed during a specific part of the cycle. During a normal menstrual cycle there is an interplay of feedback between hypothalamus, anterior pituitary, and ovaries. In the follicular phase, a low estrogen level stimulates secretion of LH and FSH by a negative feedback mechanism. The level of progesterone is also low during the follicular phase. At the end of the follicular phase both estradiol and estrogen levels are high, triggering the release of gonadotropin-releasing hormone (GnRH) from the hypothalamus; this stimulates secretion of LH from the anterior pituitary. The level of LH is highest in mid-cycle during ovulation and LH surge during mid-cycle is a good indication of ovulation. After mid-cycle, the progesterone level starts increasing in the luteal phase, reaching its highest level during the 8 days after ovulation. In the luteal phase, levels of FSH and LH decline gradually (Figure 9.1).

Certain hormone levels are elevated during stress. These include:

- Adrenocorticotropic hormone (ACTH) and cortisol
- Growth hormone (GH)
- Prolactin
- Adrenaline and noradrenaline.

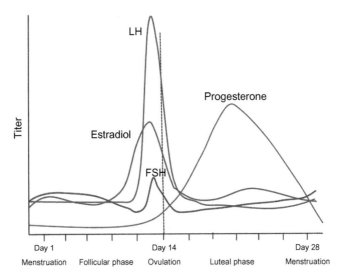

FIGURE 9.1

Titers of various hormones during menstrual cycle. This figure is reproduced in color in the color plate section. *(Courtesy of Andres Quesda, M.D., Department of Pathology and Laboratory Medicine, University of Texas-Houston Medical School.)*

Therefore it is important for a phlebotomist to wait for some time after initial venipuncture and prior to collecting blood specimens for measuring blood levels of such hormones. Certain hormone levels increase during sleep, such as growth hormone (GH) and prolactin.

This chapter is an overview of activities of the hypothalamus, pituitary, thyroid, adrenal, gonads, and pancreas in relation to their endocrine activity, as well as endocrine disorders related to these organs. Diabetes mellitus is the most prevalent endocrine disorder in the U.S., and diabetes mellitus is discussed in detail in Chapter 7. Most common endocrine disorders are listed in Table 9.1. With hormone replacement and other therapies, endocrine disorders can be treated well, and a severe consequence of endocrine disorder is rarely observed today.

9.2 HYPOTHALAMUS

The hypothalamus produces thyrotropin-releasing hormone (TRH), corticotropin hormone (CRH), gonadotropin-releasing hormone (GnRH), growth hormone-releasing hormone (GHRH), and somatostatin (growth hormone inhibitory hormone). These hormones act on the anterior pituitary, resulting in release of various other hormones, including thyroid-stimulating hormone (TSH), adrenocorticotropin (ACTH), FSH, LH, and GH. Somatostatin inhibits

Table 9.1 Most Common Endocrine Disorders

Endocrine Disorder	Cause
Diabetes mellitus	Pancreatic beta-cell dysfunction that produces insulin.
Acromegaly	Overproduction of growth hormone.
Addison's disease	Decreased production of hormones by adrenal glands.
Cushing syndrome	High cortisol produced by adrenal glands.
Graves' disease	Hyperthyroidism due to excess production of thyroid hormones.
Hashimoto's thyroiditis	Autoimmune disease related to hypothyroidism.
Hyperthyroidism	Excess production of thyroid hormones.
Hypothyroidism	Underproduction of thyroid hormones.
Prolactinoma	Overproduction of prolactin by pituitary gland.
Polycystic ovary syndrome	Excessive production of androgenic hormone.

the release of GH. Dopamine (also known as prolactin inhibitory hormone) is also a neurotransmitter which is also produced by the hypothalamus. Dopamine can inhibit GH secretion.

The supraoptic and paraventricular nuclei of the hypothalamus produce antidiuretic hormone (ADH, also known as vasopressin) and oxytocin. These hormones are stored in the posterior pituitary and act on certain body parts rather than acting on the pituitary like other tropic hormones. ADH acts on the collecting ducts of the renal tubules and causes absorption of water. ADH secretion is linked to serum osmolality, and increased serum osmolality results in increased secretion of ADH. Lesions of the hypothalamus may result in inadequate ADH secretion, also known as cranial diabetes insipidus. Failure of ADH to act on the collecting ducts results in nephrogenic diabetes insipidus. Causes of nephrogenic diabetes insipidus include hypercalcemia, hypokalemia, and lithium therapy. In both types of diabetes insipidus, polyuria with low osmolality is a common symptom. Please see Chapter 5 for more detail.

Oxytocin is a nonapeptide hormone (9 amino acids) primarily synthesized in the magnocellular neurons of paraventricular and supraoptic nuclei of the hypothalamus and most of the oxytocin produced is transported to the posterior pituitary where it is released to regulate parturition and lactation. However, production of oxytocin by cells in various parts of the brain as well as release from magnocellular dendrites provides oxytocin responsible for modulating behavior, including maternal behavior and social attachment [2]. Hormones released by the hypothalamus and their characteristics are listed in Table 9.2.

Table 9.2 Characteristics of Hormones Released by Hypothalamus

Hormone	Composition	Action
Tropic Hormones Acting on Pituitary		
Corticotropin-releasing hormone (CRH)	41 amino acids	Stimulates adrenocorticopic hormone release (ACTH).
Gonadotropin-releasing hormone (GnRH)	10 amino acids	Stimulates follicle-stimulating hormone (FSH) and luteinizing hormone (LH) release.
Growth hormone-releasing hormone (GHRH)	44 amino acids	Stimulates growth hormone (GH) release.
Thyrotropin-releasing hormone (TRH)	3 amino acids	Stimulates thyrotropin-releasing hormone (TSH) and prolactin (PRL) release.
Somatostatin	14 amino acids	Inhibits GH release.
Hormones Acting on Other Organs		
Antidiuretic hormone (ADH)	9 amino acids	Acts on kidney: causes water reabsorption.
Oxytocin	9 amino acids	Lactation, parturition, mood.

Table 9.3 Characteristics of Hormones Released by the Anterior Pituitary

Hormone	Composition	Action
Adrenocorticotropic hormone (ACTH)	39 amino acids	Stimulates glucocorticoid secretion by adrenal cortex.
Follicle-stimulating hormone (FSH)	Contains alpha and beta subunit, but alpha subunit of FSH, LH, TSH, and hCG is same (92 amino acids); the beta subunit confers uniqueness. FSH: beta chain, 117 amino acids	Stimulates development of ovarian follicle in female, and in male stimulates spermatogenesis.
Luteinizing hormone (LH)	LH: beta subunit, 121 amino acids	In female, LH surge stimulates ovulation.
Thyrotropin-releasing hormone (TSH)	TSH beta subunit, 112 amino acids	Stimulates thyroid gland to produce T3 and T4.
Growth hormone (GH)	191 amino acids	Stimulates growth mediated via IGF-1.*
Prolactin (PRL)	199 amino acids	Initiation and maintenance of lactation.

*Insulin-like growth factor 1 (IGF-1) is produced mainly in the liver.

9.3 PITUITARY GLAND

The pituitary gland is a small gland situated at the base of the skull. The gland is divided into an anterior and posterior lobe. The anterior pituitary lobe produces six hormones and the release of such hormones is under the control of the hypothalamus through various hormones produced by the hypothalamus. Hormones produced by the pituitary gland are listed in Table 9.3. It is important to note that although ADH is synthesized by the hypothalamus, it is secreted from the pituitary gland.

Growth hormone (GH, also known as somatotropin) is the most abundant hormone produced by the anterior pituitary, and it stimulates growth of cartilage, bone, and many soft tissues. GH stimulates release of insulin-like growth factor 1 (IGF-1, somatomedin C), mostly from the liver, and IGF-1 is partly responsible for the activity of GH, but this hormone has glucose lowering and other anabolic activities. Conditions that cause the deficiency of IGF-1 include Laron syndrome in children, liver cirrhosis in adults, age-related cardiovascular and neurological diseases, and intrauterine growth retardation [3]. Although IGF-1 plays important roles in adults, concentration of the similar hormone insulin-like growth factor-2 (IGF-2, somatomedin A) is high in embryonic and neonatal tissues. Both IGF-1 and IGF-2 share 45 amino acid positions and approximately 50% amino acid homology with insulin. GH is essential for proper growth of children, and it also plays an important role in adults in maintaining healthy bones, muscles, and metabolism. GH deficiency or excess is rarely encountered clinically. Another hormone similar in structure to GH is prolactin, which plays an important role in lactation.

In general, deficiency of pituitary hormones may be selective or multiple or panhypopituitarism. The deficiency of GH causes dwarfism, which can be treated with recombinant human GH replacement therapy. Clinical features are related to the reduced activity of the peripheral endocrine gland. Causes of hypopituitarism include:

- Congenital causes: For example, Kallmann syndrome (isolated GnRH deficiency causing delayed or absent puberty).
- Infections.
- Vascular causes: Sheehan's syndrome (postpartum necrosis), pituitary apoplexy, etc.
- Tumors: For example, pituitary or hypothalamic tumors, and craniopharyngioma.
- Trauma.
- Surgery.
- Infiltrative diseases such as sarcoidosis and hemochromatosis.
- Radiation.
- Empty sella syndrome.

Hyperpituitarism is most often due to pituitary tumors affecting GH-secreting cells, prolactin-secreting cells, and adrenocorticotropic hormone (ACTH)-secreting cells. GH-secreting tumors affecting individuals before closure of epiphysis result in gigantism and after closure result in acromegaly. Prolactin-secreting tumors cause hyperprolactinemia. A high level of prolactin (prolactinemia) inhibits the action of follicle-stimulating hormone (FSH) and luteinizing hormone (LH), which results in hypogonadism and infertility. ACTH-secreting tumors cause Cushing's syndrome. Proper endocrine testing for diagnosis of hypopituitarism include measuring concentrations of various hormones in serum or plasma, including thyroid-stimulating hormone (TSH), prolactin, LH, FSH, T4, and cortisol. For diagnosis of hypopituitarism, a stimulation test with gonadotropin-releasing hormone, thyrotropin-releasing hormone, and insulin-induced hypoglycemia (triple stimulation test) is useful. Following stimulation, serum or plasma levels of FSH, LH, TSH, PRL, GH, and cortisol are measured.

Endocrine tests for hyperpituitarism include measurement of hormone levels and a suppression test using glucose (oral glucose tolerance). Administration of glucose with a rise in blood glucose should suppress anterior pituitary hormones in normal individuals.

9.4 THYROID GLAND

The thyroid gland produces two hormones, thyroxine (T4) and triiodothyronine (T3). Four steps are involved in the synthesis of these hormones:

- Inorganic iodide from the circulating blood is trapped (iodide trapping).
- Iodide is oxidized to iodine (oxidation).
- Iodine is added to tyrosine to produce monoiodotyrosine and diiodotyrosine (organification).
- One monoiodotyrosine is coupled with one diiodotyrosine to yield T3 and two diiodotyrosines are coupled to yield T4 (coupling).

Both T3 and T4 are bound to thyroglobulin and stored in the colloid. Free (unbound) T4 is the primary secretory hormone from the thyroid gland, and T4 is converted in peripheral tissue (liver, kidney, and muscle) to T3 by 5′-monodeiodination. T3 is the physiologically active hormone. T4 can also be converted to reverse T3 by 3′-monodeiodination. This form of T3 is inactive. The majority (99%) of the T3 and T4 in circulation are found to be involved in thyroxine-binding globulin (TBG), albumin, and thyroxine-binding prealbumin. T3 binds to the thyroid hormone nuclear receptor on target cells to cause modified gene transcription.

In a normal individual there is a tightly coordinated feedback mechanism between hypothalamus, pituitary, and thyroid glands. Thyrotropin-releasing hormone (TRH), a tripeptide (smallest hormone molecule known) produced in the hypothalamus stimulates the pituitary to synthesize and secrete thyroid-stimulating hormone (TSH) that finally stimulates the thyroid gland to produce thyroid hormones. There is a negative feedback mechanism where a fall in blood thyroid hormone stimulates the hypothalamus to secrete TRH. Abnormalities of enzymes involved in the synthesis of thyroid hormone may cause hypothyroidism with increased TSH secretion and goiter. This is dyshormonogenetic goiter. Dyshormonogenetic goiter may be associated with nerve deafness, referred to as Pendred's syndrome. An estimated 3% of the population suffers from thyroid disorders. Thyroid disorders are also more common in women than men with an estimated 4.1 per 1,000 women developing hypothyroidism every year while the prevalence among men is 0.6 per 1,000 adults. In addition, 0.8 women per 1,000 develop hyperthyroidism every year [4]. In addition, thyroid disorders are more common in older people than younger people.

9.5 THYROID FUNCTION TESTS

The most commonly ordered thyroid function test is TSH followed by T4 (total or free), and T3 (total or free). More recently, FT4 and FT3 tests have been ordered more frequently than T4 and T3 tests. TSH is used as a screening test for thyroid status. It is elevated in primary hypothyroidism and suppressed in thyrotoxicosis. Basic interpretation of TSH, FT4 (free T4), and FT3 (free T3) in various thyroid diseases is summarized in Table 9.4. However, there are several situations where interpretation of thyroid function tests may be confusing:

- Situations where thyroid hormone-binding proteins may be low or high, causing alteration of total T3 and T4 levels. However, free T3, T4, and TSH levels should be normal. Pregnancy and oral contraceptive pills raise concentrations of thyroid-binding proteins. Hypoproteinemic states such as cirrhosis of liver, nephrotic syndrome, etc. may cause lower concentrations of thyroid-binding proteins.
- Amiodarone can reduce peripheral conversion of T4 to T3. Free T4 levels may be high, but TSH levels could be normal. Amiodarone can also cause both hypothyroidism and hyperthyroidism because amiodarone contains iodine molecules.

Table 9.4 Interpretation of Basic Thyroid Function Tests (TSH)			
Thyroid Disorder	**TSH**	**Free T4**	**Free T3**
Primary hypothyroidism	Increased	Low	Low
Secondary hypothyroidism (lack of TSH from pituitary)	Low	Low	Low
Thyrotoxicosis	Low	High	High
T3 toxicosis	Low	Normal	High

- Seriously ill patients may have reduced production of TSH, with low T4 and reduced conversion of T4 to T3 with increased conversion of T4 to reverse T3. These patients are, however, euthyroid. This is referred to as sick euthyroid syndrome.

TSH, T4, and T3 tests can be performed using automated analyzers and immunoassays. Reverse T3 analysis may also be performed under certain circumstances, but due to low volume of this test, most hospital laboratories send this test to a reference laboratory. In addition to these tests, measuring free T4 (FT4) and free T3 (FT3) is useful for diagnosis of thyroid disorders in certain patients. Although FT4 and FT3 can be detected by direct methods such as dialysis and ultrafiltration, most clinical laboratories utilize indirect immunoassay-based methods that can be adopted in automated chemistry analyzers. Both two-step immunoassays and one-step immunoassays are commercially available for determination of FT4 and FT3. An indirect way of estimating FT4 is by free thyroxine index (FT4I), which is calculated by multiplying total T4 with the value of T3 uptake. T3 uptake assay is a measure of the number of available free binding sites on thyroxine-binding globulin (TBG), and is expressed as a percentage value. Commercial kits are available for such measurement. Free thyroxine index usually correlates with FT4 concentration.

Patients with suspected thyroid disease can also be tested for thyroxine-binding globulin (which has the greatest affinity for T4) as well as antithyroid antibodies. The most common antithyroid antibody is thyroid peroxidase antibody (TPOAb). This antibody is against thyroid peroxidase (originally described as thyroid microsomal antigen), an enzyme found in the thyroid gland that plays an important role in the production of thyroid hormones. In addition, there are antibodies against thyroglobulin and TSH receptor. TSH receptor antibodies may inhibit binding of TSH to the receptor, or they stimulate the receptor. When such antibodies stimulate the TSH receptor, they may cause thyrotoxicosis (such as Graves' disease). Thyroid antibody testing is essential in establishing a diagnosis of thyroid dysfunction, which is autoimmune in nature. Enzyme-linked immunosorbent assay (ELISA) and chemiluminescent-based immunoassays are commercially available for determination of these antibodies. Other analytical methods are also available.

Various interferences have been reported in thyroid function tests. For TSH, usually third generation assays are used. These are sensitive and can detect levels as low as 0.02 mU/L; the normal range is usually defined as 0.5−5.5 mU/L; however, there are recommendations to lower the upper limit significantly. Measurement of TSH can suffer interference from heterophilic antibody and rheumatoid antibody and cause falsely elevated results. Rarely do autoantibodies to TSH develop clinically, but such autoantibodies can also falsely increase TSH results. A rare interference in the TSH assay is due to macro-TSH, an autoimmune complex between anti-TSH IgG antibody and TSH. Loh et al. reported a very high TSH level of 122 mIU/L in a patient who showed a normal FT4 level. The falsely elevated TSH value was due to macro-TSH [5]. Several strongly protein-bound drugs can affect thyroid hormone binding and cause decreases in T4 and T3 levels. Interference in T4 or T3 assays due to autoantibodies against these hormones has also been reported, but the incidence of such interference is rare. In addition, thyroglobulin autoantibodies (may be present in patients with thyroid cancer) also interfere with thyroglobulin measurement [6].

9.6 HYPOTHYROIDISM

Hypothyroidism by definition is the failure of the thyroid gland to produce sufficient hormone in order to meet daily requirements of such hormones to maintain normal metabolic functions. Hypothyroidism can be due to thyroid gland failure (primary gland failure) or insufficient stimulation of the thyroid gland due to a dysfunctional hypothalamus (producing TRH) or pituitary (producing TSH). Causes of hypothyroidism include:

- Primary hypothyroidism (primary disease of thyroid gland): Causes include autoimmune thyroiditis, Hashimoto's thyroiditis, surgery/radiation, dyshormonogenesis, antithyroid drugs, drug therapy with amiodarone, and advanced age.
- Secondary hypothyroidism (lack of TSH from the pituitary).
- Peripheral resistance to thyroid hormones.

Autoimmune thyroid disease (especially Hashimoto's thyroiditis) is the most common etiology of hypothyroidism in the U.S. In Hashimoto's thyroiditis, the principal biochemical characteristic is the presence of thyroid autoantibodies, including thyroid peroxidase antibody (TPOAb) and thyroglobulin antibody, in the sera of patients with this disease. Both antibodies are present in higher concentration in female patients than male patients, and belong to the IgG class. These antibodies have a high affinity against respective antigens, thus causing destruction of the thyroid gland. Lymphocytic infiltration with follicle formation is seen within the thyroid gland. The gland is enlarged and the patient is hypothyroidic. During the initial phase there may be transient hyperthyroidism, referred to as Hashitoxicosis. There may a genetic link to this disease [7].

Iatrogenic forms of hypothyroidism may occur after thyroid surgery, neck irradiation, or drug therapy (amiodarone, lithium, tyrosine kinase inhibitors, etc.), including radioiodine therapy. In addition, transient hypothyroidism may occur due to postpartum thyroiditis. The best laboratory assessment of thyroid function is done by measuring serum or plasma TSH levels. If TSH is elevated, FT4 should be measured. Elevated serum TSH with low FT4 indicates primary hypothyroidism, but elevated TSH with normal FT4 indicates subclinical hypothyroidism. In secondary hypothyroidism, both TSH and FT4 should be low. If FT3 is measured instead of FT4, a similar pattern should be observed (see Table 9.4). Thyroid antibody testing is useful if autoimmune hypothyroidism such as Hashimoto's thyroiditis is suspected. Thyroid hormone requirements increase during pregnancy, and such requirements last through pregnancy. Therefore, dosage must be adjusted carefully if a pregnant woman is receiving levothyroxine. Myxedema coma is a rare medical emergency due to a severe manifestation of hypothyroidism. This condition is usually observed in older women with a history of primary hypothyroidism. A change of mental status, severe hypothermia, and even psychosis may be present in a patient suffering from myxedema coma. If not treated in a timely manner, the outcome can be fatal [8].

9.7 HYPERTHYROIDISM

Clinical hyperthyroidism, also known as thyrotoxicosis, is due to excess thyroid hormones in the circulation causing various clinical symptoms. The prevalence is 2% for women and 0.2% for men, and as many as 15% of all cases of hyperthyroidism are encountered in patients over 60 years of age. Causes of hyperthyroidism include:

- Graves' disease.
- Toxic nodular (single or multiple) goiter.
- Thyroiditis (e.g. due to viral infection).
- Drugs.
- Excess TSH (e.g. due to pituitary tumor).

Graves' disease, the most common cause of hyperthyroidism, is an autoimmune disease of the thyroid where IgG antibodies bind with the TSH receptor to cause stimulation of the thyroid gland and overproduction of thyroid hormones. These antibodies are also referred to as long acting thyroid stimulators. An infiltrative ophthalmopathy (exophthalmos, lid lag and lid retraction) may be observed in about 50% of patients with Graves' disease. TSH receptor autoantibodies are usually measured in serum or plasma for diagnosis of Graves' disease because over 90% of patients have detectable levels of these antibodies. In addition, thyroid-stimulating autoantibodies may also be present in patients with Graves' disease. Graves' disease is a complex disease where genetic predisposition is modified by environmental factors. There is an association between polymorphism of human lymphocyte antigen (HLA) genes (HLA-DRB1*3 allele) and a young-age diagnosis of Graves' disease [9].

Another cause of hyperthyroidism is toxic nodular goiter, which is more common in areas of the world where iodine deficiency is common. It is more prevalent in patients older than 40 years. Toxic adenoma is due to autonomously functioning nodules. It is usually found in younger people. Thyroiditis can be acute. In thyroiditis, thyroid hormone leaks from the inflamed gland, and is most likely due to viral illness. Postpartum thyroiditis is also an acute condition that may occur three to six months after delivery, but usually resolves itself. Amiodarone-induced hyperthyroidism is due to high iodine content (37%) of amiodarone. Iodine-induced hyperthyroidism can occur due to excess iodine in the diet or exposure to radiographic contrast media. Rare causes of hyperthyroidism include metastatic thyroid cancer, ovarian tumor that produces thyroid hormone (struma ovarii), and trophoblastic tumors that produce chorionic gonadotropin and activate TSH receptors. In addition, TSH-secreting pituitary tumors can cause hyperthyroidism. The first screening test for a patient with suspected hyperthyroidism without any evidence of pituitary disease is TSH. If TSH is undetectable or very low, hyperthyroidism should be suspected. Antithyroid antibodies are elevated in Graves' disease (see Table 9.5). Radionuclide uptake and scan can differentiate high uptake in Graves' disease versus low uptake in thyroiditis [10].

CASE REPORT

A 76-year-old woman was admitted to the psychiatric division of the authors' hospital for irritability and delusion. Her medical conditions included type 2 diabetes, hypertension, dyslipidemia, hyperuricemia, and schizophrenia, but her thyroid function tests were normal four months prior to admission. Her thyroid was not swollen, but her TSH level was low (< 0.03 μIU/mL; normal: 0.54–4.25 μIU/mL). FT3 (11.70 pg/mL: normal: 2.39–4.06 pg/mL) and FT4 (3.07 ng/dL: normal: 0.7–1.5 ng/dL), however, were elevated, indicating hyperthyroidism. However, her TSH receptor autoantibody and thyroid-stimulating autoantibody tests were negative. Because slightly increased blood flow and swelling were detected by thyroid echography, a thyroid scintigraphy test was performed; it showed diffuse and remarkably elevated uptake of radioactive iodine, thus indicating Graves' disease. She was treated with methimazole and her hyperthyroidism and psychiatric symptoms were resolved. The authors commented that thyroid autoantibody-negative Graves' disease is extremely rare, but thyroid scintigraphy is useful in diagnosis of such patients [11].

Thyroid storm is a life-threatening condition that may develop if hyperthyroidism goes untreated or can be precipitated by trauma or infection in a patient with hyperthyroidism. Thyroid storm is a medical emergency and is accompanied by elevated blood pressure and high heart rate. Laboratory findings usually include very low or undetectable levels of TSH accompanied by very high thyroid hormone levels such as FT4, total T3, etc.

CASE REPORT

A 37-year-old woman was brought to the emergency room by her boyfriend because she was agitated and had an altered mental state. The patient had been under stress lately due to the start of a new job after two years of unemployment. She had no remarkable past medical history, but a physician had prescribed her lorazepam a month earlier for anxiety. Her blood pressure was 127/82 mm of Hg but her heart rate was elevated to 117 beats/min, and her respiratory rate was 22 breaths/min. She also showed sinus tachycardia, fever, respiratory alkalosis, and urine frequency, and her urinalysis was positive for benzodiazepine; no other abnormality was found. A thyroid function test showed very low TSH with highly elevated FT4 (4.08 ng/dL) and total T3 of 900 ng/dL (normal: 75–200 ng/dL), confirming the diagnosis of thyroid storm, as suspected from her symptoms [12].

Table 9.5 Interpretation of Thyroid Autoantibody Tests

Thyroid Antibody	Associated with Thyroid Disease
Thyroid peroxidase antibody	Hashimoto thyroiditis
Thyroglobulin antibody	Hashimoto thyroiditis, thyroid cancer
TSH receptor autoantibody	Graves' disease
Thyroid-stimulating autoantibody	Graves' disease

9.8 DISORDERS OF PARATHYROID GLANDS

The parathyroid gland produces parathyroid hormone (PTH), which, along with calcitonin (produced by the thyroid gland) and vitamin D, regulates calcium metabolism. PTH is an 84-amino acid hormone secreted by the chief cells of the parathyroid, and increases calcium levels in the blood by the following mechanism:

- Increased osteoclastic activity in bone.
- Increased synthesis of 1,25-dihydroxycholecalciferol (vitamin D3).
- Increased renal reabsorption of calcium.
- Increased intestinal absorption of calcium.

Calcitonin, which is secreted by the parafollicular C cells of the thyroid, essentially has the opposite action to that of PTH. The primary source of vitamin D is photoactivation of 7-dehydrocholesterol in the skin to cholecalciferol. Exposure to the UV radiation of sunlight (UVB radiation: 290–320 nm) for 15 min in midday sun is enough to produce cholecalciferol, which is then converted to 25-hydroxycholecalciferol in the liver and then to 1,25-dihydroxycholecalciferol in the kidney. Vitamin D promotes absorption of calcium from the gut and helps in bone mineralization. Deficiency or lack of 1,25-dihydroxycholecalciferol in children causes rickets and osteomalacia in adults. In chronic kidney disease, the kidney lacks the ability to convert

25-hydroxycholecalciferol to 1,25-dihydroxycholecalciferol, and vitamin deficiency may occur. Both vitamin D and PTH are responsible for calcium metabolism. The causes of hypercalcemia include:

- Primary hyperparathyroidism.
- Tertiary hyperparathyroidism.
- Malignancy.
- Excess vitamin D.
- Excessive calcium intake (milk alkali syndrome).
- Drug therapy such as thiazide diuretics therapy.

Hyperparathyroidism is a common cause of hypercalcemia. Hyperparathyroidism can be primary (due to adenomas or hyperplasia of the parathyroid glands), secondary (due to compensatory hypertrophy of parathyroid glands from hypocalcemia, as seen in chronic kidney disease), or tertiary, where, after a long period of secondary hyperparathyroidism, the parathyroid glands develop autonomous hyperplasia and hyperparathyroidism persists, even when hypocalcemia is corrected.

Various causes of hypocalcemia include:

- Hypoparathyroidism that could be congenital (DiGeorge syndrome), secondary to hypomagnesemia (magnesium is required for PTH secretion), or due to parathyroidectomy.
- Chronic kidney disease.
- Vitamin D deficiency.
- Resistance to PTH (pseudohypoparathyroidism).
- Drugs (e.g. calcitonin, bisphosphonates).
- Acute pancreatitis.
- Malabsorption.

Hypoparathyroidism refers to low levels of PTH being secreted from the parathyroid glands whereas pseudohypoparathyroidism refers to the inability of PTH to exert its function due to receptor defects. Pseudohypoparathyroidism is a hereditary disorder, and patients, in addition to hypocalcemia, also have short stature, short metacarpals, and intellectual impairment. Pseudopseudohypoparathyroidism patients actually have no abnormality of PTH or the parathyroid. Only the somatic features seen in pseudohypoparathyroidism are present, but these patients do not have cognitive impairment as seen in patients with pseudohypoparathyroidism. The molecular basis for this disease is mutation of the Gs alpha gene [13].

9.9 ADRENAL GLANDS

The adrenal glands consist of a cortex and a medulla. The cortex has three zones: zona glomerulosa, zona fasciculata, and zona reticularis. The zona glomerulosa is responsible for secreting mineralocorticoids (aldosterone), while the zona fasciculata is responsible for secreting glucocorticoids. Finally, the zona reticularis is responsible for producing sex steroids. The adrenal medulla produces catecholamines. Steroid hormones are synthesized by adrenal glands from cholesterol, while sex steroid hormones are synthesized in the gonad (Figure 9.2). Major actions of glucocorticoids include:

- Gluconeogenesis and glycogen deposition.
- Fat deposition.
- Protein catabolism.
- Sodium retention.
- Loss of potassium.
- Increase in circulating neutrophils and decrease in circulating eosinophils and lymphocytes.

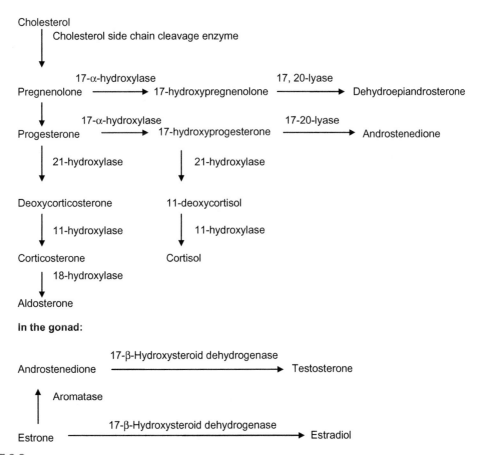

FIGURE 9.2
Steroid hormones biosynthetic pathway.

Major actions of mineralocorticoids include:

- Sodium and water retention in the distal tubule.
- Loss of potassium.

Congenital adrenal hyperplasia is most often due to the lack of 21-hydroxylase enzyme, which causes decreased production of deoxycorticosterone and aldosterone as well as reduced levels of deoxycortisol and cortisol. Adrenocorticotropic hormone (ACTH) level is also high, and as a result, 17-hydroxypregnenolone and 17-hydroxyprogesterone are produced in higher concentrations. This leads to increased production of dehydroepiandrosterone, androstenedione, and testosterone. Female children will have a virilizing effect and may also have ambiguous genitalia. Male children will have features of precocious puberty.

9.10 CUSHING'S SYNDROME

Cushing's syndrome is characterized by a high level of circulating cortisol. Major causes of Cushing's syndrome can be sub-classified under two broad categories:

- Adrenocorticotropic hormone (ACTH)-dependent disorders, which include Cushing's disease (ACTH-secreting pituitary tumor), ectopic ACTH-producing tumor (such as lung cancer), and secondary to ACTH administration.
- Non-ACTH-dependent: Adrenal tumor or secondary to glucocorticoid administration.

ACTH-dependent Cushing's syndrome is more common (70−80% of all cases) compared to non-ACTH-dependent disorders. Among ACTH-dependent disorders, Cushing's disease is observed more frequently. Diagnosis of Cushing's syndrome first requires confirmation of hypercortisolism, then finding the cause of high cortisol. Investigations that are useful for the diagnosis of Cushing's syndrome include:

- Measurement of 24-hour urinary free cortisol (values are elevated in Cushing's syndrome).
- Loss of circadian rhythm (measurement of cortisol at 9 AM and midnight should show loss of circadian rhythm as evidenced by higher midnight cortisol values compared to 9 AM values in patients with Cushing's syndrome).
- Overnight dexamethasone suppression test: patients take 1 mg of dexamethasone at bedtime and serum cortisol is measured the following morning. Cushing's syndrome patients should still show elevated levels of cortisol.

- Low and high dose dexamethasone suppression tests (discussed at the end of the chapter).
- More recently, late night or midnight salivary cortisol collection has been introduced for diagnosis of Cushing's syndrome. It has gained popularity because no venipuncture is needed and the patient can collect a specimen at home and store it because salivary cortisol is stable.

It is important to note that renal impairment can falsely lower urinary free cortisol (called urinary free cortisol because cortisol level in urine is a measure of free or unbound cortisol in circulation). A false positive test result can occur in a dexamethasone suppression test due to treatment with cytochrome P-450 liver enzyme-inducing drugs such as carbamazepine, phenytoin, phenobarbital, rifampicin, meprobamate, aminoglutethimide, methaqualone, and troglitazone. These drugs can significantly increase clearance of dexamethasone and cause a false positive result. Moreover, measurement of plasma ACTH levels is useful in differential diagnosis because ACTH levels are low in patients with autonomous adrenal diseases, normal or elevated in patients with Cushing's disease, and elevated in ectopic ACTH syndrome. In rare situations when the ACTH value is below normal, a corticotropin-releasing hormone (or desmopressin) stimulation test may be useful because desmopressin stimulates ACTH and eventually cortisol release in Cushing's disease [14].

Pseudo-Cushing's syndrome is caused by conditions (such as alcoholism, severe obesity, polycystic ovary syndrome, etc.) that can activate the hypothalamic–pituitary–adrenal axis and cause Cushing's-like syndrome. Although Cushing's syndrome is rare, pseudo-Cushing's syndrome may be observed more often; however, it may be difficult to distinguish between these two conditions because diagnostic tests can provide similar results. However, for alcoholics, pseudo-Cushing's syndrome may resolve spontaneously after cessation of alcohol consumption [15].

CASE REPORT

A 61-year-old Japanese woman who had been on hemodialysis for the past 21 years was admitted to the hospital with persistent hypokalemia and obesity. Her 8 AM serum cortisol was elevated (19.2 μg/dL, normal: 4–18.3 μg/dL) and further testing showed no circadian rhythm for cortisol. Her ACTH was low at 2 pg/mL (normal: 7.2–63.3 pg/mL). In addition a dexamethasone suppression test using 1 mg dexamethasone failed to suppress cortisol levels. Plasma ACTH and cortisol also failed to respond to corticotropin-releasing hormone (CRH) administration. A diagnosis of Cushing's syndrome was made, and further investigations revealed the disease was due to a left adrenal tumor. After surgery, her hypokalemia improved gradually without medication, and her weight gain was resolved. The authors commented that this was a rare case of adrenal Cushing's syndrome in a patient on long-term hemodialysis [16].

9.11 CONN'S SYNDROME

Conn's syndrome is most often due to an adenoma secreting aldosterone from the adrenal cortex. Clinical symptoms include hypertension (due to sodium and water retention) and hypokalemia. Therefore, it is imperative to measure serum electrolytes in a hypertensive patient if there is any suspicion of secondary hypertension. Other tests that can be helpful for diagnosis of Conn's syndrome measure the following factors:

- Aldosterone-to-renin ratio (ARR, increased).
- Renin levels (low).
- Plasma potassium and urinary potassium (hypokalemia in serum and increased loss of potassium in urine).
- Saline suppression test: Aldosterone levels are measured before and after administration of normal saline. Normal individuals should have lower aldosterone levels with the influx of sodium, but in patients with Conn's syndrome aldosterone levels may not change.

9.12 HYPOADRENALISM INCLUDING ADDISON'S DISEASE

Adrenal insufficiency may be primary, secondary, or tertiary. Primary adrenal insufficiency (hypoadrenalism) may be acute or chronic. Primary acute hypoadrenalism is most commonly due to hemorrhagic destruction of adrenal glands (Waterhouse—Friderichsen syndrome). Chronic hypoadrenalism is Addison's disease, where there is progressive dysfunction of adrenal glands due to a local disease process or systemic disorder. Secondary hypoadrenalism is due to lack of ACTH from pituitary due to hypothalamus—pituitary dysfunction. Tertiary hypoadrenalism is due to lack of corticotropin-releasing hormone (CRH). Causes of Addison's disease include:

- Congenital adrenal hyperplasia due to enzyme defect.
- Autoimmune disease.
- Post-surgery issues.
- Tuberculosis.
- Sarcoidosis.

Congenital adrenal hyperplasia is a group of inherited autosomal recessive disorders due to a defect of any of five enzymes responsible for cortisol biosynthesis. The most common cause is deficiency of 21-hydroxylase (90—95% of all cases) followed by 11-β-hydroxylase. The enzyme 21-hydroxylase converts progesterone into deoxycorticosterone, and, in addition, also converts 17-hydroxyprogesterone into 11-deoxycortisol. If this enzyme is deficient, precursors of cortisol accumulate in the blood, especially 17-hydroxyprogesterone, and, to a lesser extent, androstenedione. Peripheral conversion of androstenedione into testosterone is responsible for androgenic symptoms associated with congenital adrenal hyperplasia. Autoimmune Addison's disease is a rare disorder with symptoms typically developing over months to years due to the appearance of autoantibodies to the key enzyme, 21-hydroxylase. Usually a patient demonstrates a period of compensated or preclinical disease with elevated ACTH and renin before development of symptomatic adrenal failure. This disease may have a genetic component in man [17]. Addison's disease may be diagnosed by establishing low levels of cortisol as well as ACTH stimulation tests. These are discussed at the end of the chapter in the endocrine testing section.

CASE REPORT

A 35-year-old Caucasian man with a previous diagnosis of congenital adrenal hyperplasia in the neonatal period was admitted to the emergency department with anorexia, nausea, vomiting, and abdominal pain. He acknowledged poor adherence to his treatment regimen and irregular medical checkups. Physical examination showed marked cutaneous and gingival hyperpigmentation, hypotension, and hard nodules in both testicles. Blood tests showed mild anemia and hyponatremia, a low level of cortisol (3 μg/dL, normal: 4.3−23 μg/dL), and dehydroepiandrosterone sulfate (DHEA-sulfate: 39.4 μg/dL, normal: 80−560 μg/dL), but elevated 11-deoxycortisol and ACTH (480 pg/mL, normal: up to 46 pg/mL), and highly elevated 17-hydroxyprogesterone (20,400 ng/dL). An ACTH stimulation test failed to increase serum cortisol levels. The abdominal computed tomography showed a grossly enlarged and heterogenous adrenal gland. A bilateral adrenalectomy was performed. The patient was discharged with a prescription of hydrocortisone and fludrocortisone, and two months after surgery his cutaneous pigmentation almost disappeared and his testicular nodules were significantly smaller. The authors commented that this case showed that congenital adrenal hyperplasia due to 21-hydroxylase deficiency can progress to severe complications, and that the masses in the adrenal and testiculus of this patient resulted from chronically elevated ACTH and the growth of adrenocortical cells [18].

9.13 DYSFUNCTIONS OF GONADS

Proper function of the hypothalamic−pituitary−gonadal axis is vital for normal function of the reproductive systems of both men and women. In men, GnRH, LH, and FSH are secreted in a pulsatile pattern with higher levels in the early morning hours and lower levels at late evening. LH is also essential for men because it stimulates testosterone production by testes. Circulating testosterone is also a precursor for dihydrotestosterone and estradiol. Levels of testosterone in males may start declining after age 50. This is called andropause.

A healthy neonate female possesses approximately 400,000 primordial follicles, each containing an immature ovum, and during each menstrual cycle one ovum attains maturity approximately during mid-cycle (~Day 14). Therefore, only about 400 immature ova attain maturity during the reproductive cycle of a woman. In a normal individual, most estrogen is secreted by the ovarian follicle and the corpus luteum, but during pregnancy the placenta produces most of the estrogen. Progesterone is secreted by the corpus luteum, but during pregnancy the placenta is responsible for producing the majority of the progesterone. Progesterone is important for maintenance of pregnancy.

Hypogonadism may be broadly divided into two categories: hypergonadotropic and hypogonadotropic hypogonadism. Examples of hypergonadotropic hypogonadism include:

- Gonadal agenesis.
- Gonadal dysgenesis (e.g. Turner's syndrome, Klinefelter's syndrome).
- Steroidogenesis defect.
- Gonadal failure (e.g. mumps, radiation, chemotherapy, autoimmune diseases, granulomatous diseases).
- Chronic diseases (e.g. liver failure, renal failure).

Examples of hypogonadotropic hypogonadism include:

- Hypothalamic lesions (e.g. tumors, infections, Kallmann's syndrome).
- Pituitary lesions (e.g. adenomas, Sheehan's syndrome, sarcoidosis, hemochromatosis).

Polycystic ovary syndrome is the most common endocrinological dysfunction in women and affects 6–10% of women during the reproductive age. This syndrome has diverse clinical implications, including reproductive dysfunctions (infertility, hirsutism, hyperandrogenism) and metabolic dysfunctions (type 2 diabetes mellitus or impaired glucose tolerance, insulin resistance, adverse cardiovascular event), as well as a psychological component such as poor quality of life, depression, and anxiety. Polycystic ovary syndrome is a polygenic disease with both genetic and environmental components. Obesity-induced insulin resistance is known to exacerbate all features of polycystic ovary syndrome. Modest weight loss may improve many features of this syndrome [19].

Investigations of hypogonadism can include:

- Measurement of basal levels of testosterone, estrogens, FSH, and LH.
- Measurement of urinary FSH and LH: This can be done in males as FSH and LH levels in males may be undetectable in blood due to the pulsatile nature of secretion.
- GnRH stimulation test: In individuals with hypogonadotropic hypogonadism this will help to distinguish hypothalamic causes from

pituitary lesions. If administration of GnRH results in increased FSH and LH levels, this means the pituitary is functional.

- Clomiphene stimulation test: Clomiphene has anti-estrogenic effects. Estrogen inhibits the release of GnRh from the hypothalamus. Administration of clomiphene in normal individuals results in release of GnRH, which in turn causes FSH and LH levels to rise.
- hCG (human chorionic gonadotropin) stimulation test: This is done in males with low testosterone levels. hCG binds to LH receptors and stimulates testosterone production by the Leydig cells.

9.14 PANCREATIC ENDOCRINE DISORDERS

The most common endocrine pancreatic disorder is diabetes mellitus (discussed in Chapter 7). Other endocrine disorders of the pancreas include the following:

- Islet cell tumors.
- Non-functioning islet cell tumors.
- Insulinoma (see Chapter 7).
- Gastrinoma.
- VIPoma.
- Glucagonoma.
- Somatostatinoma.

Gastrinomas cause increased secretion of gastric acid, which results in multiple recurrent duodenal ulcers, and is referred to as Zollinger–Ellison syndrome. VIPomas produce excessive vasoactive intestinal polypeptides (VIP) that cause watery diarrhea (Verner–Morrison syndrome). Glucagonomas are rare tumors from the alpha islet cells. Features include diabetes mellitus, migratory necrolytic dermatitis, and deep vein thrombosis. Somatostatinomas are rare tumors derived from the delta islet cells. Features include diabetes mellitus and gallstones. Non-functioning tumors may produce a mass effect and biliary obstruction. Approximately 25% of islet cell tumors form part of multiple endocrine neoplasias (MEN) type 1.

9.15 MULTIPLE ENDOCRINE NEOPLASIAS

This condition is caused by the occurrence of simultaneous or metachronous tumors involving multiple endocrine glands, and includes the following subtypes:

- MEN type 1: Parathyroid adenoma or hyperplasia with pituitary adenoma and pancreatic islet cell tumor.

- MEN type 2a: Adrenal tumor with medullary carcinoma of thyroid and parathyroid hyperplasia.
- MEN type 2b: Type 2a with marfanoid habitus, intestinal, and visceral ganglioneuromas.

MEN type 1 is due to a defect in the gene menin located on chromosome 11. Menin normally suppresses a transcription factor (JunD), and lack of this suppression results in oncogenesis. Patients with MEN 1 have one defective menin gene and one wild gene. When the wild type undergoes somatic mutation it results in tumorigenesis. MEN 2a and 2b are due to mutations of the RET proto-oncogene located on chromosome 10. Pheochromocytoma is a rare tumor of the adrenal gland (usually in the adrenal medulla) tissue and may be classified as a multiple endocrine neoplasia category of disease. This tumor is responsible for the release of too much epinephrine and norepinephrine (that control heart rate, metabolism, and blood pressure), but rarely is pheochromocytoma cancerous. Surgical correction is the best therapy to cure this disease, although drug therapy may also be used depending on the judgment of the clinician.

9.16 ENDOCRINE TESTINGS: SUPPRESSION AND STIMULATION TESTS

Endocrine testing consists of measuring blood levels of various hormones as well as using suppression and stimulation tests. If a high level of hormone is observed, then the suppression test is more appropriate to see if hormone levels can be suppressed by using an appropriate agent. If the reason behind high levels is physiological, then hormone levels would be suppressed. If not, the underlying cause is pathological. Similarly, if the initial hormone level is low then the stimulation test is performed. If the underlying condition is physiological, levels should rise with the stimulation challenge. Common suppression and stimulation tests are summarized in Table 9.6.

A glucose tolerance test is performed most commonly with a 75-g oral glucose dose for diagnosis of gestational diabetes, as well as diabetes mellitus if glucose levels are in the borderline zone. However, a glucose tolerance test is also useful in diagnosis of hyperpituitarism. Most commonly this test is used for diagnosis of acromegaly. The most common cause of acromegaly is a growth hormone (GH)-secreting pituitary adenoma, and surgical removal of the tumor is the first choice for therapy. Immediate postoperative GH levels are a good indicator of successful surgery, but an oral glucose tolerance test performed a week later where the GH level is suppressed below 1 μg/L is a good predictor of long-term remission of acromegaly [20]. Although a glucose tolerance test is used most commonly for diagnosis of acromegaly, it can be used in general to diagnose

Table 9.6 Common Endocrine Tests

Endocrine Test	Analytes Measured	Interpretation
Glucose tolerance test (GTT); for hyperpituitarism.	Basal levels of FSH, LH, TSH, ACTH, cortisol, GH, and then reanalysis of these analytes after administration of oral glucose.	With true hyperpituitarism basal levels will be high and will not reduce.
Triple bolus (insulin, GnRH, TRH).	Basal levels of FSH, LH, TSH, prolactin, cortisol, GH, and reanalysis of these analytes after administration of insulin, GnRH, and TRH.	With true hypopituitarism basal levels will be low and will not rise.
Overnight dexamethasone suppression test (1 mg of dexamethasone at bedtime); for Cushing's syndrome.	Basal cortisol at 8–9 AM and then the next morning after receiving dexamethasone.	Normal patients should have cortisol below 5 μg/dL, but patients with Cushing's should not show any suppression of morning cortisol level.
Low-dose dexamethasone test (0.5 mg, q6h for 2 days); for Cushing's syndrome.	Basal cortisol level and reanalysis of cortisol level after 48 hours.	True Cushing's syndrome patient will have high levels and will not reduce.
High-dose dexamethasone suppression test (2 mg q6h for 2 days). Done after positive low dose dexamethasone suppression test to differentiate Cushing's disease from other causes.	Basal cortisol level and after 48 hours.	Cushing's disease patient will show 50% or more reduction of cortisol level; other causes of Cushing's syndrome will not.
Short ACTH (250 μg) stimulation test for hypoadrenalism.	Cortisol level before and after.	True hypoadrenalism patients will have low basal levels and will not rise.
Long ACTH (1 mg) stimulation test to differentiate primary from secondary hypoadrenalism.	Cortisol level before and after (up to 24 hours).	Patients with primary hypoadrenalism will show no rise at all here as patients with secondary hypoadrenalism will show gradual increase with time.

pituitarism by measuring base levels of any of a combination of hormones, including LH, FSH, ACTH, cortisol, and GH. Following oral administration of glucose, values of these hormones should be suppressed, but with true hyperpituitarism, basal levels will be high and will not be suppressed following administration of oral glucose.

For diagnosis of hypopituitarism, especially dysfunction of the anterior pituitary, the bolus test (also known as the dynamic pituitary function test) is used. In this test, three hormones, including insulin, gonadotropin-releasing hormone (GnRH), and thyrotropin-releasing hormone (TRH), are injected in a bolus into a patient's vein to stimulate the anterior pituitary gland. Before the bolus injection, baseline levels of cortisol, GH, prolactin, TSH, LH, and FSH are measured. After bolus administration, insulin-induced hypoglycemia should increase levels of cortisol and GH, while TRH should increase levels of TSH and prolactin, and levels of LH and FSH should increase due to administration of GnRH. The serum glucose value is also measured to ensure hypoglycemia induced by insulin. However, in a patient with hypopituitarism, levels of these hormones should stay low at baseline values despite administration of these hormones by bolus injection.

Various dexamethasone suppression tests are useful in the diagnosis of Cushing's syndrome. Dexamethasone is a potent glucocorticoid that suppresses the nocturnal rise in ACTH levels and thus suppresses 8 AM cortisol levels in a normal individual. In an overnight dexamethasone suppression test, 1 mg of dexamethasone is given at bedtime and the serum cortisol level at 8−9 AM is measured. In a normal individual the serum cortisol level should be $< 5 \, \mu g/dL$ following administration of dexamethasone; a value over $5 \, \mu g/dL$ indicates Cushing's syndrome. In a low-dose dexamethasone suppression test, 0.5 mg of dexamethasone is administered every 6 h for two days. The cortisol level is measured in the morning before and after administration of dexamethasone. In patients with Cushing's syndrome, no suppression of the cortisol level is observed following administration of dexamethasone. However, due to simplicity, a 1-mg dexamethasone suppression test is used more frequently. A high-dose dexamethasone suppression test is useful to differentiate Cushing's syndrome caused by adrenal tumors and non-endocrine ACTH-secreting tumors from Cushing's disease. This test is usually performed after a low-dose dexamethasone suppression test or a 1-mg dexamethasone suppression test. In this test, 2 mg of dexamethasone are administered every six hours for two days (an 8-mg total dosage), and serum cortisol is measured in the morning before and after administration of dexamethasone. In patients with Cushing's syndrome, no suppression of the morning cortisol level should be observed, but in patients with Cushing's disease, a 50% or more reduction of serum cortisol should be observed.

Although the glucose tolerance test is sometimes considered a gold standard for evaluating hypothalamus−pituitary−adrenal function in adrenal insufficiency, the ACTH stimulation test (also known as the cosyntropin test) is also used to determine functional capacity of adrenal glands in evaluating patients with suspected adrenal insufficiency. A normal individual should show two- to three-fold increases in serum cortisol (a gradual increase with time) within 1 h after administration of exogenous ACTH. In this test, after administration of synthetic ACTH (tetracosactrin: 1−24 amino acid sequence of human ACTH), if the serum cortisol level is not increased, it is indicative of adrenal insufficiency. In the standard ACTH stimulation test (also known as the short ACTH stimulation test), $250 \, \mu g$ of synthetic ACTH is administered intramuscularly or intravenously and a subnormal cortisol response ($< 18 \, \mu g/dL; < 500 \, nmol/L$) 30 to 60 min after the stimulation test is considered a positive test and indicates an increased possibility of primary or secondary adrenal insufficiency. A value over 20 µg/dL is considered a normal response. Sometimes a long-acting ACTH stimulation test using 1 mg of synthetic ACTH is used for differentiation between primary and secondary hypoadrenalism. More recently, a low-dose ACTH stimulation test using only 1 µg of synthetic ACTH has been introduced. This test is useful for the diagnosis of adrenal insufficiency; however, older males may have a more decreased responsiveness to this test than older females [21]. Another alternative to test the function of the hypothalamus–pituitary–adrenal axis is administration of metyrapone, an inhibitor of 11 β-hydroxylase enzyme that converts 11-deoxycortisol to cortisol. Under normal conditions, a reduced cortisol level in plasma stimulates ACTH release and the concentration of 11-deoxycortisol in serum increases significantly; a lack of response suggests primary adrenal failure.

KEY POINTS

- Endocrine activity can be classified as autocrine, paracrine, or classical endocrine activity. In autocrine activity, chemicals produced by a cell act on the cell itself. In paracrine activity chemicals produced by a cell act locally. However, in classical endocrine activity, chemicals produced by an endocrine gland act at a distant site after their release in the circulation; these chemicals are called hormones. Most classical hormones are secreted into the systemic circulation. However, hypothalamic hormones are secreted into the pituitary portal system.
- Receptors for hormones may be cell surface, membrane, or nuclear receptors.
- Hormone secretion may be continuous or intermittent. Thyroid hormone secretion is continuous. Thus, levels may be measured at any time to assess hormonal status. Secretion of follicle-stimulating hormone (FSH), luteinizing hormone (LH), and growth hormone (GH) are pulsatile. Thus, a single measurement may not reflect hormonal status. Some hormones exhibit biological rhythms. Cortisol exhibits a circadian rhythm where levels are highest in the morning and lowest during late night. The menstrual cycle is an example of a longer biological rhythm where different levels of a hormone are observed during a specific part of the cycle.
- Certain hormone levels are elevated during stress. These include adrenocorticotropic hormone (ACTH), as well as cortisol, GH, prolactin, adrenaline, and noradrenaline.
- Certain hormone levels are increased during sleep, such as GH and prolactin.
- The hypothalamus produces thyrotropin-releasing hormone (TRH), corticotropin hormone (CRH), gonadotropin-releasing hormone (GnRH), growth hormone-releasing hormone (GHRH), and somatostatin (growth hormone inhibitory hormone). These hormones act on the anterior pituitary and result in the release of various other hormones, including thyroid-stimulating hormone (TSH), ACTH, FSH, LH, and GH. Somatostatin inhibits the release of GH. Dopamine (also known as prolactin inhibitory

hormone) is also a neurotransmitter produced by the hypothalamus. Dopamine can inhibit GH secretion.

- The supraoptic and paraventricular nuclei of the hypothalamus produce antidiuretic hormone (ADH, i.e. vasopressin) and oxytocin. These hormones are stored in the posterior pituitary and act on certain body parts rather than on the pituitary like other tropic hormones do. ADH acts on the collecting ducts of the renal tubules and causes absorption of water.

- Growth hormone (GH, also known as somatotropin) is the most abundant hormone produced by the anterior pituitary, and it stimulates growth of cartilage, bone, and many soft tissues. GH stimulates release of insulin-like growth factor-1 (IGF-1, somatomedin C), mostly from the liver.

- Hyperpituitarism is most often due to pituitary tumors affecting GH-secreting cells, prolactin-secreting cells, and ACTH-secreting cells. GH-secreting tumors that affect individuals before closure of the epiphyses result in gigantism, and after closure they result in acromegaly. Prolactin-secreting tumors cause hyperprolactinemia. A high level of prolactin (prolactinemia) inhibits action of FSH and LH, which results in hypogonadism and infertility. ACTH-secreting tumors cause Cushing's syndrome.

- For diagnosis of hypopituitarism, stimulation tests with GnRH, TRH, and insulin-induced hypoglycemia (a triple stimulation test) is useful. Following stimulation, serum or plasma levels of FSH, LH, TSH, PRL, GH, and cortisol are measured.

- Endocrine tests for hyperpituitarism include measurement of hormone levels and a suppression test using glucose (oral glucose tolerance). Administration of glucose with a rise in blood glucose should suppress anterior pituitary hormones in normal individuals.

- Four steps are involved in the synthesis of thyroid hormones: (1) inorganic iodide from the circulating blood is trapped (iodide trapping), (2) iodide is oxidized to iodine, (3) iodine is added to tyrosine to produce monoiodotyrosine and diiodotyrosine (referred to as organification), and (4) one monoiodotyrosine is coupled with one diiodotyrosine to yield T3 and two diiodotyrosines are coupled to yield T4 (coupling).

- Free (unbound) T4 is the primary secretory hormone from the thyroid gland. T4 is converted in peripheral tissue (liver, kidney, and muscle) to T3 by 5'-monodeiodination. T3 is the physiologically active hormone. T4 can also be converted to reverse T3 by 3'-monodeiodination. This form of T3 is inactive. The majority (99%) of the T3 and T4 in circulation is found to be thyroxine-binding globulin (TBG), albumin, and thyroxine-binding prealbumin.

- Dyshormonogenetic goiter may be associated with nerve deafness, referred to as Pendred's syndrome.

- There are situations where thyroid hormone-binding proteins may be low or high, causing alteration of total T3 and T4 levels. However, free T3, T4, and TSH levels should be normal. Pregnancy and oral contraceptive pills raise concentrations of thyroid-binding proteins. Hypoproteinemic states such as cirrhosis of the liver, nephrotic syndrome, etc., may cause lower concentrations of thyroid-binding proteins.

- Amiodarone can reduce peripheral conversion of T4 to T3. Free T4 levels may be high, but TSH levels could be normal. Because it contains an iodine molecule, amiodarone can also cause both hypothyroidism and hyperthyroidism.

- Seriously ill patients may have reduced production of TSH with low T4 and reduced conversion of T4 to T3 with increased conversion of T4 to reverse T3. Patients are, however, euthyroid. This is referred to as sick euthyroid syndrome.

- Measurement of TSH can suffer interference from heterophilic antibody and rheumatoid antibody, causing falsely elevated results. Rarely, autoantibodies to TSH develop clinically, but such autoantibodies can also falsely increase TSH results. A rare interference in the TSH assay is due to macro-TSH, an autoimmune complex between anti-TSH IgG antibody and TSH.

- Primary hypothyroidism (primary disease of thyroid gland): Causes include autoimmune thyroiditis, Hashimoto's thyroiditis, surgery/radiation, dyshormonogenesis, antithyroid drugs, drug therapy with amiodarone, and advanced age.

- Secondary hypothyroidism can be due to lack of TSH from pituitary or peripheral resistance to thyroid hormones.

- Causes of hyperthyroidism include Graves' disease, toxic nodular (single or multiple) goiter, thyroiditis (e.g. due to viral infection), drugs, and excess TSH (e.g. due to pituitary tumor).

- Parathyroid hormone (PTH) is an 84-amino acid hormone secreted by the chief cells of the parathyroid, and it increases calcium levels in blood by increasing osteoclastic activity in bone, increasing synthesis of 1,25-dihydroxycholecalciferol (vitamin D3), increasing renal reabsorption of calcium, as well as by increasing intestinal absorption of calcium.

- Calcitonin, which is secreted by the parafollicular C cells of the thyroid, essentially has the opposite action to that of PTH.

- Hyperparathyroidism is a common cause of hypercalcemia and can be primary (due to adenomas or hyperplasia of parathyroid glands), secondary (due to compensatory hypertrophy of parathyroid glands due to hypocalcemia, as seen in chronic kidney disease), or tertiary (where after a long period of secondary hyperparathyroidism the parathyroid glands develop autonomous hyperplasia and hyperparathyroidism persists even when hypocalcemia is corrected).

- Hypoparathyroidism refers to low levels of PTH being secreted from the parathyroid glands, whereas pseudohypoparathyroidism refers to the inability of PTH to exert its function due to a receptor defect. Pseudohypoparathyroidism is a hereditary disorder, and patients, in addition to hypocalcemia, also have short stature, short metacarpals, and intellectual impairment. Pseudo-pseudohypoparathyroidism patients actually have no abnormality of PTH or parathyroid. Only the somatic features seen in pseudohypoparathyroidism are present, but these patients do not have cognitive impairment as seen in patients with pseudohypoparathyroidism.

- The adrenal glands consist of a cortex and a medulla. The cortex has three zones: zona glomerulosa, zona fasciculata, and zona reticularis. The zona glomerulosa is responsible for secreting mineralocorticoids (aldosterone), while the zona fasciculata is responsible for secreting glucocorticoids. Finally, the zona reticularis is responsible for producing sex steroids. The adrenal medulla produces catecholamines.

- Congenital adrenal hyperplasia is most often due to lack of 21-hydroxylase enzyme causing decreased production of deoxycorticosterone and aldosterone, as well as reduced levels of deoxycortisol and cortisol. ACTH level is also high, and as a result, 17-hydroxypregnenolone and 17-hydroxyprogesterone are produced in higher concentrations, which leads to increased production of dehydroepiandrosterone, androstenedione, and testosterone. Female children suffer a virilizing effect and may also have ambiguous genitalia. Male children have features of precocious puberty.

- Major causes of Cushing's syndrome can be sub-classified under two broad categories: (1) ACTH-dependent disorders, which include Cushing's disease (ACTH-secreting pituitary tumor), ectopic ACTH-producing tumor (such as lung cancer), and secondary due to ACTH administration; and (2) non-ACTH-dependent disorders (adrenal tumor or secondary due to glucocorticoid administration). ACTH-dependent Cushing's syndrome is more common (70–80% of all cases) compared to non-ACTH-dependent disorders. Among ACTH-dependent disorders, Cushing's disease is observed more frequently.

- Investigations that are useful for the diagnosis of Cushing's syndrome include measurement of 24-hour urinary free cortisol (values are elevated in Cushing's syndrome), loss of circadian rhythm (measurement of cortisol at 9 AM and midnight should show loss of circadian rhythm as evidenced by higher midnight cortisol values compared to 9 AM values in patients with Cushing's syndrome), an overnight dexamethasone suppression test (patients take 1 mg of dexamethasone at bedtime and serum cortisol is measured the following morning; Cushing syndrome patients should still show elevated levels of cortisol), as well as low- and high-dose dexamethasone suppression tests.

- Pseudo-Cushing's syndrome is caused by conditions such as alcoholism, severe obesity, polycystic ovary syndrome, etc.; these can activate the hypothalamic–pituitary–adrenal axis and cause Cushing's-like syndrome.

- Conn's syndrome is most often due to an adenoma secreting aldosterone from the adrenal cortex. Clinical symptoms include hypertension (due to sodium and water retention) and hypokalemia. Therefore, it is imperative to measure serum electrolytes in a hypertensive patient if there is any suspicion of secondary hypertension. Other tests that may be helpful for diagnosis of Conn's syndrome include aldosterone-to-renin ratio (ARR; increased in Conn's syndrome), plasma potassium and urinary potassium tests (in Conn's syndrome hypokalemia in serum and increased loss of potassium in urine is observed), and a saline suppression test (aldosterone levels are measured before and after administration of normal saline). Normal individuals should have lower aldosterone levels with the influx of sodium, but in patients with Conn's syndrome aldosterone levels may not change. Renin levels are also low in Conn's syndrome.

- Adrenal insufficiency can be primary, secondary, or tertiary. Primary adrenal insufficiency (hypoadrenalism) can be either acute or chronic. Primary acute hypoadrenalism is most commonly due to hemorrhagic destruction of adrenal glands (Waterhouse–Friderichsen syndrome). Chronic hypoadrenalism is Addison's disease, where there is progressive dysfunction of adrenal glands a local disease process or systematic disorder. Secondary hypoadrenalism is due to lack of ACTH from the pituitary because of hypothalamus–pituitary dysfunction. Tertiary hypoadrenalism is due to lack of corticotropin-releasing hormone (CRH).

- Causes of Addison's disease include congenital adrenal hyperplasia due to enzyme defect, autoimmune disease, post-surgery complications, tuberculosis, and sarcoidosis.

- Hypogonadism may be broadly divided into two categories: hypergonadotropic and hypogonadotropic hypogonadism. Examples of hypergonadotropic hypogonadism include gonadal agenesis, gonadal dysgenesis (e.g. Turner's syndrome and Klinefelter's syndrome), steroidogenesis defect, gonadal failure (e.g. mumps, radiation, chemotherapy, autoimmune diseases, granulomatous diseases), and chronic diseases (e.g. liver failure, renal failure).

- Examples of hypogonadotropic hypogonadism include hypothalamic lesions (e.g. tumors, infections, Kallmann's syndrome) and pituitary lesions (e.g. adenomas, Sheehan's syndrome, sarcoidosis, hemochromatosis).

- Gastrinomas cause increased secretion of gastric acid, which results in multiple recurrent duodenal ulcers (Zollinger–Ellison syndrome). VIPomas produce excessive vasoactive intestinal polypeptides (VIP) that cause watery diarrhea (Verner–Morrison syndrome). Glucagonomas are rare tumors from the alpha islet cells. Features include diabetes mellitus, migratory necrolytic dermatitis, and deep vein thrombosis. Somatostatinomas are rare tumors derived from the delta islet cells. Features include diabetes mellitus and gallstones. This condition is caused by the occurrence of simultaneous or metachronous tumors that involve multiple endocrine glands. The subtypes are MEN type 1 (parathyroid adenoma or hyperplasia with pituitary adenoma and pancreatic islet cell tumor), MEN type 2a (adrenal tumor with medullary carcinoma of thyroid and parathyroid hyperplasia), and MEN type 2b (type 2a with marfanoid habitus, intestinal, and visceral ganglioneuromas).

- Glucose tolerance test is useful in the diagnosis of hyperpituitarism.

- For diagnosis of hypopituitarism, a bolus test (also known as dynamic pituitary function test) is used where three hormones, including insulin, GnRH, and TRH are injected as a bolus into a patient's vein to stimulate the anterior pituitary gland. Before bolus injection, baseline levels of cortisol, GH, prolactin, TSH, LH, and FSH are measured. After bolus administration, insulin-induced hypoglycemia should increase levels of cortisol and GH, while TRH should increase levels of TSH and prolactin; finally, levels of LH and FSH should be increased due to administration of GnRH. Serum glucose value is also measured to ensure hypoglycemia induced by insulin. In a patient with hypopituitarism, levels of these hormones should stay low at baseline values despite administration of these hormones by bolus injection.
- In patients with Cushing's syndrome, no suppression of cortisol level is observed following administration of dexamethasone.
- High-dose dexamethasone suppression test is useful to differentiate Cushing's syndrome caused by adrenal tumors and non-endocrine ACTH-secreting tumors from Cushing's disease.
- Although the glucose tolerance test is sometimes considered the gold standard for evaluating hypothalamus–pituitary–adrenal function in adrenal insufficiency, the ACTH-stimulation test (i.e. the cosyntropin test) is also used to evaluate the functional capacity of adrenal glands in a patient with suspected adrenal insufficiency. Sometimes the long-acting ACTH-stimulation test using 1 mg of synthetic ACTH is used for differentiation between primary and secondary hypoadrenalism. In primary hypoadrenalism there will be no rise in serum cortisol. However, in secondary hypoadrenalism there will be a gradual increase in serum cortisol.

REFERENCES

[1] Kostoglou-Athanassiou I. Therapeutic application of melatonin. Ther Adv Endocrinol Metab 2013;4:13−24.

[2] Lee HJ, Macbeth AH, Pagani J, Young 3rd WS. Oxytocin: the great facilitator of life. Prog Neurobiol 2009;88:127−51.

[3] Pauche JE, Castilla-Cortazar I. Human conditions of insulin-like growth factor-1 (IGF-1) deficiency. J Transl Med 2012;10:224.

[4] Todd CH. Management of thyroid disorders in primary care challenges and controversies. Postgrad Med J 2009;85:655−9.

[5] Loh TP, Kao SL, Halsall DJ, Toh SA, et al. Macro-thyrotropin: a case report and review of literature. J Clin Endocrinol Metab 2012;97:1823−8.

[6] Dufour DR. Laboratory tests of thyroid function: use and limitations. Endocrinol Metab Clin N Am 2007;36:579−94.

[7] Zaletel K, Gaberscek S. Hashimoto's thyroidititis: from genes to the disease. Curr Genomics 2011;12:576−88.

[8] Gaitonde D, Rowley K, Sweeney LB. Hypothyroidism: an update. Am Fam Physician 2012;86:244−51.

[9] Jurecka-Lubieniecka B, Ploski R, Kula D, Krol A, et al. Association between age at diagnosis of Graves' disease and variants in genes involved in immune response. PLoS One 2013;8 (3):e59349.

[10] Reid JR, Wheeler SF. Hyperthyroidism: diagnosis and treatment. Am Fam Physician 2005;72:623−30.

[11] Hamasaki H, Yoshimi T, Yanai H. A patient with Graves' disease showing only psychiatric symptoms and negativity for both TSH receptor autoantibody and thyroid stimulating antibody. Thyroid Res 2012;5:19.

[12] Ravikumar D, Lin M. A 37 year old woman with altered mental status and urinary frequency. West J Emerg Med 2013;14:177−9.

[13] Mouallem M, Shaharabany M, Weintrob N, Shalitin S, et al. Cognitive impairment is prevalent in pseudohypoparathyroidism type Ia, but not in pseudopseudohypoparathyroidism: possible cerebral imprinting of Gsalpha. Clin Endocrinol (Oxf) 2008;68:233−9.

[14] Vilar L, Freitas Mda C, Faria M, Montenegro R, et al. Pitfalls in the diagnosis of Cushing's syndrome. Arq Bras Endocrinol Metabol 2007;51:1207−16.

[15] Besemer F, Pireira AM, Smit JW. Alcohol induced Cushing syndrome: hypercortisolism cause by alcohol abuse. Neth J Med 2011;69:318−23.

[16] Mise K, Ubara Y, Sumida K, Hiramatsu R, et al. Cushing's syndrome after hemodialysis for 21 years. J Clin Endocrinol Metab 2013;98:13−9.

[17] Mitchell AL, Pearce SH. Autoimmune Addison's disease: pathophysiology and genetic complexity. Nat Rev Endocrinol 2012;8:306−16.

[18] Ferreira F, Martins JM, do Vale S, Esteves R, et al. Rare and severe complications of congenital adrenal hyperplasia due to 21 hydroxylase deficiency: a case report. J Med Case Rep 2013;6:39.

[19] Teede H, Deeks A, Moran L. Polycystic ovary syndrome: a complex condition with psychological, reproductive and metabolic manifestations that impacts on health across the life span. BMC Med 2010;8:41.

[20] Kim EU, Oh MC, Lee EJ, Kim SH. Predicting long term remission by measuring immediate postoperative growth hormone levels and oral glucose tolerance test in acromegaly. Neurosurgery 2012;70:1106−13.

[21] Lekkakou L, Tzanela M, Lymberi M, Consoulas C, et al. Effects of gender and age on hypothalamic-pituitary-adrenal reactivity after pharmacological challenge with low dose 1 µg ACTH test: a prospective study in healthy adults. Clin Endocrinol (Oxf) 2013; Mar;10 [e-pub ahead of print].

CHAPTER 10

Liver Diseases and Liver Function Tests

10.1 LIVER PHYSIOLOGY

The liver is the largest internal organ of the body, approximately 1.2 to 1.5 kg in weight. The liver performs multiple functions essential to sustaining life and is the principal site for synthesis of all circulating proteins, except gamma globulins. A functioning normal liver produces 10–12 g of albumin daily; the half-life of albumin is approximately 3 weeks. When liver function is impaired over a prolonged period, albumin synthesis is severely impaired. Hypoalbuminemia is commonly found in chronic liver disease. However, a significant reduction in serum albumin levels may not be observed in patients with acute liver failure. In addition to albumin, all clotting factors (with the exception of Factor VIII) are produced in the liver. Therefore, as expected when liver function is significantly impaired, there is reduced production of clotting factors by the liver. As a result, coagulation tests such as prothrombin time (PT) is prolonged. Liver is also the site of urea production. In severe liver disease, such as fulminant hepatic failure, urea levels may be low. The liver also stores about 80 g of glycogen. Liver releases glucose into the circulation by glycogenolysis and gluconeogenesis. Again, in severe liver disease, hypoglycemia may be apparent due to depletion of the glycogen supply. Therefore, common features of significant liver dysfunction include:

- Prolonged PT, low serum glucose, and urea.
- Severe hypoalbuminemia, a common feature of chronic liver disease.

The liver also plays a major role in the synthesis of various lipoproteins, including very low density lipoprotein (VLDL) and high density lipoprotein (HDL). Hepatic lipase removes triglycerides from intermediate density lipoprotein (IDL) to produce low density lipoprotein (LDL). Liver is also a site for cholesterol synthesis. Cholesterol is esterified with fatty acids by the action of enzyme lecithin cholesterol acyl transferase (LCAT). In liver disease LCAT activity may be reduced, resulting in an increased ratio of cholesterol to cholesteryl ester.

CONTENTS

A. Dasgupta and A. Wahed: Clinical Chemistry, Immunology and Laboratory Quality Control
DOI: http://dx.doi.org/10.1016/B978-0-12-407821-5.00010-3

This may alter membrane structure with formation of target cells, as seen in liver disease. Bile acids are also synthesized in the liver from cholesterol and are excreted as bile salts. The primary bile acids (cholic acid and chenodeoxycholic acid) are converted into secondary bile acids by bacterial enzymes in the intestine. The secondary bile acids are deoxycholic and lithocholic acid. In liver diseases decreased production of bile acids may result in fat malabsorption.

Liver is the site of bilirubin metabolism. Heme, derived from the breakdown of hemoglobin, is converted to biliverdin and finally into bilirubin, which is water-soluble, unconjugated bilirubin. Unconjugated bilirubin can also bind with serum proteins, most commonly albumin. Unconjugated bilirubin is taken up by the liver, and, with the help of the enzyme UDP (uridine-5'-diphosphate) glucuronyl transferase, is converted to conjugated bilirubin (bilirubin conjugated with glucuronide). This conjugation takes place in the smooth endoplasmic reticulum of the hepatocyte. Conjugated bilirubin is water-soluble and is excreted in bile. In the clinical laboratory, conjugated bilirubin is measured as direct bilirubin, while subtracting total bilirubin from the direct bilirubin value provides the concentration of unconjugated bilirubin (also referred to as indirect bilirubin). In the intestine, bacterial enzymes hydrolyze conjugated bilirubin and release free bilirubin, which is reduced to urobilinogen. Urobilinogen bound to albumin is excreted in the urine. Some urobilinogen is converted to stercobilinogen in the intestine and is excreted in stool. Thus, in normal urine, only urobilinogen is present and in normal stool stercobilinogen is present. In obstructive (cholestatic) jaundice conjugated bilirubin regurgitates into the blood, and, because it is water-soluble, it is excreted into the urine. This is called choluria, or the presence of bile in urine. In obstructive jaundice, less conjugated bilirubin is taken by the intestine and as a result less stercobilinogen is found in the stool (pale stools). Normal individuals have mostly unconjugated bilirubin in their blood, urobilinogen in their urine, and stercobilinogen in their stool. The distribution of bilirubin, urobilinogen, and stercobilinogen in various diseases are summarized below:

- In individuals with hemolytic anemia, the excess breakdown of hemoglobin causes unconjugated hyperbilirubinemia. Urobilinogen in urine and stercobilinogen in stool may also be increased.
- In hepatocellular jaundice, uptake and conversion of unconjugated bilirubin into conjugated bilirubin is also reduced, resulting in unconjugated hyperbilirubinemia. However, amounts of urobilinogen in urine and stercobilinogen in stool are not increased.
- In cholestatic jaundice, conjugated hyperbilirubinemia is usually observed. Because conjugated bilirubin is water-soluble, it is excreted in urine (choluria). However, urobilinogen and stercobilinogen quantities are reduced.

Table 10.1 Physiological Function of the Liver

Liver Function	Comments
Protein synthesis	Low albumin is a feature of chronic liver disease; however, acute liver failure may not cause hypoalbuminemia.
Synthesis of clotting factors	Factor VIII is not produced in the liver. Significant liver dysfunction results in prolonged PT.
Urea synthesis from ammonia and carbon dioxide	Low urea level in fulminant hepatic failure.
Liver releases glucose by glycogenolysis and gluconeogenesis	Hypoglycemia in fulminant hepatic failure.
Lipid metabolism	Liver plays important role in lipoprotein and lipid metabolism.
Bilirubin synthesis	Hemolytic and hepatocellular jaundice results in unconjugated hyperbilirubinemia. Cholestatic jaundice results in increase in conjugated bilirubin with resultant choluria.
Bile acid synthesis	Liver dysfunction may result in malabsorption.

Major functions of the liver are summarized in Table 10.1.

10.2 LIVER FUNCTION TESTS AND INTERPRETATIONS

The conventional liver function test (LFT) consists of determination of serum or plasma levels of bilirubin as well as activities of various liver enzymes, including alanine aminotransferase (ALT), aspartate aminotransferase (AST), alkaline phosphatase (ALP), and sometimes gamma-glutamyltransferase (GGT). In addition, serum or plasma concentrations of total protein and albumin are useful in assessing liver function. Prolonged PT indicates significant impairment of liver function. Most of the normal bilirubin found in serum or plasma is unconjugated (indirect bilirubin). However, in hemolytic anemia, indirect bilirubin concentration may be increased. In addition, defects in the uptake of unconjugated bilirubin by hepatocytes may also increase the concentration of unconjugated bilirubin (indirect bilirubin) in serum. In contrast, an increased concentration of conjugated bilirubin (direct bilirubin) is seen in cholestatic jaundice.

The breakdown of hepatocytes results in the release of aminotransferases (also referred to as transaminases) such as ALT (alanine aminotransferase) and AST (aspartate aminotransferase) into the blood. ALT is a cytosol enzyme and is more specific for liver disease. AST is primarily a mitochondrial enzyme that is also found in the heart, muscle, kidney, and brain. ALT has a longer half-life than AST. In acute liver injury, AST levels are higher than ALT; however, after 24−48 hours ALT levels should be higher than AST. ALT is considered as a more specific marker for liver injury.

Alkaline phosphatase (ALP) is found in liver, bone, intestine, and placenta. ALP is located in the canalicular and sinusoidal membrane of the liver. Production of ALP is increased during cholestasis (intrahepatic or extrahepatic), which results in elevated activity of ALP in serum; however, it is important to determine if the source of ALP is the liver or other organs. If ALP level is raised, then the next question is whether the source of this enzyme is the liver or not. Measurement of gamma-glutamyl transpeptidase (GGT) or 5′-nucleotidase levels can be used to determine if the source of ALP is the liver or not because both GGT and 5′-nucleotidase are solely produced by the biliary epithelium. GGT is a microsomal enzyme. Therefore, the following criteria can be used to determine the origin of ALP:

- In cholestasis ALP, GGT and 5′-nucleotidase levels are raised, indicating that the liver is the source of ALP.
- If ALP activity is increased but activity of GGT or 5′-nucleotidase is normal, then it is unlikely that the liver is the source of excess ALP.

In acute liver disease both total protein and serum albumin concentrations are unaltered. In chronic liver disease total protein may be low or high. If total protein is high it is most likely due to polyclonal hypergammaglobulinemia. Serum albumin is likely to be low in chronic liver disease and serves as a guide to the severity of the liver disease. Prothrombin time (PT) reflects extrinsic and common pathway activity of the coagulation cascade. Clotting factors, with the exception of Factor VIII, are produced in the liver. Factor VII is one of the factors involved in the extrinsic pathway and has the shortest half-life among all clotting factors (4−6 hours). Therefore, within a short period after significant liver dysfunction, PT is prolonged and the magnitude of prolongation is correlated with the severity of liver dysfunction. Key points regarding interpretation of liver function tests include:

- In acute liver disease without cholestasis, levels of ALT and AST are significantly elevated and ALP is raised, but usually less than three times the upper limit of the normal value. Therefore, ALT and AST levels exceeding 500 U/L are a common finding in acute liver disease.
- In acute cholestasis, ALT and AST are raised but levels are not very high. ALP is usually more than three times the upper limit of the normal value with a parallel increase in GGT levels.

- In acute liver disease, total protein and albumin levels are unaltered.
- PT is the best test to assess the extent of liver dysfunction.
- In chronic liver disease, albumin levels are low but total protein may be elevated.
- In elevated ALP with normal bilirubin, ALT and AST may be seen in patients with hepatic metastasis or bone metastasis. Patients with hepatic metastasis may also have elevated GGT.
- In children showing only elevated ALP, it is most likely related to osteoblastic activity in their growing bones.
- Isolated elevated ALP in the elderly is a characteristic feature of Paget's disease.
- Very high isolated ALP levels can be seen in primary biliary cirrhosis.
- A normal liver function test (except elevated GGT) is a characteristic of excessive alcohol intake. Certain drugs (e.g. warfarin, anti-convulsants) may also produce similar results.

GGT is a well-established marker for alcohol abuse. In men over 40, consumption of even 8 standard drinks per week can elevate serum GGT levels, whereas those below 40 may show a significant change in GGT levels after consumption of 14 standard drinks per week [1]. Blunt abdominal trauma is a common reason for presentation in an emergency department; liver injury is common among patients with blunt abdominal injury. Modest elevation of ALT and AST are observed in blunt traumatic liver injury. Serum AST level >100 U/L and serum ALT level >80 U/L after blunt abdominal trauma may be associated with liver laceration. White blood cell (WBC) count may also be elevated [2]. Abnormal patterns of liver function tests are summarized in Table 10.2.

CASE REPORT

A 25-year-old man presented to the emergency department approximately 1.5 h after falling of his bike during his morning commute to work. During the fall he landed on the blunt end of a wooden post, with impact to his anterior/inferior right chest and abdominal right upper quadrant. His physical examination was unremarkable, but his ALT level was 249 U/L (normal < 40 U/L), AST level was 295 U/L (normal < 35 U/L), and lactate dehydrogenase level was 427 U/L (normal: 99–250 U/L). However, his GGT, ALP, and bilirubin levels were within normal range. These elevated liver function test results prompted the clinician to order a CT scan of his abdomen, which revealed a Grade III laceration of the liver and a small amount of hemoperitoneum. However, the patient continued to look and feel well and after 24 h of observation, he was discharged. He felt well when contacted after five days, and at that time his liver enzyme tests were repeated and showed results within normal range [3].

Table 10.2 Pattern of Abnormal Liver Function Tests in Various Liver Diseases

Liver Disease	Abnormal Liver Function Tests
Pre-hepatic jaundice (hemolytic anemia)	Total and unconjugated bilirubin is high. Liver enzymes (ALT, AST, ALP, and GGT) are normal. PT and proteins are normal.
Hepatocellular jaundice	Total bilirubin is high. Liver enzymes are elevated. ALT and AST may be very high (in thousands); ALP is less than three times normal. Total protein and albumin may be normal. PT may be prolonged if liver damage is significant.
Cholestatic jaundice	Total bilirubin and conjugated bilirubin is high. Liver enzymes are elevated (typically mildly). ALP is more than three times normal. PT may be prolonged due to lack of absorption of fat-soluble vitamin K.
Acute liver disease	PT is the best marker for assessment of extent of acute liver disease. Total protein and albumin are typically normal.
Chronic liver disease	Albumin level is decreased. Total protein could be elevated if hypergammaglobulinemia is present, liver enzymes may or may not be elevated.
Liver metastasis	Only ALP and GGT may be elevated.
Bone disease or metastasis	Only ALP may be elevated.
Alcoholic liver disease	Only GGT is elevated.

10.3 JAUNDICE: AN INTRODUCTION

Jaundice is defined as yellow discoloration of sclera and skin and is associated with hyperbilirubinemia (serum bilirubin > 2 mg/dL). Common causes of jaundice include:

- Congenital hyperbilirubinemia.
- Hemolytic (prehepatic) jaundice.
- Hepatocellular jaundice.
- Cholestatic jaundice (obstructive).

10.4 CONGENITAL HYPERBILIRUBINEMIA

There are three common causes of congenital hyperbilirubinemia: Gilbert's syndrome, Crigler-Najjar syndrome, and Dubin–Johnson syndrome.

Gilbert's syndrome is the most common familial hyperbilirubinemia, and affects 2–7% of the population. This disease is transmitted as autosomal dominant. The cause of Gilbert's syndrome is a mutation of the *UGT1A1* gene that codes uridine diphosphate glucuronosyltransferase (UDP-glucuronosyltransferase) enzyme, which is essential for glucuronidation of bilirubin (conjugated bilirubin) for excretion. The most common allele in Gilbert's syndrome is *UGT1A1*28*, and activity of UDP-glucuronosyltransferase is reduced by approximately 30% in patients with Gilbert's syndrome. In addition there is evidence for reduced hepatic uptake of unconjugated bilirubin. As a result, serum bilirubin is elevated due to increased concentration of unconjugated bilirubin. However, this condition is benign and often recognized during routine liver function tests with the only abnormality being elevated unconjugated bilirubin in serum. However, patients are asymptomatic. In addition, when a patient with Gilbert's syndrome recovers from viral hepatitis, serum bilirubin may be elevated for a prolonged time with values between 2–6 mg/dL. Viral illness such as influenza may also cause prolonged hyperbilirubinemia. No specific treatment is necessary except reassuring the patient that this is a benign condition. Interestingly, high bilirubin may protect against cardiovascular disease and patients with Gilbert's syndrome may be at lower risk of developing cardiovascular disease [4].

Crigler–Najjar syndrome is a rare autosomal recessive condition caused by complete absence of UDP-glucuronosyltransferase enzyme in Type I, or severe deficiency of this enzyme in Type II Crigler–Najjar disorder (also known as Arias syndrome). For Crigler–Najjar syndrome Type I, a child must receive a copy of the defective gene from both parents to develop this severe disease, which is usually life threatening and evident after birth. Fortunately Type I syndrome affects only 1 in one million newborn babies worldwide. A high level of unconjugated bilirubin is apparent at birth and leads to kernicterus, which is severe brain damage due to accumulation of unconjugated bilirubin. Few therapy options are available, including liver transplant. In Type II of the disorder, approximately 20% of UDP-glucuronosyltransferase enzyme is retained, and these patients can be treated with enzyme inducers such as phenobarbital; they usually present with milder hyperbilirubinemia than Type I. These patients may not have any brain damage and may live a normal life.

It is important to note that Gilbert's syndrome and Crigler–Najjar syndrome Types I and II are all associated with a defective gene that leads to lower than normal activity of the enzyme UDP-glucuronosyltransferase that conjugates bilirubin. Crigler–Najjar syndrome Type I is the most severe form with absence of any enzymatic activity. Crigler–Najjar syndrome Type II is the intermediate form where UDP-glucuronosyltransferase may be reduced up to 80%, while in Gilbert's syndrome only a 30% reduction in enzymatic activity is observed.

Table 10.3 Various Congenital Hyperbilirubinemias

Congenital Disease	Mode of Transmission	Comments
Gilbert's syndrome	Autosomal dominant	Reduced levels of UDP-glucuronyl transferase leading to elevated unconjugated bilirubin that increases with fasting.
Crigler–Najjar Type I	Autosomal recessive	Total absence of UDP-glucuronyl transferase; may be fatal.
Crigler–Najjar Type II	Autosomal dominant	Reduced levels of UDP-glucuronyl transferase but can be treated with enzyme inducer such as phenobarbital.
Dubin–Johnson	Autosomal dominant	Impaired excretion of conjugated bilirubin, causing conjugated hyperbilirubinemia. Melanin pigment found within hepatocytes.
Rotor	Autosomal dominant	Impaired excretion of conjugated bilirubin.

Dubin–Johnson (autosomal recessive) and Rotor's syndrome (possibly autosomal dominant) are due to impaired excretion of conjugated bilirubin from hepatocytes. Both conditions result in conjugated hyperbilirubinemia. In Dubin–Johnson syndrome, melanin pigment may be found within the hepatocytes. Various congenital hyperbilirubinemias are summarized in Table 10.3.

CASE REPORT

A 37-year-old pregnant woman had an uncomplicated pregnancy but developed gestational diabetes at 34 weeks of gestation. She was treated with insulin. The patient was treated previously with phenobarbital for hyperbilirubinemia due to Crigler–Najjar syndrome Type II disorder, but her obstetrician decided to lower the phenobarbital dosage from 100–150 mg/day to 25 mg/day to reduce the risk of teratogenicity. However, her serum bilirubin concentrations were stable between 4–6 mg/dL and never exceeded 9 mg/dL. Ultrasound performed in the 34th week of gestation showed no abnormality, and so no liver biopsy was performed. She gave birth to the baby spontaneously through vaginal delivery at 40.2 weeks of gestation, and no abnormality was detected in the newborn [5].

10.5 HEMOLYTIC (PREHEPATIC) JAUNDICE

Hemolytic jaundice is due to hemolytic anemias. In hemolytic jaundice, increased unconjugated hyperbilirubinemia and reticulocytosis are accompanied by normal liver enzyme levels. However, lactate dehydrogenase (LDH) levels may be high due to increased destruction of red blood cells. For the same reason, serum haptoglobins may be low because haptoglobins bind to free hemoglobin released due to hemolysis. Clinical conditions that result in increased in vivo hemolysis are due to constitutive changes in the erythrocytes, such as glucose-6-phosphate dehydrogenase deficiency or hereditary spherocytosis. Hemolytic jaundice may be observed due to hematoma reabsorption, or after blood transfusion. In addition, conditions such as thalassemia, cobalamin, and folate or iron deficiency that produce ineffective erythrocytes may also cause early destruction of erythrocytes, thus causing hemolytic anemia.

10.6 HEPATOCELLULAR JAUNDICE

Hepatocellular jaundice may be seen in patients with acute hepatitis or chronic liver disease. Common causes of acute hepatitis include viral hepatitis, drugs, and alcohol. Important issues regarding viral hepatitis include:

- Hepatitis B virus is a DNA virus but hepatitis A, C, D, and E are all RNA viruses.
- Hepatitis A and E are transmitted by the fecal−oral route. Hepatitis B, C, and D are transmitted parenterally, vertically, and during sex.
- Hepatitis B, C, and D can cause chronic liver disease. However, even with acute infections in hepatitis A and E, most patients recover; the mortality rate with hepatitis A is less than 1% while with hepatitis E it is 1−2%. However, the mortality rate increases to 10−20% in pregnant patients with hepatitis E.
- Approximately 90% or more of patients with acute infection with hepatitis B eventually clear the virus and achieve immunity. The remainder are at risk of developing chronic liver disease and possibly hepatocellular carcinoma. In contrast, 50−70% of patients with hepatitis C infection fail to clear the virus and these patients are at risk for chronic liver disease and possibly hepatocellular carcinoma. Hepatitis D infection occurs either as a co-infection with hepatitis B or as a superinfection in a hepatitis B-infected, hepatitis B surface antigen (HBsAg)-positive patient. However, concurrent infection of hepatitis B along with hepatitis D results in a poor outcome of hepatitis B infection in a patient. Testing for hepatitis viral infection in patients is an important function of the clinical laboratory. Important laboratory parameters to test in a patient suspected with a hepatitis virus infection are described in detail in Chapter 23.

10.7 CHRONIC LIVER DISEASE

Common causes of chronic liver disease include chronic alcohol abuse, chronic infection with hepatitis B and D, or hepatitis B or C alone. Other causes include autoimmune liver disease, primary biliary cirrhosis, hemochromatosis, Wilson's disease, and alpha-1 antitrypsin deficiency. In chronic liver disease, moderate to severe hypoalbuminemia is commonly observed, but other liver function tests may be normal or abnormal depending on the severity of illness. However, serum gammaglobulins may be increased along with increased IgA levels. This feature is manifested in serum protein electrophoresis as polyclonal hypergammaglobulinemia with beta gamma bridging, as IgA travels at the junction of the beta and gamma bands.

In autoimmune liver disease, anti-nuclear antibody (ANA), anti-smooth muscle actin, anti-soluble liver antigen, and anti-liver/kidney microsomal antibodies may be present. IgG levels may be raised. However, primary biliary cirrhosis is mostly seen in women. Anti-mitochondrial antibodies are found in most patients. IgM levels are often raised. Pruritus preceding other features is a characteristic finding as well as secondary hyperlipidemia. Primary or hereditary hemochromatosis is transmitted as autosomal recessive. Hemochromatosis is a multisystemic disease with bronze discoloration of the skin (due to melanin deposition) and diabetes mellitus. This disease is also referred to as bronze diabetes. Hypogonadism due to pituitary dysfunction is the most common endocrine feature. There may be cardiomyopathy with heart failure and cardiac arrhythmias. Pseudo gout due to deposition of calcium pyrophosphate dihydrate (CPPD) crystals is a common feature. Iron stores as well as liver iron content may also be increased. Other causes of increased liver iron content include alcohol intake and iron overload due to chronic transfusion.

Wilson's disease is transmitted as an autosomal recessive disorder. Normally copper is incorporated into apo-ceruloplasmin to form ceruloplasmin, but this process is defective in Wilson's disease. The unbound free copper may be secreted in the urine (urinary copper is high) and may also be deposited in certain tissues including the liver, basal ganglia, and cornea. Features of chronic liver disease, as well as extrapyramidal features, are thus easily explained. Quantification of liver copper can be done and should be high. Other causes of increased liver copper include chronic cholestasis. Kayser—Fleischer ring is a greenish brown pigment at the sclera corneal junction due to deposition of copper in the Descemet's membrane in the cornea. Wilson's disease is an important cause of acute liver disease in young people.

Alpha-1 antitrypsin deficiency can cause liver disease as well as panacinar emphysema. Serum alpha-1 antitrypsin levels are typically low, which can be evident in serum protein electrophoresis. Genetic variants of alpha-1 antitrypsin are characterized by their electrophoretic mobilities as medium (M), slow (S), and very slow (Z). Normal individuals are MM. Homozygotes are ZZ and heterozygotes are either MZ or SZ. Various chronic liver diseases are summarized in Table 10.4.

Table 10.4 Chronic Liver Diseases

Chronic Liver Disease	Comments
Autoimmune liver	Associated with other autoimmune diseases. Extrahepatic features such as polyarthritis, pleurisy, and glomerulonephritis may be present. In addition, ANA, anti-smooth muscle actin, anti-soluble liver antigen, and anti-liver/kidney microsomal antibodies may be present. Polyclonal hypergammaglobulinemia may be present due to increased IgG.
Alcoholic liver disease	Alcoholic fatty liver may be present in many chronic drinkers, but alcoholic hepatitis, and especially liver cirrhosis, are serious medical conditions. A classic case is micronodular cirrhosis, and IgA level may be elevated.
Primary biliary cirrhosis	Seen primarily in women where pruritus precedes other features and secondary hyperlipidemia may be observed. Anti-mitochondrial antibody is found in most patients and IgM levels are often high.
Hemochromatosis	Primary (hereditary) disease is transmitted as autosomal recessive, also called bronze diabetes with endocrine dysfunction, cardiomyopathy, and arthropathy. Liver iron content is increased.
Wilson's disease	Transmitted as autosomal recessive where copper is deposited in the liver. Observed at young age and Kayser–Fleischer ring is seen as greenish brown pigment at the sclera cornea junction. Low serum ceruloplasmin.
Alpha-1 antitrypsin	This deficiency can cause liver disease and also panacinar emphysema. Serum alpha-1antitrypsin level is low. Genetic variants are characterized by their electrophoretic mobility as medium (M), slow (S), and very slow (Z). Normal people are MM, homozygotes are ZZ, while heterozygotes may be MZ or SZ.

10.8 CHOLESTATIC JAUNDICE

Cholestatic jaundice can be classified into two broad categories: intrahepatic and extrahepatic. Intrahepatic cholestatic jaundice is due to impaired hepatobiliary production and excretion of bile causing bile components to enter the circulation. The concentration of conjugated bilirubin in serum is elevated in cholestatic jaundice. Intrahepatic cholestasis may be due to primary biliary cirrhosis, hepatocellular disease such as acute viral hepatitis infection, Dubin–Johnson syndrome, Rotor syndrome, or cholestatic disease of pregnancy. Wilson's disease may also lead to intrahepatic cholestasis due to copper deposition into liver parenchyma, hepatocellular dysfunction, and jaundice. Extrahepatic cholestasis may be related to gallstones, a malignancy such as a pancreatic tumor, pancreatitis, or can be secondary to surgical procedure [6].

10.9 ALCOHOL- AND DRUG-INDUCED LIVER DISEASE

Alcohol abuse can produce a spectrum of liver diseases that include fatty changes in the liver, alcoholic hepatitis, and eventually cirrhosis of the liver. Fatty change is reversible. In alcoholic hepatitis, hepatocellular necrosis takes place. Cytoplasmic inclusions called Mallory bodies may be seen. Alcoholic cirrhosis may be complicated by hepatocellular carcinoma. Heavy drinking for even only a few days can produce fatty changes in the liver (steatosis), which can be reversed after abstinence. However, drinking heavily for a longer period may cause more severe alcohol-related liver injuries such as alcoholic hepatitis and cirrhosis of the liver. In general, women are more susceptible to alcoholic liver diseases than men. The diagnosis of alcoholic hepatitis is a serious medical condition because approximately 70% of such patients may progress to liver cirrhosis, a major cause of death worldwide. In the United States it is estimated that there are 2 million people suffering from alcohol-related liver diseases. Liver cirrhosis is the seventh leading cause of death among young and middle-aged adults, and approximately 10,000 to 24,000 deaths from liver cirrhosis annually may be attributable to alcohol abuse [7]. However, if a patient with alcoholic hepatitis practices complete abstinence, this condition may be reversible.

The liver is the major site of drug metabolism. Drugs are converted into more water-soluble forms through drug metabolism so that drug metabolites can be excreted in bile or urine. Drugs that cause liver damage may do so in a dose-dependent or dose-independent manner. An example of a drug that causes dose-dependent hepatotoxicity is acetaminophen. When ingested, a large proportion of acetaminophen undergoes conjugation with glucuronide and sulfate. The remainder is metabolized by microsomal enzymes to produce toxic derivatives. These are detoxified by conjugation with glutathione. Ingestion of large amounts of acetaminophen will result in excess toxic derivatives and saturation of glutathione that result in liver damage. Interestingly, alcoholics may experience acetaminophen toxicity from a therapeutic dose of acetaminophen [8]. However, acetaminophen toxicity can be treated with N-acetylcysteine that can restore the liver glutathione supply. Other drugs may also cause liver damage, such as azathioprine, methotrexate, chlorpromazine, erythromycin, and even statins in some individuals. Reye's syndrome is a potentially fatal disease observed mostly in children. If aspirin is given to an infant or a child, it could cause Reye's syndrome due to inhibition of beta-oxidation of fatty acids in the mitochondria and uncoupling of oxidative phosphorylation. There is diffuse microvesicular fatty infiltration of the liver. Mortality rate is high.

10.10 LIVER DISEASE IN PREGNANCY

Hyperemesis gravidarum is a condition during pregnancy (usually observed in the first trimester) that is associated with nausea, vomiting, and dehydration (morning sickness). The cause of this condition is still controversial, but it may be related to hormonal changes during pregnancy, most likely an elevated concentration of human chorionic gonadotropin. This condition may also be accompanied by mild jaundice and a mild elevation of liver enzymes.

Intrahepatic cholestasis of pregnancy is a liver-specific disorder characterized by maternal pruritus (itching) observed (usually) in the third trimester. The etiology of this disease is not fully elucidated, but may occur due to cholestatic effects of reproductive hormones such as estrogen. The mechanism by which this condition leads to fetal complication is not understood. Severe pruritus may or may not be accompanied by jaundice. This condition may resolve after pregnancy, but can reappear during subsequent pregnancies [9].

Acute fatty liver of pregnancy is a rare but potentially life-threatening condition that usually occurs in the third trimester with a mean gestational age of 35–36 weeks (the range is 28–40 weeks) or may also be observed in the early postpartum period. Although exact etiology is not known, this disease may be linked to an abnormality in fetal fatty acid metabolism. However, diagnosis is challenging because this condition can appear similar to conditions encountered in preeclampsia, viral hepatitis, or cholestasis of pregnancy. Supportive care and expeditious delivery are required to minimize adverse maternal and fetal outcomes [10]. Fulminant hepatic failure in late pregnancy is also a very serious condition that can be potentially fatal.

HELLP syndrome (H, hemolysis; EL, elevated liver enzyme; LP, low platelet count) is a serious complication of pregnancy that occurs most commonly in patients with severe preeclampsia or eclampsia. Unconjugated hyperbilirubinemia without encephalopathy may also be observed in HELLP syndrome. HELLP syndrome usually develops around 37 weeks of gestation or following delivery.

10.11 LIVER DISEASE IN NEONATES AND CHILDREN

Physiological jaundice is observed in neonates due to decreased activity of UDP-glucuronosyltransferase enzyme leading to unconjugated hyperbilirubinemia. It affects approximately 65% of newborns in the first week of life [11]. Breast milk may have inhibitors to this enzyme, causing unconjugated hyperbilirubinemia (also referred as breast milk jaundice). Physiological jaundice can be treated with phototherapy, but if it persists for more than two weeks after birth, a possible pathological cause of such jaundice must be investigated. Neonatal hepatitis can occur due to infection with cytomegaly virus (CMV), rubella, or toxoplasma. Biliary atresia can also cause jaundice in infants. Metabolic disorders such as tyrosinemia and galactosemia can also cause jaundice. Progressive familial intrahepatic cholestasis (PFIC) refers to a group of familial cholestatic conditions caused by defective biliary epithelial transporters. The clinical features usually appear first in childhood with progressive cholestasis and hepatic failure. A patient may eventually need a liver transplant. These heterogenous groups of conditions are inherited in autosomal recessive fashion. Alagille syndrome is a genetic disorder affecting the liver, heart, kidney, and other organs, and problems associated with this disorder first appear in infancy or early childhood. The disorder is inherited in an autosomal dominant manner with an estimated prevalence of 1 in every 100,000 live births. Facial dysmorphism, cardiac abnormalities, and cholestasis are common features of this disorder.

10.12 MACRO LIVER ENZYMES

On rare occasions, isolated and unexplained elevated levels of liver enzymes such as AST are observed which are due to AST binding with serum IgG. This bound enzyme is referred as macro AST. Binding of ALT with IgG has also been reported. Another enzyme that can bind to IgG is creatinine kinase (CK), giving rise to macro CK. This phenomenon of macro enzymes is seen more frequently in the elderly population, but macro AST may also be seen in children. There is also an association between macro enzymes and autoimmune diseases. Macro AST may also be detected in patients with chronic hepatitis or malignancy. Laboratory detection of macro enzyme can be done by gel filtration chromatography, ultracentrifugation, or polyethylene glycol precipitation.

CASE REPORT

A 27-year-old female was admitted to the gastroenterology department of the hospital because of isolated elevated AST for more than 1.5 years. Physical examination revealed a healthy woman with no apparent abnormality. Her abdomen was normal without hepatomegaly, tenderness, or abnormal peristalsis. Her AST was elevated to 740 U/L, but ALT (23 U/L), LFH (424 U/L), alkaline phosphatase (49 U/L), and GGT (17 U/L) were within normal limits. No hemolysis was detected and the presence of macro AST was suspected by the gastroenterologist. Precipitation of serum with polyethylene glycol followed by measurement of the supernatant where no AST was detectable suggested the presence of macro AST. The clinician decided that her isolated elevated AST was due to macro AST [12].

10.13 LABORATORY MEASUREMENT OF BILIRUBIN AND OTHER TESTS

For determination of bilirubin, it is important to protect the specimen from light because conjugated and unconjugated bilirubin is photooxidized. If the specimen is stored in the refrigerator, bilirubin is stable up to 3 days, and if stored at $-70°C$ in the dark, the specimen may be stable up to 3 months. For measuring conjugated bilirubin (direct bilirubin), serum or plasma is acidified with hydrochloric acid and then mixed with diazotized sulfanilic acid to produce azobilirubin. Then the reaction is stopped with ascorbic acid and the solution is made alkaline; azobilirubin produces a more intense blue color, which is measured colorimetrically. This is called direct bilirubin, which is the concentration of conjugated bilirubin in serum or plasma. For determination of total bilirubin, caffeine is added to serum or plasma in order for less reactive unconjugated bilirubin to react with diazotized sulfanilic acid. The solution after incubation is made alkaline for colorimetric measurement. This is referred to as "total bilirubin" and subtracting total bilirubin from direct bilirubin provides the value of unconjugated bilirubin ("indirect bilirubin"). This method is referred to as the Jendrassik and Grof method. It is usually assumed that direct bilirubin measures mostly conjugated bilirubin species, mono and di-conjugated bilirubin, as well as delta bilirubin (bilirubin tightly bound to albumin), while the total bilirubin method measures both conjugated and unconjugated bilirubin.

In neonates, heel puncture is painful and distressing. Therefore, bilirubin can be monitored with a transcutaneous bilirubin analyzer such as BiliChek, which is a handheld fiberoptic device that measures three wavelengths by spectral reflectance to measure bilirubin, melanin, and hemoglobin. This method accounts for differences in skin pigment.

Liver biopsy is often done to establish a diagnosis of cirrhosis. Liver biopsy is a procedure not without risks. There is significant interest in development of tests that allow clinicians to avoid performing a liver biopsy. One such test measures levels of procollagen type (III) peptide (PIIINP) in blood. Levels are increased in cirrhosis. However, levels can also be increased in inflammation and necrosis.

KEY POINTS

- Hypoalbuminemia is commonly found in chronic liver disease. In addition to albumin, all clotting factors, with the exception of Factor VIII, are produced in the liver. In significant liver disease, such as fulminant hepatic failure, urea levels may also be low. The liver releases glucose into the circulation by glycogenolysis and gluconeogenesis. In significant liver disease hypoglycemia may be apparent due to depletion of the glycogen supply.
- The liver is the site for bilirubin metabolism. Heme, derived from the breakdown of hemoglobin, is converted to biliverdin and finally into bilirubin, which is water-insoluble unconjugated bilirubin. Unconjugated bilirubin can also bind with serum proteins, most commonly albumin. Unconjugated bilirubin is also taken up by the

liver, and, with the help of UDP (uridine-5′ diphosphate)glucuronyl transferase, is converted to conjugated bilirubin (bilirubin conjugated with glucuronide). This conjugation takes place in the smooth endoplasmic reticulum of the hepatocyte. Conjugated bilirubin is water-soluble and is excreted in bile. It is measured in the clinical laboratory as direct bilirubin. Subtracting total bilirubin from the direct bilirubin value provides the concentration of unconjugated bilirubin (also referred to as indirect bilirubin). In the intestine, bacterial enzymes hydrolyze conjugated bilirubin and release free bilirubin, which is reduced to urobilinogen. Urobilinogen bound to albumin is excreted in the urine. Some urobilinogen is converted to stercobilinogen in the intestine and is excreted in stool. Thus, in normal urine, urobilinogen is only present, and in normal stool, stercobilinogen is present. In obstructive (cholestatic) jaundice, conjugated bilirubin regurgitates into blood and, as it is water-soluble, passes into the urine. This is called choluria or the "presence of bile in urine." In obstructive jaundice, less conjugated bilirubin is taken up by the intestine and, as a result, a smaller amount of stercobilinogen is found in the stool (pale stool). Normal individuals have mostly unconjugated bilirubin in their blood, urobilinogen in their urine, and stercobilinogen in their stool.

- Breakdown of hepatocytes results in the release of aminotransferases (also referred to as transaminases) such as ALT and AST into the blood. ALT is a cytosol enzyme and more specific for liver disease. AST is primarily a mitochondrial enzyme that is also found in the heart, muscle, kidney, and brain.
- Alkaline phosphatase (ALP) is found in liver, bone, intestine, and placenta. ALP is located in the canalicular and sinusoidal membrane of the liver. Production of alkaline phosphatase is increased during cholestasis (intrahepatic or extrahepatic), resulting in elevated activity of ALP in serum; however, it is important to determine if the source of ALP is the liver or other organs. If alkaline phosphatase is raised, and the question is whether the source of this enzyme is from the liver or not, measurement of GGT or 5′-nucleotidase levels can be used to determine if the source of ALP is liver or not because both GGT and 5′-nucleotidase are solely produced by the biliary epithelium. Gamma glutamyl transferase (or gamma-glutamyl transpeptidase, GGT) is a microsomal enzyme.
- In acute liver disease without cholestasis, levels of ALT and AST are significantly elevated and ALP is raised but usually less than three times normal. Therefore, ALT and AST levels exceeding 500 U/L are a common finding in acute liver disease.
- In acute cholestasis, ALT and AST are raised but levels are not very high. ALP is usually more than three times normal with a parallel increase in GGT levels.
- In acute liver disease, total protein and albumin levels are unaltered.
- PT is the best test to assess the extent of liver dysfunction.
- In chronic liver disease, albumin levels are low, but total protein may be elevated.
- Elevated ALP with normal bilirubin, ALT, and AST may be seen in patients with hepatic metastasis or bone metastasis. Patients with hepatic metastasis may also have elevated GGT.

- Elevated ALP in children is most likely related to osteoblastic activity in their growing bones.
- Isolated elevated ALP in the elderly is a characteristic feature of Paget's disease.
- Very high isolated ALP levels can be seen in primary biliary cirrhosis.
- Normal liver function tests (except elevated GGT) are a characteristic of excessive alcohol intake. Certain drugs (e.g. warfarin, anti-convulsants) may also produce similar observation.
- Gilbert's syndrome is the most common familial hyperbilirubinemia and affects 2–7% of the population; this disease is transmitted as autosomal dominant. The cause of Gilbert's syndrome is mutation of the *UGT1A1* gene that codes the uridine diphosphate glucuronosyltransferase (UDP-glucuronosyltransferase) enzyme essential for glucuronidation of bilirubin (conjugated bilirubin) for excretion. In addition, there is evidence for reduced hepatic uptake of unconjugated bilirubin. As a result, serum bilirubin is elevated due to increased concentration of unconjugated bilirubin.
- Crigler–Najjar syndrome is a rare autosomal recessive condition caused by complete absence of UDP-glucuronosyltransferase enzyme in Type I, or severe deficiency of this enzyme in Type II.
- Dubin–Johnson (autosomal recessive) and Rotor's syndrome (possibly autosomal dominant) are due to impaired excretion of conjugated bilirubin from the hepatocytes. Both conditions result in conjugated hyperbilirubinemia.
- Hepatitis B virus is a DNA virus but hepatitis A, C, D, and E are all RNA viruses.
- Hepatitis A and E are transmitted by the fecal–oral route. Hepatitis B, C, and D are transmitted parenterally, vertically, and during sex.
- Hepatitis B, C, and D can cause chronic liver disease. However, with acute infection with hepatitis A and E, most patients recover; the mortality rate with hepatitis A is less than 1%, while with hepatitis E it is 1–2%. However, the mortality rate increases to 10–20% in pregnant patients with hepatitis E.
- Approximately 90% or more of patients with acute infection of hepatitis B eventually clear the virus and achieve immunity. The rest are at risk of developing chronic liver disease and possibly hepatocellular carcinoma. In contrast, 50–70% of patients with hepatitis C infection fail to clear the virus and these patients are at risk for chronic liver disease and possibly hepatocellular carcinoma. Hepatitis D infection occurs either as a co-infection with hepatitis B, or as a superinfection in a hepatitis B-infected, hepatitis B surface antigen (HBsAg)-positive patient. However, concurrent infection of hepatitis B and hepatitis D results in a poor outcome of hepatitis B infection in a patient.
- In chronic liver disease, moderate to severe hypoalbuminemia is commonly observed, but other liver function tests may be normal or abnormal depending on the severity of illness. However, serum gammaglobulins may be increased along with increased IgA levels. This feature is manifested in serum protein electrophoresis as polyclonal hypergammaglobulinemia with beta gamma bridging, as IgA travels at the junction of the beta and gamma bands.

- In autoimmune liver disease, anti-nuclear antibody (ANA), anti-smooth muscle actin, anti-soluble liver antigen, and anti-liver/kidney microsomal antibodies may be present. IgG levels may be raised.
- Primary biliary cirrhosis is mostly seen in women. Anti-mitochondrial antibodies are found in most patients. IgM levels are often raised. Pruritus preceding other features is a characteristic finding as well as secondary hyperlipidemia.
- Primary or hereditary hemochromatosis is transmitted as autosomal recessive. Hemochromatosis is a multisystemic disease with bronze discoloration of skin (due to melanin deposition) and diabetes mellitus. This disease is also referred to as bronze diabetes.
- Hypogonadism due to pituitary dysfunction is the most common endocrine feature. It can cause cardiomyopathy with heart failure and cardiac arrhythmias. Pseudo gout due to deposition of calcium pyrophosphate dihydrate (CPPD) crystals is a common feature. Iron stores, as well as liver iron content, may also be increased. Other causes of increased liver iron content include alcohol intake and iron overload due to chronic transfusion.
- Wilson's disease is transmitted as an autosomal recessive disorder. Normally copper is incorporated into apo-ceruloplasmin to form ceruloplasmin, but this process is defective in Wilson's disease. The unbound free copper may be secreted in the urine (urinary copper is high) and may also be deposited in certain tissues, including the liver, basal ganglia, and cornea. Quantification of liver copper may be done and should be high. Other causes of increased liver copper include chronic cholestasis. Kayser—Fleischer ring is a greenish brown pigment at the sclera corneal junction due to deposition of copper in the Descemet's membrane in the cornea. Wilson's disease is an important cause of acute liver disease in young people.
- Alpha-1 antitrypsin deficiency can cause liver disease as well as panacinar emphysema. Serum alpha-1 antitrypsin levels are low, which could be evident in serum protein electrophoresis. Genetic variants of alpha-1 antitrypsin are characterized by their electrophoretic mobilities as medium (M), slow (S), and very slow (Z). Normal individuals are MM. Homozygotes are ZZ and heterozygotes are either MZ or SZ.
- The liver is the major site of drug metabolism. Drugs are converted into more water-soluble forms through drug metabolism so that drug metabolites can be excreted in bile or urine. Drugs that cause liver damage may do so in a dose-dependent or dose-independent manner. An example of a drug causing dose-dependent hepatotoxicity is acetaminophen. Interestingly, alcoholics can experience acetaminophen toxicity from a therapeutic dose of acetaminophen. However, acetaminophen toxicity can be treated with N-acetylcysteine that can restore the liver glutathione supply.
- HELLP syndrome is accompanied by hemolysis, elevated liver enzymes, and low platelet count, and is a serious complication of pregnancy that occurs most commonly in patients with severe preeclampsia or eclampsia. Unconjugated

hyperbilirubinemia without encephalopathy may also be observed in HELLP syndrome. HELLP syndrome usually develops around 37 weeks of gestation or following delivery.

■ On rare occasions, isolated and unexplained elevated levels of liver enzymes such as AST are observed. This is due to binding of AST with serum IgG. This bound enzyme is referred to as macro AST. Binding of ALT with IgG has also been reported. Another enzyme that may bind to IgG is creatinine kinase (CK), giving rise to macro CK. This phenomenon of macro enzymes is seen more frequently in the elderly population, but macro AST may also be seen in children. There is an association between macro enzyme and autoimmune diseases. Macro AST may also be detected in patients with chronic hepatitis or malignancy.

REFERENCES

[1] Tynjala J, Kangastupa P, Laatikainen T, Aalto M, et al. Effect of age and gender on the relationship between alcohol consumption and serum GGT: time to recalibrate goals for normal ranges. Alcohol Alcohol 2012;47:558−62.

[2] Lee WC, Kuo LC, Cheng YC, Chen CW, et al. Combination of white blood cell count with liver enzymes in the diagnosis of blunt liver laceration. Am J Emerg Med 2010;28:1024−9.

[3] Ritchie AH, Willscroft DM. Elevated liver enzymes as a predictor of liver injury in stable blunt abdominal trauma patients: case report and systematic review of the literature. Can J Rural Med 2006;11:283−7.

[4] Schwertner HA, Vitek L. Gilbert syndrome UGT1A1*28 allele and cardiovascular disease risk: possible protective effects and therapeutic applications of bilirubin. Atherosclerosis 2008;198:1−11.

[5] Passuello V, Puhl AG, Wirth S, Steiner E, et al. Pregnancy outcome in maternal Crigler-Najjar syndrome Type II: a case report and systematic review of literature. Fetal Diagn Ther 2009;26:121−6.

[6] Winger J, Mchelfelder A. Diagnostic approach to the patients with jaundice. Prim Care Clin Office Pract 2011;38:469−82.

[7] DeBarkey SF, Stinson FS, Grant BF, Dufour MC. Surveillance report #41. Liver cirrhosis mortality in the United States 1970−1993. National Institute of Alcohol Abuse and Alcoholism, 1996.

[8] Prescott LF. Paracetamol, alcohol and the liver. Br J Clin Pharmacol 2000;49:291−301.

[9] Greenes V, Willamson C. Intrahepatic cholestasis of pregnancy. World J Gastroenterol 2009;15:2049−66.

[10] Ko HH, Yoshida E. Acute fatty liver of pregnancy. Can J Gastroenterol 2006;20:25−30.

[11] Jangaard KA, Curtis H, Goldbloom RB. Estimation of bilirubin using BiliChek, a transcutaneous bilirubin measurement device: effect of gestational age and use of phototherapy. Paediatr Child Health 2006;11:79−83.

[12] Szmuness W, Stevens CE, Harley EJ, Zang EA, et al. Hepatitis B vaccine in medical staff of hemodialysis units: efficacy and subtype cross-protection. N Engl J Med 1982;307 (24):1481.

CHAPTER 11

Renal Function Tests

11.1 BASIC FUNCTIONS OF KIDNEYS

Kidneys are a paired organ system located in the retroperitoneal space. They weigh approximately 150 g each. Renal blood supply represents roughly 25% of cardiac output. The functional unit of the kidney is the nephron, and the components of each nephron include the glomerulus, proximal tubule, loop of Henlé, distal tubule, and the collecting duct. Kidneys have three very important physiological roles:

- Excretory Function: Removal of undesirable end products of metabolism, excess inorganic ions ingested in the diet, and drugs and toxins from the body through urine formation.
- Regulatory Function: Maintaining proper acid−base balance and homeostasis.
- Endocrine Function: The kidney can be regarded as an endocrine organ that produces certain hormones and is also responsible for activation of several hormones.

Kidneys are responsible for urine formation and secretion of undesired end products of metabolism from the body, including urea formed from protein catabolism and uric acid produced from nucleic acid metabolism. The glomerulus is the site of filtration. The basement membrane of the capillaries serves as a barrier to passage of large proteins into the glomerular filtrate. Molecules with a weight of more than 15 kilodaltons (kDa) are not found in the glomerular filtrate. Approximately two-thirds of the filtrate volume is reabsorbed in the proximal tubule. Ninety percent of hydrogen ion secretion by the kidney takes place at the proximal tubule. Further reabsorption of water and solutes takes place in the more distal parts of the nephron. Typically the volume of the glomerular filtrate in one day ranges from 150−200 liters. This volume is reduced to 1−2 liters of urine per day. The loop of Henlé is the site where urine is concentrated. At the distal tubule,

CONTENTS

A. Dasgupta and A. Wahed: Clinical Chemistry, Immunology and Laboratory Quality Control
DOI: http://dx.doi.org/10.1016/B978-0-12-407821-5.00011-5

sodium and chloride are reabsorbed while potassium and hydrogen ions are excreted. Proper function of the distal tubule is essential in maintaining plasma acid–base and electrolyte homeostasis. The collecting duct is the site of further water reabsorption, which occurs under the influence of antidiuretic hormone (ADH).

The kidneys also produce two hormones: erythropoietin and renin. Erythropoietin is produced in response to renal hypoxia and acts on the bone marrow to stimulate erythropoiesis. Renin is produced by the juxtaglomerular apparatus. Renin converts angiotensinogen released by the liver into angiotensin I, which is then converted into angiotensin II in the lungs by angiotensin-converting enzyme (ACE). Angiotensin II is a vasoconstrictor and also stimulates release of aldosterone from the adrenal cortex. This is defined as the "renin-angiotensin-aldosterone" system. Aldosterone, a mineralocorticoid, acts on the distal tubules and collecting ducts of the nephron and causes retention of water and sodium as well as excretion of potassium and hydrogen ions.

The kidneys are responsible for producing the active form of vitamin D, a fat-soluble vitamin essential for absorption of calcium. Vitamin D deficiency can cause osteomalacia in adults and rickets in children. Human skin is capable of synthesizing an inactive form of vitamin D (cholecalciferol, vitamin D3) from 7-dehydrocholesterol in the presence of sunlight (solar ultraviolet radiation: 290–315 nm; reaches earth between 10 AM and 3 PM). This is why vitamin D is also called the "sunshine vitamin." Very few foods naturally contain vitamin D and as a result many foods are fortified with an inactive form of vitamin D. The inactive form of vitamin D obtained either from skin exposure to the sun or food is first converted into 25-hydroxyvitamin D (25-hydroxycholecalciferol) in the liver by the action of enzyme vitamin D-25 hydroxylase. Finally, the kidneys (proximal tubular epithelial cells) convert this form of vitamin D into the active form, which is called 1,25-dihydroxyvitamin D (1,25-dihydroxycholecalciferol) by the action of the enzyme 25-hydroxylcholecalciferol-1α-hydroxylase. The biologically active form of vitamin D is 1,25-dihydroxycholecalciferol, which plays an important role in absorption of calcium from the gastrointestinal tract. The enzyme 25-hydroxylcholecalciferol-1α-hydroxylase is stimulated by parathyroid hormone (PTH) and inhibited by high blood levels of calcium and phosphate. Although 1,25-dihydroxyvitamin D is the bioactive form of vitamin D, the best laboratory parameter to monitor vitamin D status of a patient is to measure 25-hydroxyvitamin D. A serum 25-hydroxyvitamin D level of 30 ng/mL or greater is considered an adequate level [1]. Prostaglandins are synthesized by the action of cyclooxygenase enzyme acting on arachidonic acid. This enzyme is present in many organs, including kidneys. The kidney is also a site of degradation of hormones such as insulin and aldosterone.

11.2 GLOMERULAR FILTRATION RATE

Glomerular filtration is one of the major functions of the kidney. Neutral molecules show much higher glomerular permeability than highly negatively charged molecules. Glomerular filtration rate (GFR) is a measure of the functional capacity of the kidney and is an important parameter to assess kidney function. GFR can be estimated by using the formula in Equation 11.1:

$$GFR = (Ua \times V)/Pa \qquad (11.1)$$

Here, Ua is the concentration of a solute in urine, V is the volume of urine in mL/minute, and Pa is the concentration of the same solute in plasma. However, this formula is often corrected to take into account body surface area (Equation 11.2):

$$GFR = (Ua \times V)/Pa \times 1.73/A \qquad (11.2)$$

Here, A is the body surface area in square meters. Standard body surface is 1.73 m^2.

The body surface area of most adults is between 1.6 and 1.9. The formula for calculating GFR using any analyte that is freely filtered through the glomerula is only valid if the solute is in stable concentration in plasma and is inert (neutral charge), freely filtered at the glomerulus. This compound must not be secreted, reabsorbed, synthesized, or metabolized by the kidneys. Estimation of GFR by insulin clearance is considered to be the gold standard. However, in routine clinical practice creatinine clearance is more practical. More recently, cystatin C was introduced as an alternative to creatinine clearance. For example, if serum creatinine is 1.0 mg/dL (0.01 mg/mL), urine creatinine is 1 mg/mL, the volume of urine is 60 mL in 1 hour, and body surface area is 1.70. Then GFR for this patient using creatinine clearance should be as follows (Equation 11.3):

$$GFR \ (Creatinine \ Clearance) = 1.0 \ mg/mL \times 1 \ mL/min/0.01 \ mg/mL$$
$$\times \ 1.73/1.70 = 101.7 \ mL/min/1.73 \ m^2 \qquad (11.3)$$

In order to calculate GFR based on creatinine clearance, a 24-h urine collection is recommended which should be from one morning void to the next day's morning void. This is difficult in real practice and usually GFR is estimated using a formula.

11.3 CREATININE CLEARANCES

Creatine is synthesized in the kidneys, liver, and pancreas, and then transported in blood to other organs, especially the brain and muscles, where it is phosphorylated to phosphocreatine. Phosphocreatine is a high-energy compound and interconversion of phosphocreatine to creatinine is important for muscular functions. Creatinine is the waste product derived from creatine and phosphocreatine, and creatinine production is related to the muscle mass of an individual. Women usually excrete 1.2 g of creatinine per day while men excrete 1.5 g/day. Dietary intake of meat also affects the amount of creatinine produced daily. Therefore, serum creatinine levels are affected by gender, age, weight, lean body mass, and dietary protein intake. Creatinine is produced in the body at a constant rate and is freely filtered and not reabsorbed, although a small amount of creatinine is secreted by the tubules. Thus it is a convenient marker for estimation of glomerular filtration rate (GFR). Because collection of 24-h urine is difficult, estimation of GFR can be done using values of plasma creatinine concentration and relevant formulas. However, it is also important to take into account the age, sex, and race of the patient when performing such calculations. The Cockroft–Gault formula is widely used for calculating GFR (Equation 11.4):

$$\text{Creatinine clearance} = \frac{(140 - \text{Age in years}) \times \text{Weight in kg}}{\text{Serum creatinine } (\mu\text{mol/L})} \times 1.23 \text{ if male (or 1.04 if female)}$$

$$(11.4)$$

Another version of the formula that is also commonly used and which produces the same results is shown in Equation 11.5:

$$\text{Creatinine clearance} = \frac{(140 - \text{Age in years}) \times \text{Weight in kg}}{0.814 \times \text{Serum creatinine } (\mu\text{mol/L})} \times 0.85 \text{ (if female)} \qquad (11.5)$$

If serum creatinine is given in mg/dL, as is often the case with U.S. laboratories, then this equation can be further modified so that GFR can be calculated directly using creatinine concentration expressed in mg/dL (Equation 11.6):

$$\text{Creatinine clearance} = \frac{(140 - \text{Age in years}) \times \text{Weight in kg}}{72 \times \text{Serum creatinine } (\text{mg/dL})} \times 0.85 \text{ (if female)} \qquad (11.6)$$

The conversion factor for converting serum creatinine given in mg/dL into μmol/L is 88.4. Therefore, 88.4×0.814 is 71.9, which can be rounded up to 72, as used in the modified equation. Alternatively, if serum creatinine is expressed in μmol/L, it can be multiplied by 0.011 to get the creatinine concentration in mg/dL.

The Cockroft–Gault formula was modified to the MDRD formula by Modification of Diet in Renal Disease Study Group as follows (Equation 11.7):

$$\text{Estimated GFR (mL/min/1.73 m}^2) = 186 \times (\text{plasma creatinine in mg/dL})^{-1.154}$$
$$\times \text{Age}^{-0.203} \times F$$

$$(11.7)$$

Here, F is 0.742 for females, and 1.21 for African-Americans.

In children, the Schwartz formula is usually used to estimate GFR (Equation 11.8):

$$\text{Estimated GFR} = \frac{k \times \text{Height (cm)}}{\text{Serum creatinine (mg/dL)}} \qquad (11.8)$$

For a pre-term baby, the value of k is 0.33 in the first year of life, but for full term infants it is 0.45. For infants and children up to age 12, the value of k is assumed to be 0.55.

11.4 CHRONIC KIDNEY DISEASE

Kidney function depends on age. The GFR is low at birth, reaching adult levels at approximately two years of age. Renal function declines after the age of 40 and declines even further after the age of 65. Approximately 19 million Americans older than 20 years have chronic kidney disease. In addition, an estimated 435,000 individuals have end-stage renal disease. Early diagnosis of chronic kidney disease may delay or even prevent end-stage renal disease. The Kidney Disease Outcomes Quality Initiative from the National Kidney Foundation has developed guidelines for detection and evaluation of chronic kidney disease [2]. The following criteria are adopted to define chronic kidney disease:

- If creatinine clearance is above 90 mL/min/1.73 m^2 (preferable above 100) in the absence of any abnormal finding (imaging study or laboratory-based tests), normal kidney function can be assumed.
- If kidney damage is present for three months (structural damage of kidney based on imaging studies or functional damage based on laboratory tests), then it is usually assumed that chronic kidney disease is present. However, if GFR is below 60 mL per minute for three or more months, it can also be assumed that the patient has chronic kidney disease.
- End-stage renal disease or end-stage renal failure is defined as a GFR below 15 mL per minute, or if the patient is dependent on dialysis.

Table 11.1 Stages of Renal Disease Based on National Kidney Foundation Guidelines

Glomerular Filtration Rate	Stage	
>90 mL/min/1.73 m^2	0	If no proteinuria/hematuria present, no risk factor.
>90 mL/min/1.73 m^2	1	With proteinuria or hematuria.
60–89 mL/min/1.73 m^2	2	Mild disease with decreased GFR.
30–59 mL/min/1.73 m^2	3	Moderate chronic kidney disease.
15–29 mL/min/1.73 m^2	4	Severe chronic kidney disease.
<15 mL/min/1.73 m^2	5	End-stage renal disease or end-stage renal failure.

However, creatinine clearance may not always reflect the true nature of chronic kidney disease [3]. Various stages of renal disease based on the National Kidney Foundation Guidelines are listed in Table 11.1 [4].

Fractional excretion of sodium, a measure of percent of filtered sodium that is excreted in urine, is also useful in evaluating renal function. The formula is shown in Equation 11.9:

$$\text{Fractional excretion of sodium} = \frac{\text{Urine Sodium} \times \text{Serum Creatinine}}{\text{Serum Sodium} \times \text{Urine Creatinine}} \times 100$$

(11.9)

A value less than 1% is indicative of pre-renal disease (pre-renal azotemia), where reabsorption of almost all filtered sodium is an appropriate response to renal hypoperfusion. However, a value over 3% (some authors suggest a value over 2%) is indicative of acute tubular narcosis or urinary tract obstruction. Values between 1−2% may be observed in either disease.

11.5 CYSTATIN C

There are many sources of errors when assessing kidney function using creatinine collection. Collection of 24-h urine (even accurately timed urine collection) is cumbersome. In addition, proper hydration of a patient is important with at least a urinary flow of 2 mL/min. Cystatin C is a low-molecular-weight protein (13.3 kDa) that can be used for calculating GFR. In contrast to creatinine, plasma concentrations of cystatin C are unaffected by sex, diet, or muscle mass. It is considered to be a superior marker for estimation of GFR compared to creatinine. In addition, in the pediatric population, it has been documented that plasma cystatin C is a better marker of kidney function than creatinine. The reference value of plasma cystatin C is 0.5−1.0 mg/L.

Many authors have compared calculated GFR using creatinine (Cockroft–Gault formula), MDRD formula, and by using calculated GFR using serum cystatin concentration. Several formulas have been proposed to estimate GFR based on serum cystatin C levels. Three common formulas (Equations 11.10–11.12) are provided in this chapter but for an in-depth discussion, the review published by Rosenthal is useful [5].

$$\text{Formula 1: GFR} = (78/\text{Cystatin C, mg/L}) + 4 \tag{11.10}$$

$$\text{Formula 2: GFR} = (80.35/\text{Cystatin C, mg/L}) - 4.32 \tag{11.11}$$

$$\text{Formula 3: GFR} = 77.24/\text{Cystatin C}^{-1.263} \quad (\text{Cystatin C in mg/L}) \tag{11.12}$$

Hojs et al. commented that estimating GFR based on serum cystatin C concentration is superior to calculating GFR based on serum creatinine concentration in renally compromised patients with estimated GFR around 60 mL/min/1.73 m^2 [6].

CASE REPORT

A 44-year-old previously healthy African-American was presented to the emergency department with complaints of nausea for four days. Two months before his admission, his serum creatinine was 114 μmol/L (1.6 mg/dL) and his serum cystatin C was 0.99 mg/L. GFR calculated by using both creatinine and cystatin C was 78 mL/min/1.73 m^2. On admission, his physical examination was unremarkable but his serum creatinine was highly elevated to 999 μmol/L (11 mg/dL) and his serum cystatin C level was also elevated to 2.71 mg/L. His calculated GFR based on creatinine was 5 mL/min/1.73 m^2, which indicated severe renal failure, but GFR calculated using his serum cystatin C level was 22 mL/min/1.73 m^2. Urinalysis showed moderate blood (2+ RBC) and a urine drug test was positive for cocaine. His creatinine phosphokinase was highly elevated to 25,099 U/L, and a diagnosis of rhabdomyolysis was made. The following day, the patient's serum creatinine was further elevated to 1,022 μmol/L (11.2 mg/dL), but his serum cystatin C was reduced to 2.37 mg/L. However, the patient started producing urine and his nausea was significantly reduced. On hospital Day 6, the patient was discharged with a serum creatinine value of 458 μmol/L (5.0 mg/dL) and serum cystatin C of 1.69 mg/L. His estimated GFR was 16 mL/min/1.73 m^2 using serum creatinine concentration, but his estimated GFR was 40 mL/min/1.73 m^2 using his serum cystatin C concentration. Moreover, the patient was producing over 2 liters of urine during the last few days of his hospital stay. The authors commented that cystatin C is a better marker to evaluate renal function and resolution of acute kidney injury in patients with rhabdomyolysis than serum creatinine [7].

11.6 UREA (BLOOD UREA NITROGEN) AND URIC ACID

Although serum creatinine and cystatin C are more commonly used to evaluate renal function, urea (often called blood urea nitrogen) and uric acid measurements also have some clinical value. Urea is the result of catabolism of proteins and amino acids. This takes place in the liver. First, ammonia is formed, and then it is eventually converted into urea. The kidneys are the primary route for excretion of urea, and account for over 90% of urea excretion. Minor loss of urea takes place through the gastrointestinal tract and skin. Urea is freely filtered at the glomerulus and is subsequently not reabsorbed or secreted at the tubules. However, measurement of urea levels is inferior when assessing renal function (as compared to creatinine levels) because serum or plasma concentration of urea may be increased in the following situations:

- Dehydration.
- Hypoperfusion of the kidneys.
- High-protein diet.
- Protein catabolism.
- Steroid administration.

Under a similar situation, serum creatinine is not elevated (normal range: 0.5−1.2 mg/dL). However, measuring the urea level along with creatinine is of clinical relevance. The urea level in blood is usually measured as blood urea nitrogen (BUN), with a normal level between 6 and 20 mg/dL. The following criteria are usually used to interpret the BUN/creatinine ratio:

- The BUN/creatinine ratio for normal individuals is usually from 12:1 to 20:1. For example, if BUN is 15 mg/dL and creatinine is 1.1 mg/dL, then BUN/creatinine ratio is 13.6.
- A BUN/creatinine ratio below 10:1 may indicate intrinsic renal disease. A BUN/creatinine ratio above 20:1 may be indicative of hypoperfusion of the kidney, including pre-renal failure.

Acute kidney injury can be the result of pre-renal, renal, and post-renal causes. In critically ill patients with renal hypoperfusion but intact tubular function (pre-renal azotemia), BUN concentration may increase out of proportion to serum creatinine concentration; the BUN/creatinine ratio may exceed 20:1. However, critically ill patients are also prone to accelerated protein catabolism, which can also increase the BUN/creatinine ratio without pre-renal azotemia [8]. The BUN/creatinine ratio is not a precise test because the ratio can be altered under many conditions other than kidney diseases. The increase of serum creatinine is a better indicator of declining renal function.

In humans, purines break down into xanthine and hypoxanthine, and then xanthine oxidase transforms these compounds into uric acid (which is excreted in the urine). The normal uric acid level in serum is 2.6 to 6.0 in females and 3.5 mg/dL to 7.2 mg/dL in males. Eating purine-rich foods such as liver, anchovies, mackerel, dried beans, and peas, as well as drinking alcohol, can elevate serum uric acid levels. Some drugs, such as diuretics, can also increase uric acid levels in serum or plasma (uric acid is an antioxidant). However, an increased uric acid level in the blood can be associated with gout and can cause the formation of renal stones. However, the serum level of uric acid may also be elevated due to decreased renal function as observed in patients with renal failure. Lesch—Nyhan syndrome is a rare genetic disease associated with high serum uric acid due to deficiency of the hypo-xanthine—guanine phosphoribosyl transferase enzyme.

11.7 PROTEIN IN URINE AND PROTEINURIA

Molecules less than 15 kDa pass freely into urine through glomerular filtration whereas a selected few proteins with molecular weights between 16 and 69 kDa can also be filtered by the kidney. The molecular weight of albumin, the major protein found in serum, is 67 kDa, and, as expected, a very small amount of albumin is also found in the urine of normal individuals. Glomerular filtration of a protein depends on several factors, including the molecular weight of the protein, its concentration in serum, its charge, and its hydrostatic pressure. Although 90% of these proteins are reabsorbed (smaller proteins are effectively absorbed by the renal tubule), the following proteins can pass through the glomerular filtration process:

- Albumin.
- Alpha-1 acid glycoprotein (orosomucoid).
- Alpha-1-microglobulin.
- Beta-2-microglobulin.
- Gamma trace protein.
- Retinol binding protein.

Normally, total urinary protein is <150 mg/24 h and consists of mostly albumin and Tamm—Horsfall protein (secreted from the ascending limb of the Loop of Henlé). The extent of proteinuria can be assessed by quantifying the amount of proteinuria as well as by expressing it as the protein-to-creatinine ratio. A normal ratio is as follows:

- Adults: < 0.2
- Children 6 months to 2 years <0.5; and older than 2 years <0.25.

Proteinuria with minor injury (typically only albumin is lost in the urine) can be due to vigorous physical exercise, congestive heart failure, pregnancy, certain drug therapies, high fever, and alcohol abuse. Proteinuria can be classified into glomerular, tubular, and combined proteinurias. Glomerular proteinuria can be sub-classified as selective (albumin and transferrin in urine) and non-selective (all proteins are present). In glomerular proteinuria, albumin is always the major protein.

In mild glomerular proteinuria, total protein concentration is usually within 1,500 mg/24 h of urine, but with moderate glomerular proteinuria, the total protein level can be 1,500—3,000 mg/24 h. In the case of non-selective proteinuria, total protein in urine often exceeds 3,000 mg/24 h.

Total protein/creatinine ratio is also useful in grading proteinuria:

- Low-grade proteinuria: 0.2—1.0
- Moderate proteinuria: 1.0—5.0
- Non-selective proteinuria: > 5.0.

The major difference between glomerular and tubular proteinuria is the difference between the molecular weight ranges of protein found in urine. In glomerular proteinuria, albumin is the major component found in urine, while in tubular proteinuria, albumin is a minor component; proteins with smaller molecular weight, such as alpha-1 microglobulin and beta-2 microglobulin, are the major proteins found in urine. In mixed-type proteinuria, both albumin and low-molecular-weight proteins such as alpha-1 microglobulin and beta-2 microglobulin are present [9].

11.8 OTHER RENAL DISEASES

Acute and chronic renal failure, acute nephritis, and nephrotic syndrome represent commonly observed renal diseases. Drug-induced renal injury also represents a frequent clinical entity. The most common drugs encountered in renal failure include vancomycin, aminoglycosides, amphotericin B, cyclosporine, and radiographic contrast agents. Various renal diseases are summarized in Table 11.2.

CASE REPORT

A 48-year-old woman with a diagnosis of hypertension for 2 years and hyperlipidemia for 10 months showed a steadily increased creatinine level from 0.7 mg/dL to 1.8 mg/dL over a period of 8 months. Her medications included hydrochlorothiazide (12.5 mg/day) for hypertension and fenofibrate (200 mg/day) for reducing cholesterol. Because of the increasing creatinine related to fenofibrate therapy, the drug was discontinued and her creatinine level returned to normal in a few months [10]. The precise mechanism by which fenofibrate therapy results in increased levels of creatinine (which is reversible upon discontinuation of therapy) is not fully understood. However, it has been speculated that fenofibrate therapy may impair GFR in certain patients.

Table 11.2 Various Renal Disorders

Disease	Comments
Acute renal failure (ARF)	ARF is the sudden deterioration of renal function that can be broadly divided into pre-renal, renal, and post-renal subtypes. The pre-renal subtype is associated with hypoperfusion of the kidneys. Renal causes include glomerulonephritis and interstitial nephritis. Post-renal causes are related to obstructive uropathy.
Acute interstitial nephritis	Tubulointerstitium is damaged due to various agents such as drugs, infections, and immunological injuries.
Acute tubular necrosis (ATN)	Necrosis of renal tubules may be related to hypoperfusion and hypoxia. Patient undergoes three phases: oliguric phase, polyuric phase, and, finally, phase of recovery.
Chronic renal failure (CRF)/Chronic kidney disease (CKD)	Defined as chronic and progressive loss of renal function. Based on the GFR it is divided into 5 stages (stage 5 with the lowest GFR, see Table 11.1).
Nephrotic syndrome	Defined as proteinuria (>3 g/day), hypoalbuminemia, hypercholesterolemia, and edema. Most common cause of nephritic syndrome in adults is membranous glomerulonephritis, and common cause of nephrotic syndrome in children is minimal lesion.
Nephritic syndrome	Defined as oliguria, hematuria with hypertension, and edema. Acute diffuse glomerulonephritis is the leading cause of nephritic syndrome.
Renal tubular acidosis	A group of disorders characterized by normal anion gap metabolic acidosis with inappropriately high urine pH (>5.5 in early morning urine). Type I (distal) is associated with decreased hydrogen ion secretion at the distal tubule. Type II is associated with increased loss of bicarbonate from the proximal tubule. In type IV (type III is discontinued) there is hyporeninemic hypoaldosteronism.

11.9 LABORATORY MEASUREMENTS OF CREATININE AND RELATED TESTS

Plasma creatinine may be measured using either chemical or enzymatic methods. Most chemical methods utilize the Jaffe reaction. In this method creatinine reacts with picrate ion in an alkaline medium to produce an orange–red complex. The Jaffe reaction is not entirely specific for creatinine. Substances such as ascorbic acid, high glucose, cephalosporins, and ketone bodies can interfere with this method. High bilirubin (both conjugated and unconjugated) may falsely lower the creatinine value (negative interference) as measured by the Jaffe reaction. Enzymatic methods are also available for serum creatinine determination. Enzymes commonly used for creatinine determination are creatininase (also called creatinine deaminase) and creatinine hydrolase (also called creatinine aminohydrolase). Although enzymatic creatinine methods are subject to less interference than the Jaffe method, interferences in enzymatic methods have nevertheless been reported. The reference method for creatinine measurement is isotope dilution mass spectrometry. Liu et al. reported that although enzymatic methods are less affected than creatinine determination using the Jaffe reaction in patients undergoing hemodialysis, the gold standard for creatinine determination is isotope dilution mass spectrometry, which is free from interferences [11].

CASE REPORT

A healthy 3-year, 4-month-old boy was brought into the pediatric emergency department by his parents after presumed ingestion of model car fuel (Dynamite Blue Thunder, Horizon Hobby, Inc., Champaign, IN). This car fuel contains nitromethane and methanol, and it was estimated that the boy ingested only 5 mL of fuel. When examined in the emergency department, the patient was found to be alert and responsive with no abnormality in respiration. His serum electrolytes, glucose, and venous blood gas parameters were within normal ranges. However, serum creatinine (measured by the Jaffe reaction) was highly elevated (926 μmol/L, or 10.2 mg/dL), indicating acute renal failure; but the patient was not as unwell as expected from such high serum creatinine. His urea, however, was only slightly elevated (4.7 mmol/L, or 28.3 mg/dL). The initial methanol level was 4.2 mmol/L. The patient continued to do well, and three hours later his creatinine level dropped to 817 μmol/L (9.0 mg/dL) and serum methanol dropped to 2.2 mmol/L. The patient received intravenous fluid and supportive therapy. Considering that creatinine values were falsely elevated, specimens were sent to another hospital laboratory and the specimen that showed creatinine of 926 μmol/L (10.2 mg/dL) showed a normal creatinine level of 29 μmol/L (0.3 mg/dL) using an enzymatic method. Other elevated creatinine levels when reanalyzed by an enzymatic method showed normal creatinine values. The authors concluded that falsely elevated creatinine as measured by the Jaffe reaction was due to interference of nitromethane present in the model car fuel [12].

Blood urea can be measured by both chemical and enzymatic methods. Most chemical methods are based on the "Fearon Reaction," where urea reacts with diacetyl-forming diazine (which absorbs at 540 nm). Enzymatic methods are based on hydrolysis of ureas by the enzyme urease and these reactions generate ammonia. Ammonia can be measured using the Berthelot method or another enzymatic method such as with glutamate dehydrogenase. Ammonia can also be measured by conductometry.

Measurement of uric acid can be done by either a chemical or enzymatic method. A commonly used colorimetric method employs phosphotungstic acid, which is reduced by uric acid in alkaline medium to produce a blue color (tungsten blue) that can be measured spectrophotometrically. However, this method is subject to interferences, including interference from endogenous compounds such as high glucose and ascorbic acid (vitamin C). The enzymatic method based on uricase is more specific.

11.10 URINE DIPSTICK ANALYSIS

Urinalysis is a good screening tool for diagnosis of urological conditions such as urinary tract infection, as well as sub-clinical kidney disease. Urine dipstick analysis is usually the first test performed during urinalysis, followed by microscopic examination. Urine dipsticks are inexpensive paper or plastic devices with various segments (reaction pads) capable of color change if a particular substance of interest is present; such change in color can be compared to a color chart provided by the manufacturer for interpretation of results. Usually test strips can detect the presence of glucose, bilirubin, ketones, blood, protein, urobilinogen, nitrite, and leukocytes in the urine. Specific gravity of urine and pH can also be roughly estimated using a dipstick. Normal specific gravity of urine is between 1.002 and 1.035, and pH is between 4.5 and 8.0. On a typical Western diet, urine pH is around 6.0 [13]. The urine dipstick is very sensitive to the presence of red blood cells and free hemoglobin. Negative or trace protein in urine is normal, but a value of 1+ should be investigated further. Typically glucose does not appear in urine unless plasma glucose is over 180 mg/dL to 200 mg/dL. A positive nitrite test is indicative of bacteria in urine, and a urine culture is recommended. In addition, a positive test for leukocyte esterase indicates the presence of neutrophils (neutrophils produce leukocyte esterase) due to infection or inflammation. However, both false positive and false negative test results may be encountered with urine dipstick analysis. Major interferences include:

- A protein reaction pad of urine dipstick detects albumin in urine but cannot detect Bence–Jones proteins. If urine is alkaline, a false positive protein test result may occur.
- A hemoglobin test pad can show a false positive result if myoglobin is present.
- A ketone reaction pad based on sodium nitroprusside can detect only acetoacetic acid and is weakly sensitive to acetone, but cannot detect beta-hydroxybutyric acid.
- The presence of ascorbic acid (vitamin C) in urine can cause a false negative dipstick test with glucose and hemoglobin. Such interference may occur after taking vitamin C supplements or even fruit juice enriched with vitamin C [14]. Most glucose test strips use a glucose oxidase-based method where ascorbic acid can cause falsely lower values (negative interference). However, in a glucometer that uses glucose dehydrogenase, ascorbic acid can cause a false positive result (see Chapter 7).

KEY POINTS

- The kidney has three important functions: excretory, regulatory, and endocrine functions.
- The kidney produces two important hormones: erythropoietin and renin. Erythropoietin is produced in response to renal hypoxia and acts on the bone marrow to stimulate erythropoiesis. Renin is produced by the juxtaglomerular

apparatus. Renin converts angiotensinogen released by the liver into angiotensin I, which is then converted into angiotensin II in the lungs by angiotensin-converting enzyme (ACE). Angiotensin II is a vasoconstrictor and also stimulates release of aldosterone from the adrenal cortex.

- The kidney also produces an active form of vitamin D (1,25-dihydroxyvitamin D or 1,25-dihydroxycholecalciferol). A serum vitamin D level over 30 ng/mL is considered adequate.
- The basement membrane of capillaries serves as a barrier to passage of large proteins into the glomerular filtrate. Molecules with a weight of more than 15 kilodaltons (kDa) are not found in the glomerular filtrate. The loop of Henlé is the site where urine is concentrated.
- GFR can be estimated with the formula: $GFR = (Ua \times V)/Pa$. Ua is the concentration of a solute in urine, V is the volume of urine in mL/minute, and Pa is the concentration of the same solute in plasma.
- Serum creatinine levels are affected by gender, age, weight, lean body mass, and dietary protein intake (mol/L). Cystatin C is a low-molecular-weight protein (13.3 kDa) that can be used for calculating GFR. In contrast to creatinine, plasma concentrations of cystatin C are unaffected by sex, diet, or muscle mass.
- Both creatinine clearance and cystatin C clearance may be used to evaluate glomerular filtration rate, but cystatin C may be slightly superior to creatinine.
- The Cockroft–Gault formula is widely used for calculating GFR.
- The Cockroft–Gault formula: Creatinine Clearance = ((140 − Age in years) × (Weight in kg))/(Serum Creatinine in μmol/L) × (1.23 if male or 1.04 if female).
- However, in the U.S., creatinine concentration is expressed in mg/dL, and this formula can be modified into: Creatinine Clearance = ((140 − Age in years) × (Weight in kg))/(72 × (Serum Creatinine in mg/dL)) × (0.85 if female).
- The Cockroft–Gault formula was modified to the MDRD formula by Modification of Diet in Renal Disease Study Group as follows: Estimated GFR (mL/min/ 1.73 m^2) = 186 × (plasma creatinine in mg/dL)$^{-1.154}$ × Age$^{-0.203}$ × F.
- In chronic renal disease, creatinine clearance is usually less than 60 mL/min/ 1.73 m^2, but a value below 15 mL/min is indicative of end-stage renal disease.
- Fractional excretion of sodium over 3% may indicate acute tubular necrosis, but less than 1% may indicate hypoperfusion of the kidney.
- The BUN/creatinine ratio for normal individuals is usually from 12:1 to 20:1. A BUN/creatinine ratio below 10:1 may indicate intrinsic renal disease. A BUN/ creatinine ratio above 20:1 may be because of hypoperfusion of the kidney, including pre-renal failure.
- Normally, total urinary protein is <150 mg/24 h and consists of mostly albumin and Tamm–Horsfall protein (secreted from the ascending limb of the Loop of Henlé).
- Proteinuria can be classified into glomerular proteinuria, tubular proteinuria, and combined proteinuria. Glomerular proteinuria can be sub-classified as: selective

(albumin and transferrin in urine) and non-selective (all proteins are present). In glomerular proteinuria the major protein present is always albumin.

- In tubular proteinuria, albumin is a minor component, but proteins with smaller molecular weight such as alpha-1 microglobulin and beta-2 microglobulin are the major proteins found in the urine. In mixed-type proteinuria both albumin and low-molecular-weight proteins such as alpha-1 microglobulin and beta-2 microglobulin are present.
- Plasma creatinine can be measured using chemical or enzymatic methods. Most chemical methods utilize the Jaffe reaction. In this method creatinine reacts with picrate ion in an alkaline medium to produce an orange–red complex.
- The Jaffe reaction is not entirely specific for creatinine. Substances such as ascorbic acid, high glucose, cephalosporins, and ketone bodies can interfere with this method. High bilirubin (both conjugated and unconjugated) can falsely lower the creatinine value (negative interference) measured by using the Jaffe reaction.
- Usually test strips can detect the presence of glucose, bilirubin, ketones, blood, protein, urobilinogen, nitrite, and leukocytes in the urine. The specific gravity of urine and pH can also be roughly estimated using a dipstick.
- Typically glucose does not appear in urine unless plasma glucose is over 180 mg/dL to 200 mg/dL. A positive nitrite test is indicative of bacteria in urine and urine culture is recommended. In addition, a positive test for leukocyte esterase indicates the presence of neutrophils (neutrophils produce leukocyte esterase) due to infection or inflammation.
- A protein reaction pad of urine dipstick detects albumin in urine but cannot detect Bence–Jones proteins. If urine is alkaline, a false positive protein test result may occur.
- A hemoglobin test pad can show a false positive result if myoglobin is present.
- A ketone reaction pad based on sodium nitroprusside can detect only acetoacetic acid and is weakly sensitive to acetone, but cannot detect beta-hydroxybutyric acid.
- The presence of ascorbic acid (vitamin C) in urine can cause a false negative dipstick test with glucose and hemoglobin. Such interference may occur after taking a vitamin C supplement or even fruit juice enriched with vitamin C. Most glucose test strips use glucose oxidase-based methods where ascorbic acid can cause falsely lower values (negative interference).

REFERENCES

[1] Khan KA, Akram J, Fazal M. Hormonal cations of vitamin D and its role beyond just a vitamin: a review article. Int J Med Mol Med 2011;3:65−72.

[2] Snyder S, Pendergraph B. Detection and evaluation of chronic kidney disease. Am Fam Physician 2005;72:1723−32.

[3] Stevens L, Coresh J, Greene T, Levey AS. Assessing kidney function: measured and estimated glomerular filtration rate. N Eng J Med 2006;345:2473−83.

[4] National Kidney Foundation. K/DQQI clinical practice guidelines for chronic kidney disease: evaluation, classification and stratification. Am J Kidney Dis 2002;39(Suppl. 2): S1–266.

[5] Rosenthal SH, Bokenkamp A, Hoffmann W. How to estimate GDR serum creatinine, serum cystatin C or equation? Clin Biochem 2007;40:153–61.

[6] Hojs R, Bevc S, Ekhart R, Gorenjak M, et al. Serum cystatin C based equation compared to serum creatinine based equations for estimation of glomerular filtration rate in patients with chronic kidney disease. Clin Nephrol 2008;70:10–7.

[7] Yap L, Lamarche J, Peguero A, Courville C. Serum cystatin C verus serum creatinine in the estimation of glomerular filtration rate in rhabdomyolysis. J Ren Care 2011;37:155–7.

[8] Rachoin JS, Dahar R, Moussallem C, Milcarek B, et al. The fallacy of the BUN: creatinine ratio in critically ill patients. Nephrol Dial Transplant 2012;27:2248–54.

[9] Lillehoj EP, Poulik MD. Normal and abnormal aspects of proteinuria: Part I: Mechanisms, characteristics and analyses of urinary protein. Part II: Clinical considerations. Exp Pathol 1986;29:1–28.

[10] Samara M, Abcar AC. False estimate of elevated creatinine. Perm J 2012;16:51–2.

[11] Liu WS, Chung YT, Yang CY, Lin CC, et al. Serum creatinine determined by Jaffe, enzymatic methods and isotope dilution liquid chromatography-mass spectrometry in patients under hemodialysis. J Clin Lab Anal 2012;26:206–14.

[12] Killorn E, Lim RK, Rieder M. Apparent elevated creatinine after ingestion of nitromethane: interference with the Jaffe reaction. Ther Drug Monit 2011;33:1–2.

[13] Patel H. The abnormal urinalysis. Pediatr Clin N Am 2006;53:325–7.

[14] Brigden ML, Edgell D, McPherson M, Leadbeater A, et al. High incidence of significant urinary ascorbic acid concentrations in west coast population-implications for routine urinalysis. Clin Chem 1992;38:426–31.

CHAPTER 12

Inborn Errors of Metabolism

12.1 OVERVIEW OF INBORN ERRORS OF METABOLISM

Congenital metabolic disorders are a class of genetic diseases that result from lack of (or abnormality of) an enzyme or its cofactor that is responsible for a clinically significant block in a metabolic pathway. As a result, abnormal accumulation of a substrate or deficit of the product is observed. In the majority of cases this is due to a single gene defect that encodes a particular enzyme important in the metabolic pathway. All inborn errors of metabolism are genetically transmitted, typically in an autosomal recessive or X-linked recessive fashion. Although individual inborn errors of metabolism are rare genetic disorders, over 500 human diseases related to inborn errors of metabolism have been reported. Therefore, collectively inborn errors of metabolism affect more than one baby out of 1,000 live births [1]. Children with inherited metabolic disorders most likely appear normal at birth because metabolic intermediates responsible for the disorder are usually small molecules that can be transported by the placenta and then eliminated by the mother's metabolism. However, symptoms usually appear due to accumulation of metabolites days, weeks, or months after birth, and very rarely a few years after birth. Although clinical presentation may vary, infants with metabolic disorders typically present with lethargy, decreased feeding, vomiting, tachypnea (related to acidosis), decreased perfusion, and seizure. With progression of the disease, infants may be presented to the hospital with stupor or coma. Metabolic screening must be initiated in any infant suspected of inborn errors of metabolism; elevated plasma ammonia level, hypoglycemia, and metabolic acidosis are indications of inborn errors of metabolism. Therefore, presenting clinical features of inborn errors of metabolism, although variable, may include:

- Failure to thrive, weight loss, delayed puberty, precocious puberty.
- Recurrent vomiting, diarrhea, abdominal pain.

CONTENTS

213

A. Dasgupta and A. Wahed: Clinical Chemistry, Immunology and Laboratory Quality Control
DOI: http://dx.doi.org/10.1016/B978-0-12-407821-5.00012-7

- Neurologic features such as seizures and stroke.
- Organomegaly such as lymphadenopathy and hepatosplenomegaly.
- Dysmorphic features.
- Cytopenias.
- Heart failure.
- Immunodeficiency.

Currently, newborn screenings are performed in many states to potentially identify any of 40 of the most commonly encountered inborn errors of metabolism, preferably using the new technology of tandem mass spectrometry. Common inborn errors of metabolism are listed in Table 12.1.

12.2 AMINO ACID DISORDERS

Amino acids are an integral part of proteins and may also act as substrates for gluconeogenesis. Out of twenty amino acids, nine of them are essential because they cannot be synthesized by the human body. In a patient with amino acid disorders, accumulation of amino acids in the blood is a common feature, and, as expected, increased excretion of amino acids is observed in urine. Common amino acid disorders are phenylketonuria and maple syrup urine disease.

12.2.1 Phenylketonuria

Phenylketonuria is due to deficiency of phenylalanine hydroxylase enzyme, which converts phenylalanine into tyrosine. As a result, phenylalanine accumulates in the circulation and is then converted to phenylpyruvate, a phenyl ketone that is eventually excreted in the urine. Phenylketonuria is an autosomal recessive disorder caused by a mutation in the gene that is responsible for coding of phenylalanine hydroxylase. A sustained phenylalanine concentration greater than 20 mg/dL (1,211 µmol/L) correlates with classical symptoms of phenylketonuria such as mental retardation, impaired head circumference growth, poor cognitive function, and lighter skin pigmentation. The disease is mild if phenylalanine concentration is in the range of 9.9−19.9 mg/dL (600−1,200 µmol/L). The phenylalanine-to-tyrosine ratio is also used for diagnosis of phenylketonuria; this ratio is helpful in reducing false positive rates. Treatment consists of a phenylalanine-restricted diet.

12.2.2 Maple Syrup Urine Disease (MSUD)

Maple syrup urine disease is a metabolic disorder caused by a deficiency of the branched-chain alpha-keto acid dehydrogenase complex that results in accumulation of branched-chain amino acids including leucine, isoleucine, and valine. The urine of such patients has an odor like maple syrup, thus the name

Table 12.1 Common Inborn Errors of Metabolism

Disorder	Enzyme Defect
Amino Acid Metabolism Disorders	
Phenylketonuria	Phenylalanine hydroxylase
Maple syrup disease	Branched-chain alpha-keto acid dehydrogenase complex
Tyrosinemia type I	Fumarylacetoacetate hydrolase
Tyrosinemia type II	Tyrosine aminotransferase
Homocystinuria	Cystathionine beta-synthase
Carbohydrate Metabolism Disorders	
Galactosemia	Galactose-1-phosphate uridyl transferase (most common cause: other enzyme defect)
GSD Type I (Von Gierke's disease)	Glucose-6-phosphatase
GSD Type II (Pompe's disease)	Acid alpha-glucosidase
GSD Type V (McArdle disease)	Muscle glycogen phosphorylase
Hereditary fructose intolerance	Aldolase B
Fructose intolerance (benign)	Fructose kinase
Lactose intolerance	Lactase
Urea Cycle Defect	
Most common cause	Ornithine transcarbamylase or carbamoyl synthase
Organic Aciduria	
Methylmalonic acidemia	Methylmalonyl-CoA mutase
Propionic aciduria	Propionyl-CoA carboxylase
Isovaleric aciduria	Isovaleryl-CoA dehydrogenase
Glutaric aciduria type I	Glutaryl-CoA dehydrogenase
Fatty Acid Oxidation Disorders	
MCAD deficiency (most common)	Medium-chain acyl coenzyme A dehydrogenase (MCAD)
SCAD deficiency	Short-chain acetyl-CoA dehydrogenase deficiency (SCAD)
LCAD deficiency	Long-chain acetyl-CoA dehydrogenase deficiency (LCAD)
VLCAD deficiency	Very-long chain acetyl-CoA dehydrogenase deficiency (VLCAD)
CPT-I deficiency	Carnitine palmitoyl transferase type I (CPT-I)
CPT-II deficiency	Carnitine palmitoyl transferase type II (CPT-II)
CACT deficiency	Carnitine acylcarnitine translocase (CACT)
Mitochondrial Disorders	
Kearns–Sayre syndrome (KSS)	Mitochondrial DNA abnormality
Peroxisomal Disorders	
Zellweger syndrome	Peroxisome membrane protein
Lysosomal Storage Disorders	
Hunter syndrome	Iduronate sulfatase

Continued...

Table 12.1 Common Inborn Errors of Metabolism *Continued*

Disorder	Enzyme Defect
Hurler syndrome	Alpha-L-iduronidase
Gaucher's disease	Beta-glucocerebrosidase
Tay-Sachs disease	Hexosaminidase A
Fabry's disease	Alpha-galactosidase A
Niemann–Pick disease Type A and B	Sphingomyelinase
Purine or Pyrimidine Metabolic Disorders	
Lesch-Nyhan syndrome	Hypoxanthine guanine phosphoribosyltransferase

CASE REPORT

A female (1,890 g) was born after 34 weeks of gestation and was admitted to the neonatal intensive care unit due to tachypnea and grunting that occurred over the first 2 h of delivery. Total parenteral nutrition (TPN) was initiated at approximately 28 to 29 h of life starting at 2 g amino acids per kilogram of body weight. A heelstick blood specimen obtained for newborn screening at 66 h (day 3) showed an abnormally high level of phenylalanine (1,420 µmol/L). Phenylalanine was restricted and the infant was started with breast milk on Day 7. A follow-up specimen on Day 7 showed a phenylalanine level of 4,164 µmol/L. Phenylalanine was completely eliminated from the diet on Days 12 to 16, and a reduced phenylalanine diet containing 40 mg of phenylalanine per kilogram of body weight was started on Day 17. The phenylalanine level of the infant was 304 µmol/L on discharge on Day 20. The median phenylalanine level was 287 µmol/L in the first year of life and the child was put on a restricted phenylalanine diet. However, by Month 13 the child could walk and showed normal development [2].

maple syrup urine disease. Elevated leucine is responsible for brain injury and neurological symptoms in these patients. This disease is inherited in an autosomal recessive manner, and newborn screening involves plasma amino acid analysis for diagnosis; this can be conducted using a dried blood spot.

12.2.3 Other Amino Acid Disorders

Tyrosinemia type I is caused by a deficiency of fumarylacetoacetate hydrolase and affected patients may present in childhood to their physicians with acute hepatic failure, coagulopathy, renal dysfunction, growth retardation, and possibly peripheral nerve involvement. Tyrosinemia type II is caused by a deficiency of tyrosine aminotransferase and is an oculocutaneous form of the disease that causes corneal lesions and skin involvement. Treatment involves a low tyrosine/phenylalanine diet and may also include nitisinone. Newborn screening involving only tyrosine has certain limitations in the diagnosis of tyrosinemia. Analysis of succinylacetone using tandem mass spectrometry may be helpful [3].

Homocystinuria is due to cystathionine beta-synthase deficiency. Patients have developmental delay and may present with ocular, skeletal, vascular, and central nervous system abnormalities. Newborn diagnosis is based on high methionine and high homocysteine levels.

12.3 CARBOHYDRATE METABOLISM DISORDERS

Carbohydrate disorders may include deficiencies of enzymes involved in the metabolism of glycogen, galactose, and fructose. These diseases can be broadly subclassified as diseases causing liver dysfunction, diseases affecting muscle and liver, and diseases affecting only muscle. Galactosemia, glycogen storage diseases, hereditary fructose intolerance, and fructose 1,6-diphosphate deficiency are common examples of carbohydrate metabolism disorders.

12.3.1 Galactosemia

Three enzymes are involved in the metabolism of galactose, and deficiency of any three can cause galactosemia. The most common form of galactosemia is caused by lack of the enzyme galactose-1-phosphate uridyl transferase, and affects an estimated 1 in every 55,000 newborns. In these patients, galactose-1-phosphate accumulates and is then degraded to galactonate and galactitol, causing early cataract formation in the eyes. If untreated, these children may develop an intellectual disability, speech problems, and other dysfunctions. Diagnosis can be established by measuring galactose-1-phosphate uridyl transferase activity in a dried blood spot.

12.3.2 Glycogen Storage Disease

There are multiple types of glycogen storage diseases depending on the exact deficient enzyme. They are numbered according to their discovery, though numbers are not useful in separating these disorders according to clinical symptoms. It is noteworthy to mention Type I (Von Gierke's disease), which is due to the absence of glucose-6-phosphatase, Type II (Pompe's disease) due to absence of acid alpha-glucosidase, and Type V (McArdle disease) due to the absence of muscle glycogen phosphorylase. In general, patients with glycogen storage disease I, III, VI, and IX present with hepatomegaly and hypoglycemia, while patients with glycogen storage disease IV often experience liver failure prior to symptoms of hypoglycemia. Patients with glycogen storage disease II, V, and VII primarily have muscle dysfunction. The onset of glycogen storage II disease may be during early childhood, but in type V and

VII patients it often presents to clinics during adolescence with complaints of exercise intolerance accompanied (usually) by myoglobinuria.

12.3.3 Fructose Intolerance

Hereditary fructose intolerance is a rare (1 in 20,000 births) recessive inherited disorder of carbohydrate metabolism due to catalytic deficiency of aldolase B (fructose biphosphate or liver aldolase). These patients show impaired fructose metabolism when exposed to fructose or sucrose during infancy through diet. Persistent ingestion of fructose and sucrose can cause severe liver and kidney damage that may be associated with seizure, coma, and even death. Early diagnosis is essential for good prognosis because these individuals can live a normal life by avoiding fruits and sweets containing fructose. The diagnosis can be confirmed by measuring a particular enzyme activity on a liver biopsy. DNA analysis is also available for diagnosis [4]. The most common mutation in fructose metabolism is due to lack of fructokinase, which is the first step in the metabolism of dietary fructose. This condition, however, is asymptomatic, and excess fructose is excreted in urine (fructosuria).

12.3.4 Lactose Intolerance

Lactose intolerance, also called lactase deficiency, is due to an insufficient level of lactase, which hydrolyzes lactose into glucose and galactose. Therefore, ingestion of milk and dairy products results in bloating, abdominal cramps, diarrhea, and related symptoms. This is a common condition that may develop later in life or may be manifested in early childhood. These individuals can live a normal life by either avoiding milk or dairy products, by consuming lactose-free milk, or by taking a lactase enzyme as supplement. Lactose intolerance is transmitted either in autosomal recessive (Caucasian population) or autosomal dominant (Asian population) fashion.

12.4 UREA CYCLE DISORDERS

Degradation of amino acids results in formation of ammonia as a waste product. This waste then enters the urea cycle in the liver, and in the first step ammonia combines with carbon dioxide to form carbamoyl phosphate. Finally, in the urea cycle, ammonia is converted into urea for excretion by kidneys. There are six known enzymes involved in the urea cycle, and a defect in any one of these enzymes can cause these disorders. Ornithine transcarbamylase deficiency, an X-linked disorder (occurring most commonly in males) is the most common urea cycle defect. Newborns with urea cycle disorder develop high levels of ammonia after a protein feed. Only patients with arginase deficiency, a defect in the last step of the urea cycle, do not present with hyperammonemia; instead, they present with neurological dysfunctions. It is often difficult to diagnose a urea cycle defect, although hyperammonemia in a sick neonate is an indication. Determination of orotic acid concentration can help differentiate ornithine transcarbamylase deficiency (elevated orotic acid) from carbamyl phosphatase synthetase deficiency (normal or low orotic acid level). Urea cycle disorders are treatable causes of hyperammonemia in infants and the pediatric age group. Presentation in adolescence or adult life of urea cycle disorder is rare.

CASE REPORT

An 18-year-old, previously healthy male who consumed energy drinks (12 cans, 14 g protein per bottle) after elective wisdom tooth extraction presented to the hospital with decreased consciousness and encephalopathy. His ammonia level was 300 µmol/L (normal: 10−40 µmol/L), but the rest of his blood and cerebrospinal fluid analytes were within acceptable levels. He underwent hemodialysis in the community hospital and was later transferred to a local tertiary care hospital where a diagnosis of late-onset urea cycle disorder was made based on his history of intolerance to a high-protein diet and presentation with hyperammonemia. However, he showed no clinical or biochemical evidence of liver disease or metabolic acidosis. His diagnosis was confirmed by genetic analysis showing a genetic defect that caused deficiency in ornithine transcarbamylase enzyme. His hyperammonemia was managed with nitrogen scavenging medications (sodium benzoate, phenyl acetate). He was discharged on Day 11 with a normal ammonia level of 35 µmol/L [5].

12.5 ORGANIC ACID DISORDERS (ORGANIC ACIDURIA)

Organic acid disorders are a group of inborn errors of metabolism due to enzyme deficiency in the amino acid degradation pathways including defects in metabolism of branched-chain amino acids (leucine, isoleucine, and valine) as well as other amino acids including homocysteine, tyrosine, methionine, threonine, lysine, and tryptophan. As a result, toxic organic acids accumulate in circulation and are eventually excreted in urine to cause organic aciduria. More than 25 disorders are known. However, division of organic acid disorders and amino acid metabolism disorders are somehow arbitrary because phenylketonuria and maple syrup urine disease also cause organic aciduria. Amino acid disorders are traditionally diagnosed by amino acid analysis in blood, while organic acidurias are traditionally diagnosed by urine organic acid analysis by gas chromatography/mass spectrometry or tandem mass spectrometry. Alternatively, a dried blood spot in a newborn can be analyzed to establish diagnosis of organic aciduria.

Methylmalonic acidemia is due to a defect in methylmalonyl CoA mutase enzyme, which is involved in metabolism of branched-chain amino acids. Vitamin B_{12} is also required for this conversion. Mutations leading to defects in vitamin B_{12} metabolism or in its transport frequently result in the development of methylmalonic acidemia.

In healthy individuals, the enzyme propionyl CoA carboxylase converts propionyl CoA to methylmalonyl CoA. In individuals with propionic acidemia this pathway is blocked, causing conversion or excessive propionyl CoA to propionic acid, thus leading to propionic acidemia. Isovaleric acidemia is caused by a deficiency of isovaleryl-CoA dehydrogenase, which is involved in the metabolism of leucine. Glutaric aciduria type I is a rare organic aciduria due to deficiency of glutaryl-CoA dehydrogenase, which is involved in catabolism of lysine, hydroxylysine, and tryptophan [6].

12.6 FATTY ACID OXIDATION DISORDERS

Mitochondrial fatty acid oxidation is a major pathway for energy production during fasting and strenuous exercise that may cause a hypoglycemic condition. However, fatty acids must be transported to mitochondria prior to oxidation, and either a transport defect of fatty acid or a defect in any enzyme involved in the fatty acid oxidation pathway can cause fatty acid oxidation disorders (inherited in an autosomal recessive pattern). Patients with fatty acid oxidation disorders usually have features of hypoglycemia without ketosis during episodes of decreased carbohydrate intake. The most common fatty acid oxidation disorder is due to the deficiency of medium-chain acyl coenzyme A dehydrogenase (MCAD) deficiency, but long-chain acetyl-CoA dehydrogenase deficiency (LCAD), very-long-chain acetyl-CoA dehydrogenase deficiency (VLCAD), and short-chain acetyl CoA dehydrogenase deficiency (SCAD) have also been reported.

Carnitine is essential for transporting long-chain fatty acids because long-chain fatty acids cannot pass through the mitochondrial membrane, although short- and medium-chain fatty acids can. Carnitine palmitoyl transferase type I (CPT-I) is responsible for attaching carnitine to long-chain fatty acid molecules, then carnitine acylcarnitine translocase (CACT) transports the resulting molecule into the mitochondria, and finally palmitoyl transferase type II (CPT-II) removes the carnitine and releases fatty acids for beta-oxidation inside the mitochondria (which produces energy). The free carnitine is transported back into circulation for binding with more fatty acids. A defect in any of these carnitine transport enzymes can also cause a fatty acid oxidation defect in an individual. Newborn screening can identify different types of fatty acid oxidation disorders and treatment of fatty acid oxidation is primarily aimed at maintaining blood glucose by feeding at regular intervals and a diet high in carbohydrates but low in fat. Some patients may also need carnitine supplements.

12.7 MITOCHONDRIAL DISORDERS

Organic acids, fatty acids, and amino acids are metabolized to acetyl-CoA within the mitochondria. Acetyl-CoA combines with oxaloacetate to form citric acid, which is oxidized in the Krebs cycle (also known as the citric acid cycle). If there is a defect in the energy-producing pathway, especially during oxidative phosphorylation, these abnormalities are called mitochondrial disorders or diseases. Patients with mitochondrial disorders may present with hypoglycemia with ketosis. Mitochondrial disorders may affect muscle alone. Multiple organ involvement such as brain, heart, kidney, and liver may also be seen. Mitochondrial diseases are due to a mutation of mitochondrial DNA (mitochondria have their own DNA) and all mitochondrial DNA is derived from the ovum (i.e. these diseases are maternally inherited). Examples of mitochondrial disorders include cytochrome c oxidase deficiency and Kearns—Sayre syndrome (KSS). KSS is a syndrome characterized by isolated involvement of the muscles that control eyelid movement (levator palpebrae, orbicularis oculi) and those controlling eye movement (extraocular muscles). This results in ptosis and ophthalmoplegia, respectively. KSS involves a triad of the eye, changes with bilateral pigmentary retinopathy, and cardiac conduction abnormalities.

12.8 PEROXISOMAL DISORDERS

Peroxisomes are cellular organelles that play an important role in beta-oxidation of very-long-chain fatty acids, degradation of phytanic acid by alpha-oxidation, degradation of hydrogen peroxide, as well as synthesis of bile acids and plasmalogen (an important component of cell membranes and myelin). Examples of peroxisomal disorders are Zellweger syndrome and adrenoleukodystrophy. Zellweger syndrome is due to a biogenesis defect, and as a result all peroxisomal enzymes are deficient, making it a very severe disorder. However, neonatal adrenoleukodystrophy is milder than Zellweger syndrome.

12.9 LYSOSOMAL STORAGE DISORDERS

Lysosomes are cellular organelles that contain more than 30 acid hydrolases that can degrade unwanted complex molecules such as mucopolysaccharides, sphingolipids, glycoproteins, etc., into molecules which could be used by the body again. Therefore, lysosomes can be regarded as the recycling centers of the body. Lysosomal storage diseases are a heterogenous group of more than 50 disorders due to defects in lysosomal enzymes, enzyme receptors, membrane proteins, activator proteins, or transporters. In lysosomal storage disorders, accumulation of a few complex lipids that should be degraded in a normal person results in coprecipitation of other hydrophobic compounds in the endolysosome system and impairs lysosomal function such as delivery of nutrients. As a result, cellular starvation finally causes organ dysfunction. All lysosomal storage diseases are inherited in autosomal recessive manner except Hunter syndrome, Danon disease, and Fabry disease, which are inherited in an X-linked manner. Common examples of lysosomal storage disorders include:

- Mucopolysaccharidoses (e.g. Hunter syndrome, Hurler syndrome, Sanfilippo syndrome, and Scheie syndrome).
- Sphingolipidoses (e.g. Gaucher's disease, Tay-Sachs, Fabry disease, Niemann—Pick disease).
- Glycoproteinoses (e.g. mannosidosis).
- Mucolipidosis.

Mucopolysaccharidoses are a group of metabolic disorders due to deficiency of lysosomal enzymes responsible for the breakdown of polysaccharide chains (glycosaminoglycan). Gaucher's disease is the most common form of lysosomal storage disease, and is due to a deficiency of the enzyme glucocerebrosidase, leading to accumulation of glucocerebroside. Gaucher's disease has three common clinical subtypes:

- Type I (or non-neuropathic) is the most common form of the disease. It is seen most often in Ashkenazi Jews. Features are apparent early in life or in adulthood and include hepatosplenomegaly. Neurological features are not seen. Depending on disease onset and severity, Type I patients may live well into adulthood. Many individuals have a mild form of the disease or may not show any symptoms at all.
- Type II (or acute infantile neuropathic Gaucher's disease) typically begins within 6 months of birth. Neurological features are prominent and most children die at a very early age.
- Type III (the chronic neuropathic form) can begin at any time in childhood or even in adulthood. It is characterized by slowly progressive but milder neurologic symptoms compared to the acute or type II version.

Tay—Sachs disease has a higher frequency in Ashkenazi Jews and is caused by a deficiency of hexosaminidase. Niemann—Pick Type A disease is a fatal disorder of infancy (life expectancy: 2—3 years) due to accumulation of sphingomyelin as a result of a mutation in the sphingomyelin phosphodiesterase 1 gene encoding enzyme acid sphingomyelinase. In Type A disease, activity of this enzyme is almost completely absent. In Type B disease, some activity of this enzyme is preserved. However, in Type C disease, accumulation of non-esterified cholesterol takes place [7].

12.10 PURINE OR PYRIMIDINE METABOLIC DISORDERS

Purine and pyrimidine nucleotides are part of DNA, RNA, ATP, and nicotinamide adenine dinucleotide (NAD). Examples of purine and pyrimidine disorders include Lesch–Nyhan disease or syndrome and adenosine deaminase deficiency.

Lesch–Nyhan disease is a rare monogenic disorder that is transmitted in an X-linked recessive fashion. These patients have a high risk of developing gout due to overproduction of uric acid as a result of deficiency of the enzyme hypoxanthine guanine phosphoribosyltransferase. The patients with classical phenotype present with overproduction of uric acid, severe motor dysfunction resembling patients with dystonic cerebral palsy, intellectual deficiency, and self-injurious behavior. The mildest form of this disease includes only overproduction of uric acid. In between the classical (extreme) and mild forms, there is an intermediate form of this disease where patients experience some motor and cognitive dysfunction, but no self-injurious behavior [8].

12.11 DISORDERS OF PORPHYRIN METABOLISM

The various porphyrias are due to abnormalities in the enzymes involved in the synthesis of heme, resulting in accumulation of intermediate compounds. Porphyrins consist of four pyrrole rings, and the precursors in the formation of the pyrrole rings are glycine and succinyl-CoA, which combine in the presence of delta-aminolevulinate synthase to form delta-aminolevulinic acid inside the mitochondria. It is then transported into the cytosol for further transformation. Then two molecules of delta-aminolevulinic acid condense to form one pyrrole ring (mono-pyrrole porphobilinogen). Next, porphobilinogen molecules are cyclized to form hydroxymethylbilane, which is eventually converted into coproporphyrinogen III and enters into the mitochondria for further transformation into protoporphyrinogen IX by the action of coproporphyrinogen oxidase enzyme. Finally, protoporphyrinogen IX is converted into protoporphyrin and ferrous iron is incorporated into the molecule by the action of ferrochelatase to form a heme molecule inside the mitochondria. In porphyria the heme intermediates accumulate due to partial deficiency in certain enzymes involved in heme biosynthesis. Porphyrias can be sub-classified under two broad categories:

■ Porphyrias involving skin lesions and photosensitivity.

Table 12.2 Various Types of Porphyrias

Enzyme	Function of Enzyme	Type of Porphyria for Missing Enzyme
ALA (delta-aminolevulinate) synthase	Forms ALA from glycine and succinyl-CoA	Not known
ALA dehydratase	Converts ALA to porphobilinogen (PBG)	ALA dehydrase deficiency porphyria
PBG deaminase	Converts PBG to hydroxymethylbilane	Acute intermittent porphyria (AIP)
Uroporphyrinogen synthase	Converts hydroxy-methylbilane to uro-porphyrinogen III	Congenital erythropoietic porphyria
Uroporphyrinogen decarboxylase	Converts uroporphyrinogen III to coproporphyrinogen III	Porphyria cutanea tarda and hepatoerythropoietic porphyria
Coproporphyrinogen oxidase	Converts coproporphyrinogen III to protoporphyrinogen	Hereditary coproporphyria (HCP)
Protoporphyrinogen oxidase	Converts protoporphyrinogen to protoporphyrin	Variegate porphyria (VP)
Ferrochelatase	Adds iron to protoporphyrin to form heme	Erythropoietic protoporphyria (EPP)

- Porphyrias with neurovisceral dysfunction such as neuropathy, convulsions, psychiatric disorders, acute abdomen, hypertension, tachycardia, etc.

Acute porphyrias such as acute intermittent porphyria, variegate porphyria, and hereditary coproporphyria are inherited in autosomal dominant fashion. Acute life-threatening neurovisceral attacks seen in these three porphyrias are similar in nature. Non-acute porphyrias include congenital erythropoietic porphyria, porphyria cutanea tarda, and erythropoietic protoporphyria. Various types of porphyrias are listed in Table 12.2.

12.12 NEWBORN SCREENING AND EVALUATION

Newborn screening tests are routinely performed to identify approximately 40 disorders. The tests and methods may vary from state to state and country to country. False positive and false negative screening tests can occur. Clinical evaluation includes a detailed history, including family history, physical examination, and laboratory evaluation. Laboratory evaluation may include initial tests and specialized tests. Initial tests can include complete blood count, serum levels of glucose, ammonia, creatinine, urea, uric acid, electrolytes, muscle enzymes such as creatinine kinase, aldolase, as well as liver function tests. Urinalysis is also helpful. However, specialized tests are needed for diagnosis of inborn errors of metabolism. These tests include:

- Quantitative plasma amino acid profile (for urea cycle disorders and disorders of amino acid metabolism).
- Urine organic acids (for diagnosis of various acidurias).
- Serum pyruvate and lactate (lactic acidosis is seen in mitochondrial disorders as well as disorders of carbohydrate metabolism and glycogen storage diseases).
- Acylcarnitine profile (used for fatty acid oxidation disorders).

More recently, new platform technology such as tandem mass spectrometry (MS/MS) has been widely used in developed countries for newborn screening. New developments in tandem mass spectrometry coupled with electrospray detection (ESI) allow rapid and high-throughput screening for a large number of inborn errors of metabolism from a single dried blood spot specimen after extraction [9].

KEY POINTS

- Phenylketonuria is due to a deficiency of phenylalanine hydroxylase, which converts phenylalanine into tyrosine. As a result, phenylalanine accumulates in the circulation and is converted to phenylpyruvate, a phenyl ketone that is eventually excreted in urine (hence the name phenylketonuria). Phenylketonuria is an autosomal recessive disorder.
- Maple syrup urine disease is a metabolic disorder caused by a deficiency of the branched-chain alpha-keto acid dehydrogenase complex that results in the accumulation of branched-chain amino acids, including leucine, isoleucine, and valine. The urine of such patients has the same odor as maple syrup.
- The most common form of galactosemia is caused by lack of the enzyme galactose-1-phosphate uridyl transferase, and affects an estimated 1 in every 55,000 newborns. In these patients galactose-1-phosphate accumulates.
- The most common mutation in fructose metabolism (due to lack of fructokinase) affects the first step in the metabolism of dietary fructose. This condition, however, is asymptomatic, and excess fructose is excreted in urine (fructosuria).
- Lactose intolerance (also called lactase deficiency) is due to an insufficient level of lactase, which hydrolyzes lactose into glucose and galactose. Therefore, ingestion of milk and dairy products results in bloating, abdominal cramps, diarrhea, and related symptoms.
- Ornithine transcarbamylase deficiency, an X-linked disorder (occurring most commonly in males) is the most common urea cycle defect. Newborns with urea cycle disorder develop high levels of ammonia after a protein feed.
- Organic acid disorders are a group of inborn errors of metabolism due to enzyme deficiency in the amino acid degradation pathways, including defects in metabolism of branched-chain amino acids (leucine, isoleucine, and valine) as well

as other amino acids, including homocysteine, tyrosine, methionine, threonine, lysine, and tryptophan. As a result, toxic organic acids accumulate in circulation and are eventually excreted in urine, causing organic aciduria.

- The most common fatty acid oxidation disorder is due to deficiency of medium-chain acyl coenzyme A dehydrogenase (MCAD) deficiency.
- Organic acids, fatty acids, and amino acids are metabolized to acetyl-CoA within the mitochondria. Acetyl-CoA combines with oxaloacetate to form citric acid, which is oxidized in the Krebs cycle (also known as the citric acid cycle). If there is a defect in the energy-producing pathway (especially during oxidative phosphorylation) it causes a disease called a mitochondrial disorder or mitochondrial disease. Patients with mitochondrial disorders may present with hypoglycemia with ketosis. Although mitochondrial disorders may affect muscle alone, multiple organ involvement such as brain, heart, kidney, and liver may also be seen. Mitochondrial diseases are due to the mutation of mitochondrial DNA (mitochondria have their own DNA); all mitochondrial DNA are derived from the ovum, so these diseases are maternally inherited.
- Peroxisomes are cellular organelles that play an important role in beta-oxidation of very-long-chain fatty acids, degradation of phytanic acid by alpha-oxidation, degradation of hydrogen peroxide, as well as in the synthesis of bile acids and plasmalogen (an important component of cell membranes and myelin). Examples of peroxisomal disorders are Zellweger syndrome and adrenoleukodystrophy.
- Lysosomal storage diseases are a heterogenous group of more than 50 disorders that are due to defects in lysosomal enzymes, enzyme receptors, membrane proteins, activator proteins, or transporters. Common examples of lysosomal storage disorders include: mucopolysaccharidoses (e.g. Hunter syndrome, Hurler syndrome, Sanfilippo syndrome, and Scheie syndrome), sphingolipidoses (e.g. Gaucher's disease, Tay-Sachs, Fabry disease, and Niemann—Pick disease), glycoproteinoses (e.g. mannosidosis), and mucolipidosis.
- Gaucher's disease is the most common form of lysosomal storage disease, and is due to a deficiency of the enzyme glucocerebrosidase, which leads to accumulation of glucocerebroside. Gaucher's disease has three common clinical subtypes:
  - Type I (or non-neuropathic type) is the most common form of the disease. It is seen most often in Ashkenazi Jews. Features are apparent early in life or in adulthood and include hepatosplenomegaly. Neurological features are not seen. Depending on disease onset and severity, Type I patients may live well into adulthood. Many individuals have a mild form of the disease or may not show any symptoms at all.
  - Type II (or acute infantile neuropathic Gaucher's disease) typically begins within 6 months of birth. Neurological features are prominent, and most children die at a very early age.

- Type III (the chronic neuropathic form) can begin at any time in childhood, or even in adulthood. It is characterized by slowly progressive, but milder neurologic, symptoms compared to the acute or Type II version.
- Tay—Sachs disease has a higher frequency in Ashkenazi Jews and is caused by a deficiency of hexosaminidase. Niemann—Pick type A disease is a fatal disorder of infancy (life expectancy: 2—3 years) due to accumulation of sphingomyelin as a result of mutation in the sphingomyelin phosphodiesterase 1 gene encoding enzyme acid sphingomyelinase.
- Examples of purine and pyrimidine disorders include Lesch—Nyhan disease or syndrome and adenosine deaminase deficiency. Lesch—Nyhan disease is a rare monogenic disorder, and is transmitted in an X-linked recessive fashion. These patients have a high risk of developing gout due to overproduction of uric acid as a result of deficiency of the enzyme hypoxanthine guanine phosphoribosyltransferase.
- Acute porphyrias, such as acute intermittent porphyria, variegate porphyria, and hereditary coproporphyria, are inherited in an autosomal dominant fashion. Acute life-threatening neurovisceral attacks seen in these three porphyrias are similar in nature. Non-acute porphyrias include congenital erythropoietic porphyria, porphyria cutanea tarda, and erythropoietic protoporphyria.
- Newborn screening tests are routinely performed to identify approximately 40 disorders. Tests and methods may vary from state to state and country to country. False positive and false negative screening tests can occur. Clinical evaluation includes a detailed history, including family history, physical examination, and laboratory evaluation. Laboratory evaluation may include initial tests and specialized tests. Initial tests may include: complete blood count, serum levels of glucose, ammonia, creatinine, urea, uric acid, electrolytes, muscle enzymes such as creatinine kinase, aldolase, and also liver function tests. Urinalysis is also helpful. However, specialized tests are needed for diagnosis of inborn errors of metabolism. These tests include: quantitative plasma amino acid profile (for urea cycle disorders and disorders of amino acid metabolism), urine organic acids (for diagnosis of various acidurias), serum pyruvate and lactate (lactic acidosis is seen in mitochondrial disorders as well as disorders of carbohydrate metabolism and glycogen storage diseases), and acylcarnitine profile (used for fatty acid oxidation disorders).

REFERENCES

[1] Alfadhel M, Al-Thihli K, Moubayed H, Eyaid W, Zytkovicz TH. Drug treatment of inborn errors of metabolism: a systematic review. Arch Dis Child 2013;Mar:26 [e-pub ahead of print].

[2] Lin HJ, Kwong AM, Carter JM, Ferreira BF, et al. Extremely high phenylalanine levels in a newborn on parenteral nutrition: phenylketonuria in a neonatal intensive care unit. J Perinatol 2011;31:507—10.

[3] Allard P, Greiner A, Korson MS, Zytkovicz TH. Newborn screening for hepatorenal tyrosinemia by tandem mass spectrometry: analysis of succinylacetone extracted from dried blood spot. Clin Biochem 2004;37:1010—5.

[4] Ferri L, Caciotti A, Cavicchi C, Rigoldi M, et al. Integration of PCR sequencing analysis with multiplex ligation dependent probe amplification for diagnosis of hereditary fructose intolerance. JIMD Reports 2012. Available from: http://dx.doi.org/10.1007/8904_2012_125.

[5] Iyer H, Sen M, Prasad C, Rupar CA, et al. Coma, hyperammonemia, metabolic acidosis, and mutation: lessons learned in the acute management of late onset urea cycle disorders. Hemodial Int 2012;16:95−100.

[6] Kolker S, Christensen E, Leonard J, Greenberg C, et al. Diagnosis and management of glutaric aciduria type I-revised recommendations. J Inherit Metab Dis 2011;34:677−94.

[7] Schulze H, Sandhoff K. Lysosomal lipid storage disease. Cold Spring Harb Perspect Biol 2011;3 [pii a004804].

[8] Torres RJ, Puig JG, Jinnah HA. Update on the phenotypic spectrum of Lesch-Nyhan disease and its attenuated variants. Curr Rheumatol Rep 2012;14:189−94.

[9] Ozben T. Expanded newborn screening and confirmatory follow-up testing for inborn errors of metabolism detected by tandem mass spectrometry. Clin Chem Lab Med 2013;51:157−76.

CHAPTER 13

Tumor Markers

13.1 INTRODUCTION TO TUMOR MARKERS

Most tumor markers are produced by normal cells as well as by cancer cells, but in the process of developing cancer, concentrations of these markers are elevated many fold compared to very low concentrations of these markers observed in blood under non-cancerous conditions. In addition to blood, these markers are also found in urine, stool, or bodily fluids of patients with cancer. So far, more than 20 different tumor markers have been characterized and are used clinically for diagnosis and monitoring of treatment. Some tumor markers are elevated with only one type of cancer, whereas others are associated with two or more cancer types. Although theoretically any type of biological molecule can act as a tumor marker, in practice, most markers are either proteins or glycoproteins. However, low-molecular-weight substances (e.g. vanillylmandelic acid and homovanillic acid) are used as markers for diagnosis of neuroblastoma. More recently, patterns of gene expression and changes to DNA are also under intense investigation for potential use as tumor markers. These types of markers are measured specifically in tumor tissues. Most of the traditionally used markers are probably not involved in tumorigenesis, but are likely to be by-products of malignant transformation.

13.2 CLINICAL USES OF TUMOR MARKERS AND COMMON TUMOR MARKERS

Tumor markers can be used for one of five purposes:

- Screening a healthy population or a high-risk population for the probable presence of cancer.
- Diagnosis of cancer or a specific type of cancer.
- Evaluating prognosis in a patient.

CONTENTS

A. Dasgupta and A. Wahed: Clinical Chemistry, Immunology and Laboratory Quality Control
DOI: http://dx.doi.org/10.1016/B978-0-12-407821-5.00013-9

- Predicting potential response of a patient to therapy.
- Monitoring recovery of a patient receiving surgery, radiation, or chemotherapy.

Tumor markers were first developed to test for cancer in people without symptoms, but very few markers are effective in achieving this goal. Today, the most widely used tumor marker in the clinical setting is prostate-specific antigen (PSA). In addition, only a few markers that are now available have clinically useful predictive values for cancer at an early stage, and only when patients at high risk are tested. Tumor markers are not the gold standard for diagnosis of a cancer. In most cases, a suspected cancer can only be diagnosed by a biopsy. Alpha-fetoprotein (AFP) is an example of a tumor marker that can be used to aid in diagnosis of cancer, especially hepatocellular carcinoma (HCC). However, the level of AFP can also be increased in some liver diseases, although when it reaches a certain threshold it is usually indicative of hepatocellular carcinoma.

Some types of cancer grow and spread faster than others, while some cancers also respond well to various therapies. Sometimes the level of a tumor marker can be useful in predicting the behavior and outcome for certain cancers. For example, in testicular cancer, very high levels of a tumor marker such as human chorionic gonadotropin (hCG) or AFP may indicate an aggressive cancer with poor survival outcome. Patients with these high levels may require very aggressive therapy even at the initiation of cancer therapy. Certain markers found in cancer cells can be used to predict whether a treatment is likely to produce a favorable outcome or not. For example, in breast and stomach cancers, if the cells have too much of a protein called human epidermal growth factor receptor 2 (HER2), drugs such as trastuzumab (Herceptin®) can be helpful if used during chemotherapy. However, with normal expression of HER2, these drugs may not produce the expected therapeutic benefits. Tumor markers are also used to identify the recurrence of certain tumors after successful therapy. Certain tumor markers may be useful for further evaluation of a patient after completion of the treatment when there is no obvious sign of cancer in the body. Commonly measured tumor markers in clinical laboratories include:

- Prostate-specific antigen (PSA) for prostate cancer.
- Human chorionic gonadotropin (hCG) for gestational trophoblastic tumors and some germ cell tumors.
- Alpha-fetoprotein (AFP) for certain germ cell tumors and HCC.
- CA-125 (carbohydrate antigen or cancer antigen 125) for ovarian cancer.
- CA 19-9 (carbohydrate antigen 19-9) for pancreatic and gastrointestinal cancers.
- Carcinoembryonic antigen (CEA) for colon and rectal cancers.

Less commonly monitored tumor markers include:

- Cancer antigen (CA) 15-3, a marker for breast cancer.
- Cancer antigen (CA) 72-4, a marker for colorectal cancer.
- Cytokeratin fragment (CYFRA) 21-1, a marker for lung cancer.
- Squamous cell carcinoma antigen, a marker for squamous cell lung cancer.
- Neuron-specific enolase, a marker for lung cancer.
- Chromogranin A, a marker for neuroendocrine tumor.

Clinical utilities of common tumor markers are listed in Table 13.1. Common causes of elevated levels of tumor markers in the absence of cancer are listed in Table 13.2.

Table 13.1 Commonly Used Tumor Markers

Tumor Marker	Use
Prostatic-specific antigen (PSA)	Prostate carcinoma
Carbohydrate antigen-125 (Cancer antigen-125)	Ovarian and Fallopian carcinoma
Alpha-fetoprotein	Hepatocellular carcinoma and germ cell tumors
Carcinoembryonic antigen (CEA)	Colorectal, gastric, pancreatic, lung, and breast carcinomas
CA-19-9 (carbohydrate antigen 19-9)	Pancreatic carcinoma, cholangiocarcinoma
Beta-2-microglobulin	Multiple myeloma and lymphoma
Beta-human chorionic gonadotropin (β-hCG)	Choriocarcinoma and testicular carcinoma

Table 13.2 Common Causes of Elevation of Tumor Marker Levels (In the Absence of Neoplasia)

Tumor Marker	Common Causes of Elevated Levels
Prostate-specific antigen	Prostatitis/benign prostatic hyperplasia, cyst, heterophilic antibody
Alpha-fetoprotein	Hepatobiliary disease, pneumonia, pregnancy, autoimmune disease, heterophilic antibody
CA-125	Hepatobiliary disease, pulmonary disease, renal failure, hypothyroidism, endometriosis, pregnancy, autoimmune disease, skin disease, cardiovascular disease, heterophilic Antibody
Carcinoembryonic antigen (CEA)	Hepatobiliary disease, renal failure, hypothyroidism, gastrointestinal disease, pancreatitis, Endometriosis, autoimmune disease, heterophilic antibody
CA-19-9	Hepatobiliary disease, renal failure, pulmonary disease, pancreatitis, gastrointestinal disease, endometriosis, heterophilic antibody
CA-15-3	Vitamin B12 deficiency, renal failure
Beta-2-microglobulin	Renal failure, autoimmune disease, cerebral lesion
hCG or beta-hCG	Renal failure, pregnancy, autoimmune disease, heterophilic hntibody
CA-72-4	Hepatobiliary disease, renal failure, pancreatitis, gastrointestinal disorder
CYFRA-21-1	Hepatobiliary disease, renal failure, pulmonary disease
Chromogranin A	Cardiovascular disease, viral Infection, prostatitis/benign prostatic hyperplasia, gastrointestinal disease, heterophilic antibody

13.3 PROSTATE-SPECIFIC ANTIGEN (PSA)

Prostate-specific antigen (PSA) is a serine protease belonging to the kallikrein family. PSA is a single-chain glycoprotein containing 237 amino acids and four carbohydrate side chains (molecular weight: 28,430). PSA is expressed by both normal and neoplastic prostate tissue. Under normal conditions, PSA is produced as a proantigen (proPSA) by the secretory cells that line the prostate glands, and is secreted into the lumen, where the propeptide moiety is removed to generate active PSA. The active PSA can then undergo proteolysis to generate inactive PSA, of which a small portion then enters the bloodstream and circulates in an unbound state (free PSA). Alternatively, active PSA can diffuse directly into the circulation where it is rapidly bound to alpha-1-antichymotrypsin (ACT) and alpha-2-macroglobulin [1]. In men with a normal prostate, the majority of free PSA in the serum reflects the mature protein that has been inactivated by internal proteolytic cleavage. In contrast, free PSA is relatively decreased in patients with prostate cancer. Thus, the percentage of free or unbound PSA is lower in the serum of men with prostate cancer. Therefore, the ratio of free to total PSA or complexed PSA (cPSA) is a means of distinguishing between prostate cancer and benign prostatic hyperplasia (BPH) in a patient with elevated PSA. Causes of elevated PSA include:

- Benign prostatic hyperplasia (BPH).
- Prostate cancer.
- Prostatic inflammation/infection.
- Perineal trauma.

Studies in the 1980s confirmed that serum total PSA could be used as a screening tool to identify men with prostate cancer because elevated serum PSA is clearly a more sensitive marker than digital rectal examination. However, between 20 and 50 percent of men with newly diagnosed prostate cancers may have serum PSA values below 4.0 ng/mL (the upper end of normal is usually 4.0 ng/mL), indicating that PSA lacks specificity as a tumor marker. In general, patients with a PSA below 4.0 ng/mL are more likely to have prostate cancer that is confined to the organ. These patients have a better prognosis than patients with prostate cancer who show levels above 4.0 ng/mL [2].

Prostatitis with or without active infection is an important cause of an elevated PSA, and levels as high as 75 ng/mL have been reported in the literature. Thus, many physicians often initially treat a man with an isolated elevated serum PSA with antibiotics for a presumed diagnosis of prostatitis, and then obtain a repeat serum PSA for further clinical evaluation. The percent free PSA may be less affected by the presence of inflammation, particularly when the total serum PSA is less than 10 ng/mL. However, the free-to-

total ratio of serum PSA may be unable to distinguish chronic inflammation from prostate cancer, as both conditions can lower the percentage of free PSA. This would be expected because inflammation leads to elevated serum PSA in a similar fashion as prostate cancer (i.e. through disruption of the basal membrane and increased leakage of "immature" PSA into the blood stream).

Any perineal trauma can also increase the serum PSA. Prostate massage and digital rectal examination may cause minor transient elevations that can be clinically insignificant. Mechanical manipulation of the prostate by cystoscopy, prostate biopsy, or transurethral resection of the prostate (TURP) can more significantly affect the serum PSA. Vigorous bicycle riding has been reported to cause substantial elevations in serum PSA, but this is not a consistent finding. Sexual activity can minimally elevate the PSA (usually in the 0.4 to 0.5 ng/mL range) for approximately 48 to 72 hours after ejaculation.

Emerging concepts regarding PSA testing that may help refine the interpretation of an elevated concentration include PSA density, and PSA velocity of free versus complexed or bound PSA. These modifications would presumably be most useful for prostate cancer screening when the total PSA is between 2.5 and 10.0 ng/mL, the range in which decisions regarding further diagnostic testing are most difficult. To more directly compensate for BPH and prostate size, transrectal ultrasound (TRUS) has been used to measure prostate volume. Serum PSA is then divided by prostate volume to obtain PSA density, with higher PSA values (greater than 0.15 ng/mL/cc) being more suggestive of prostate cancer and lower values more suggestive of BPH. Another approach has been to assess the rate of PSA change over time (the PSA velocity). An elevated serum PSA that continues to rise over time is more likely to reflect prostate cancer than one that is consistently stable. For practical purposes, the clinical usefulness of PSA velocity is in part limited by intra-patient variability in the serum PSA; at least three consecutive measurements should be performed. A longer time over which values are continuously measured can be useful in reducing the general variation in the PSA measurements.

Prostate cancer is associated with a lower percentage of free PSA in the serum compared to PSA values observed in benign conditions. The percentage of free PSA has been used to improve the sensitivity of cancer detection when total PSA is in the normal range (<4 ng/mL) and also to increase the specificity of cancer detection when total PSA is in the "gray zone" (4.1 to 10 ng/mL). In this latter group (PSA between 4.1 and 10 ng/mL) the lower the value of free PSA, the greater the likelihood that an elevated PSA represents cancer rather than BPH. As with PSA, there is no absolute free/total cutoff that can completely differentiate prostate cancer from BPH. The optimal cutoff value for free PSA is unclear and depends upon whether optimal sensitivity or specificity is sought. The higher the cutoff value, the greater the sensitivity but the lower the specificity. Free PSA could be useful for risk stratification in men with prostate cancer. A lower percentage of free-to-total PSA may be associated with a more aggressive form of prostate cancer.

Assays for alpha-1-antichymotrypsin (ACT)-complexed PSA (cPSA) have recently been implemented that could theoretically provide a similar enhanced degree of specificity compared to the free-to-total PSA ratio. Most but not all reports suggest that cPSA outperforms both total PSA and the ratio of free to total PSA with similar sensitivity but a higher specificity. According to one study, for men with total PSA in the diagnostic gray zone (4.0 and 10.0 ng/mL), the use of cPSA alone would have missed only one of the 36 men with cancer who would be diagnosed with prostate cancer using both total PSA and biopsies. Interestingly, free-to-total PSA alone would also have also missed one cancer, but eliminated biopsy in only 20 men compared to 34 men where biopsy could be eliminated using cPSA assay alone. The utility of cPSA in men with a lower total PSA (2 to 4 ng/mL) is under investigation as there are conflicting data as to whether cPSA improves specificity compared with the free-to-total PSA ratio [3]. Complexed PSA has been approved for the monitoring of men with prostatic carcinoma. The utility of complexed PSA for screening is uncertain and is not routinely used at this time in clinical practice on a regular basis.

PSA is initially produced as proPSA and this form can preferentially leak into the blood stream in men with prostate cancer. One specific isoform of proPSA is [−2]proPSA, which is unbound and potentially higher in concentration in men with prostate cancer. Based upon this observation, there has been growing interest in using the ratio of [−2]proPSA-to-free PSA (expressed as percent [−2]proPSA or %[−2]proPSA) for screening of prostate cancer [4]. Percent [−2]proPSA is currently approved by the European Union for prostate cancer detection and is being evaluated in the United States by the Food and Drug Administration (FDA).

13.4 FALSE POSITIVE AND UNEXPECTED PSA RESULTS

False positive PSA test results may be encountered and can cause confusion regarding the diagnosis of prostate cancer. Kilpelainen et al., based on the screening of 61,604 men in Europe, observed 17.8% false positive PSA results. However, men who tested false positive with one PSA screening test were more prone to be diagnosed with prostate cancer in the future [5].

Nevertheless, the major cause of false elevation of PSA is the presence of heterophilic antibody in the serum. In fact the presence of heterophilic antibody in the specimen not only may cause false elevation of PSA, but also false elevation of other tumor markers. Falsely elevated PSA due to interference of heterophilic antibody can result in inappropriate and unnecessary treatment for prostate cancer. Morgan and Tarter commented that human anti-mouse IgG heterophilic antibody, if present in a patient's serum, can interfere with a serum PSA assay. If PSA is detectable after radical prostatectomy and the likelihood of incomplete resection or systematic disease is low, an unexpected PSA result due to the presence of heterophilic antibody must be considered [6].

CASE REPORT

A 58-year-old man without any personal history of familial risk for prostate cancer had a serum PSA level of 83 ng/mL (Access Hybritech PSA, Hybritech Inc., San Diego, CA) and was referred to an urologist. However, digital rectal examination was normal and a prostate biopsy did not indicate presence of prostate cancer. Subsequent PSA analysis 1 and 2 months apart showed a similar result and the patient was started on androgen deprivation therapy with goserelin and bicalutamide. At 3 months the patient was still asymptomatic but his serum PSA remained high in the absence of radiographic evidence of any advanced cancer as suggested by the PSA value. At that time a repeat of the PSA assay by a different method (Immulite PSA assay) was conducted and the value was undetectable (<0.03 ng/mL). The treating physician suspected that the initial PSA results were falsely elevated due to interference, and treating high PSA samples with heterophilic antibody blocking agent resulted in no detectable PSA levels by the original Hybritech assay. This confirmed a falsely elevated PSA result due to the presence of heterophilic antibody. At 1 year after his testosterone level was normalized, his PSA was 1.09 ng/mL and prostatic dynamic MRI showed no sign of tumor in the prostate or in the pelvis [7].

13.5 CANCER ANTIGEN 125 (CARBOHYDRATE ANTIGEN 125: CA-125)

CA-125 (also known as mucin 16 or MUC16) is a glycoprotein. CA-125 in humans is encoded by the *MUC16* gene. CA-125 is used as a tumor marker because CA-125 concentrations may be elevated in the blood of some patients with ovarian cancer, but also in some benign conditions. CA-125 levels in serum are elevated in approximately 50 percent of women with the early stage the disease, and in over 80 percent of women with advanced ovarian cancer. Monitoring CA-125 serum levels is also useful for determining the response of a patient to ovarian cancer therapy as well as for predicting a patient's prognosis after treatment. In general, persistence of high levels of CA-125 during therapy is associated with poor survival rates in patients. Also, an increase in CA-125 levels in a patient during remission is a strong predictor of recurrence of ovarian cancer. The specificity of CA-125 is limited because CA-125 levels are elevated in approximately 1% of healthy women and fluctuate during the menstrual cycle. CA-125 is also increased in a variety of benign and malignant conditions, including:

- Endometriosis.
- Uterine leiomyoma.
- Cirrhosis with or without ascites.
- Pelvic inflammatory disease.
- Cancers of the endometrium, breast, lung, and pancreas.
- Pleural or peritoneal fluid inflammation due to any cause.

13.6 FALSE POSITIVE CA-125

Meigs' syndrome (association of ovarian fibroma, pleural effusion, and ascites) may also cause marked elevation of CA-125. Abnormally high values of both CA-125 and CA-19-9 (false positive) have also been reported in women with benign tumors. Sometimes F(ab')2 fragments of the murine monoclonal antibody OC-125 are administered to patients with ovarian cancer because OC-125 is directed against the CA-125 antigen present on the surface of human ovarian cancers. Exposure to such antibodies may lead to development of an immune response that causes the presence of human anti-mouse monoclonal antibody (HAMA; also broadly termed as heterophilic antibody), which may interfere in an unpredictable manner with the determination of CA-125 using serum specimens in such patients.

Measurable CA-125 concentrations can also be observed in patients without any cancer. CA-125 concentrations are known to rise in patients with severe congestive heart failure, and the elevations correlate with the severity of disease and elevations of a specific marker of heart failure, for example, B-type natriuretic peptide (BNP). In the menstrual phase of the cycle in women, CA-125 values may be elevated, causing false positive test results. CA-125 may also increase after abdominal surgery, chronic obstructive pulmonary disease, active tuberculosis, and lupus erythematosis. During pregnancy CA-125 concentrations increase 10 weeks after gestation and remain high throughout the pregnancy. During the terminal phase of pregnancy, the CA-125 concentration may be as high as twice the upper limit of the reference range.

13.7 ALPHA-FETAL PROTEIN

AFP, sometimes called alpha-1-fetoprotein or alpha-fetoglobulin, is a protein encoded in humans by the *AFP* gene which is located on the q arm of chromosome 4 (4q25). AFP is a major plasma protein produced by the yolk sac and the liver during fetal development and is considered to be the fetal form of albumin. The half-life of AFP is approximately five to seven days.

Following effective cancer therapy, normalization of the serum AFP concentration over 25 to 30 days is indicative of an appropriate decline. However, it is essentially undetectable in the serum in normal men. The upper limit of normal serum AFP concentration is less than $10-15\,\mu g/L$. Many tissues regain the ability to produce this oncofetal protein while undergoing malignant degeneration, but serum AFP concentrations above $10,000\,\mu g/L$ are most commonly observed in patients with non-seminomatous germ cell tumors (NSGCTs) or hepatocellular carcinoma. In men with NSGCTs, AFP is produced by yolk sac (endodermal sinus) tumors, and, less often, by embryonal carcinomas. As with beta-human chorionic gonadotropin (β-hCG), the frequency of elevated serum AFP increases with the advancing clinical stage of the tumor, from $10-20\%$ in men with stage I tumors to $40-60\%$ of those with disseminated NSGCTs. By definition, pure seminomas do not cause an elevated serum AFP. However, molecular studies have demonstrated AFP mRNA in minute quantities in pure seminoma, and several case reports have documented pure seminoma with borderline elevations in serum AFP (10.4 to 16 ng/mL). Higher serum AFP concentrations are considered diagnostic of a non-seminomatous component of the tumor (especially yolk sac elements) or hepatic metastases. If the presence of an elevated serum AFP is confirmed, patients should be treated as if they had an NSGCT.

Serum AFP is the most commonly used marker for diagnosis of hepatocellular carcinoma (HCC). Serum levels of AFP do not correlate well with other clinical features of HCC, such as size, stage, or prognosis. Elevated serum AFP may also be seen in patients without HCC such as acute or chronic viral hepatitis. AFP may be slightly elevated in patients with liver cirrhosis due to chronic hepatitis C infection. A significant rise in serum AFP in a patient with cirrhosis should raise concerns that HCC may have developed. It is generally accepted that serum levels greater than $500\,\mu g/L$ (upper limit of normal in most laboratories is between 10 and $20\,\mu g/L$) in a high-risk patient is diagnostic of HCC. However, HCC is often diagnosed at a lower AFP level in patients undergoing screening. Not all tumors secrete AFP, and serum concentrations could be normal in up to 40 percent of patients with small HCCs. In a study of 357 patients with hepatitis C and without HCC, 23 percent had an AFP $>10.0\,\mu g/L$. Elevated levels were associated with the presence of stage III or IV fibrosis, an elevated international normalized ratio, and an elevated serum aspartate aminotransferase level [8]. AFP levels are, however, normal in the majority of patients with fibrolamellar carcinoma, a variant of HCC. Despite the issues inherent in using AFP for the diagnosis of HCC, it has emerged as an important prognostic marker, especially in patients being considered for liver transplantation. Patients with AFP levels $>1,000\,\mu g/L$ have an extremely high risk of recurrent disease following the transplant, irrespective of the tumor size seen on imaging.

13.8 FALSE POSITIVE AFP

False positive elevations of serum AFP can occur from tumors of the gastrointestinal tract, particularly hepatocellular carcinoma, or from liver damage (e.g. cirrhosis, hepatitis, or drug or alcohol abuse). Lysis of tumor cells during the initiation of chemotherapy may result in a transient increase in serum AFP. Elevated serum AFP occurs in pregnancy with tumors of gonadal origin (both germ cell and non-germ cell) and in a variety of other malignancies, of which gastric cancer is the most common. As expected, heterophilic antibodies, if present in the specimen, can also cause a falsely elevated alpha-fetoprotein concentration.

13.9 CARCINOEMBRYONIC ANTIGEN (CEA)

CEA is a glycoprotein involved in cell adhesion that is normally produced during fetal development. The production of CEA stops before birth. Therefore, it is not usually present in the blood of healthy adults, although levels are raised in heavy smokers. CEA is a glycosyl phosphatidyl inositol (GPI)-cell surface-anchored glycoprotein whose specialized sialofucosylated glycoforms serve as functional colon carcinoma L-selectin and E-selectin ligands, which may be critical to the metastatic dissemination of colon carcinoma cells. It is found in the sera of patients with colorectal carcinoma (CRC), gastric carcinoma, pancreatic carcinoma, lung carcinoma and breast carcinoma. Patients with medullary thyroid carcinoma also have higher levels of CEA compared to healthy individuals (above 2.5 ng/mL). However, a CEA blood test is not reliable for diagnosing cancer or as a screening test for early detection of cancer. Most types of cancer do not produce a high level of CEA. Elevated CEA levels should return to normal after successful surgical resection or within 6 weeks of starting treatment if cancer treatment is successful. However, due to lack of both sensitivity and specificity, serum CEA is not a useful screening tool for CRC. In patients with established disease, the absolute level of the serum CEA correlates with disease burden and is of prognostic value. Furthermore, elevated preoperative levels of CEA should return to baseline after complete resection; residual disease should be suspected if they do not. Serum levels of the tumor marker CEA should be routinely measured preoperatively in patients undergoing potentially curative resections for CRC for two reasons:

- Elevated preoperative CEA levels that do not normalize following surgical resection imply the presence of persistent disease and the need for further evaluation.
- Preoperative CEA values are of prognostic significance.
  CEA levels ≥ 5.0 ng/mL are associated with an adverse impact on survival that is independent of tumor stage.

As a single analyte, serum levels of CEA are neither sufficiently sensitive nor specific to diagnose cholangiocarcinoma. Many conditions other than cholangiocarcinoma can increase serum levels of CEA. Non-cancer-related causes of elevated CEA include gastritis, peptic ulcer disease, diverticulitis, liver disease, chronic obstructive pulmonary disease, diabetes, and any acute or chronic inflammatory state.

13.10 FALSE POSITIVE CEA

As expected, false positive CEA test results can occur due to the presence of heterophilic antibodies in the specimen. However, CEA concentrations can also be elevated in non-neoplastic conditions. Renal failure and fulminant hepatitis can falsely increase CEA values. CEA concentrations may be also elevated in patients receiving hemodialysis. Patients with hypothyroidism may also show elevated levels of CEA correlated with the duration of hypothyroidism. CEA levels may also be raised in some non-neoplastic conditions like ulcerative colitis, pancreatitis, cirrhosis, chronic obstructive pulmonary disease (COPD), Crohn's disease, as well as in smokers.

13.11 CANCER ANTIGEN-19-9

CA-19-9, also called cancer antigen-19-9 or sialylated Lewis (a) antigen, is a tumor marker used primarily in the management of pancreatic cancer. Guidelines from the American Society of Clinical Oncology discourage the use of CA-19-9 as a screening test for cancer, particularly pancreatic cancer, because the test may be falsely negative in many cases, or abnormally elevated in people with no cancer at all (false positive). However, in individuals with pancreatic masses, CA-19-9 can be useful in distinguishing between cancer and other pathologies of the gland. The reported sensitivity and specificity of CA-19-9 for pancreatic cancer are 80% and 90%, respectively; these values are closely related to tumor size. The accuracy of CA-19-9 for identification of patients with small surgically resectable cancers is limited. The specificity of CA-19-9 is limited because CA-19-9 is frequently elevated in patients with cancers other than pancreatic cancer, and various benign pancreaticobiliary disorders. As a result of all of these issues, CA-19-9 is not recommended as a screening test for pancreatic cancer.

The degree of elevation of CA-19-9 (both at initial presentation and in the postoperative setting) is associated with long-term prognosis. Furthermore, in patients who appear to have potentially resectable disease, the magnitude of the CA-19-9 level can also be useful in predicting the presence of radiographically occult metastatic disease. The rates of unresectable disease among all patients with a CA-19-9 level $\geq$130 units/mL versus $<$130 units/mL were

26% and 11%, respectively. Among patients with tumors in the body/tail of the pancreas, more than one-third of those who had a CA-19-9 level $\geq$130 units/mL had unresectable disease.

Serial monitoring of CA-19-9 levels (once every one to three months) is useful for further monitoring of patients after potentially curative surgery and for those who are receiving chemotherapy for advanced disease. Elevated CA-19-9 levels usually precede the radiographic appearance of recurrent disease, but confirmation of disease progression should be pursued with imaging studies and/or biopsy. CA-19-9 can be elevated in many types of gastrointestinal cancer, such as colorectal cancer, esophageal cancer, and hepatocellular carcinoma. Apart from cancer, elevated levels may also occur in pancreatitis, cirrhosis, and diseases of the bile ducts. It can be elevated in people with obstruction of the bile duct. In patients who lack the Lewis antigen (a blood type protein on red blood cells), which is about 10% of the Caucasian population, CA-19-9 is not expressed even in those with large tumors. This is due to deficiency of the fucosyltransferase enzyme that is needed to produce CA-19-9 as well as the Lewis antigen. The use of a combined index of serum CA-19-9 and CEA (CA-19-9 + [CEA $\times$ 40]) has also been proposed for screening of cholangiocarcinoma.

Interference of heterophilic antibodies causing false positive CA-19-9 results has been documented, and usually treating the specimen with heterophilic antibody-blocking agents can eliminate such interference. Patients with acute or chronic pancreatitis may also have elevated levels of CA-19-9. In addition, pulmonary diseases may also elevate CA-19-9 levels. Liver cirrhosis, Crohn's disease, and benign gastrointestinal diseases can also increase CA-19-9 levels.

13.12 β_2-MICROGLOBULIN

Beta-2-microglobulin (β_2-microglobulin) is a component of major histocompatibility complex (MHC) class I molecules that is present on all nucleated cells (excluding red blood cells). In humans, the β_2-microglobulin protein is encoded by the *B2M* gene. For the diagnosis of multiple myeloma, the serum β_2-microglobulin level is one of the prognostic factors incorporated into the International Staging System. The serum β_2-microglobulin level is elevated (>2.7 mg/L) in 75% of patients at the time of diagnosis. Patients with high values have inferior survival. The prognostic value of serum β_2-microglobulin levels in myeloma is probably due to two factors:

- High levels are associated with greater tumor burden.
- High levels are also associated with renal failure, which carries an unfavorable prognosis.

In lymphoma, β_2-microglobulin levels usually correlate with the disease stage and tumor burden in patients with chronic lymphocytic leukemia (CLL); increasing levels are associated with a poorer prognosis. Beta-2-microglobulin may be regulated, at least in part, by exogenous cytokines. The source of these elevated cytokines in CLL is unclear, although IL-6, which inhibits apoptosis in CLL cells, may be released from vascular endothelium. However, β_2-microglobulin levels also rise with worsening renal dysfunction, leading some investigators to suggest a measure of β_2-microglobulin adjusted for the glomerular filtration rate (GFR) [9]. This GFR-adjusted B2M requires validation in prospective confirmatory studies. The plasma β_2-microglobulin concentration is increased in dialyzed patients, with a level ranging from 30 to 50 mg/L, much higher than the normal value of 0.8 to 3.0 mg/L. Infection with the AIDS virus, hepatitis, and active tuberculosis may also elevate levels of β_2-microglobulin.

13.13 HUMAN CHORIONIC GONADOTROPIN (HCG)

Human chorionic gonadotropin (hCG) is a hormone composed of alpha and beta subunits; the beta subunit is specific for hCG (beta-hCG) and provides functional specificity. Beta-hCG is synthesized in large amounts by placental trophoblastic tissue and in much smaller amounts by the hypophysis and other organs such as testicles, liver, and colon. Therefore, elevated levels of beta-hCG are observed during pregnancy, produced by the developing placenta after conception, and later by the placental component syncytiotrophoblast. Laboratory tests for hCG are essentially very sensitive and specific for diagnosis of trophoblast-related conditions, including pregnancy and the gestational trophoblastic diseases. Rarely, very low levels of hCG are detected in the absence of one of these conditions. However, hCG exists in many forms in serum, including intact molecules, beta-hCG, a hyper-glycated form, and other forms such as a C-terminal peptide. Therefore, an assay capable of measuring all forms of hCG is desirable to resolve low values to ensure it is indeed a low value. Although all assays detect regular hCG, they do not necessarily detect all hCG variants. For example, many over-the-counter pregnancy tests do not measure hyperglycosylated hCG, which accounts for most of the total hCG at the time of missed menses. Clinical tests for pregnancy may only detect total hCG levels ≥ 20 mIU/mL. Therefore, when following hCG levels to negative (<1 mIU/mL) in women with gestational trophoblastic disease, it is important to use a sensitive hCG test that detects both regular and other forms of hCG.

At levels of hCG above 500,000 mIU/mL, a "hook effect" can occur, which results in an artifactually low value for hCG (i.e. 1 to 100 mIU/mL). This is because the sensitivity of most hCG tests is set to the pregnancy hCG range (i.e. 27,300 to 233,000 mIU/mL at 8 to 11 weeks of gestation); therefore, when an extremely high hCG concentration is present, both the capture and tracer antibodies used in assays become saturated, which prevents the binding of the two to create a sandwich. For this reason, a suspected diagnosis of gestational trophoblastic disease must be communicated to the laboratory so that the hCG assay can also be performed at 1:1,000 dilution to eliminate any hook effect.

Molar pregnancy (hydatidiform mole) is a non-malignant tumor that arises from the trophoblast in early pregnancy after an embryo fails to develop. Molar pregnancy is known to produce high amounts of beta-hCG. However, in a urine pregnancy test, a false negative beta-hCG result may be observed due to the hook effect because very large amounts of beta-hCG may be present. Dilution of the specimen is essential to further investigate such false negative results. Another approach is to perform a serum beta-hCG test.

CASE REPORT

A 47-year-old woman presented to the emergency department with a two-month history of abdominal bloating, mild epigastric discomfort, and loss of appetite. She had a regular menstrual cycle, but her last menstrual bleeding was one month before her presentation in the emergency room; she did have vaginal spotting for the previous month. On examination, a large non-tender abdominal mass corresponding to 22 weeks of gestation was noticed, and she also had moderate vaginal bleeding during examination. The urine pregnancy test was negative and her hemoglobin was 7.8 g/dL.

Therefore, based on negative beta-hCG in her urine, a diagnosis of molar pregnancy was missed. Ultrasonography revealed an enlarged uterus with a heterogenous mass. At that time a serum beta-hCG test was performed and the value was > 1,000,000 U/L. Ultrasonography was repeated and a provisional diagnosis of molar pregnancy was made. An emergency evacuation of her uterus was made and the patient was discharged two days postoperatively. The authors concluded that negative urine beta-hCG was due to the hook effect [10].

13.14 CAUSES AND EVALUATION OF PERSISTENT LOW LEVELS OF HCG

Determining the clinical value of a low level of hCG can be challenging. It is important to determine if the hCG represents an actual early pregnancy (intrauterine or ectopic), active gestational trophoblastic disease (complete or partial mole, invasive mole, choriocarcinoma), quiescent gestational trophoblastic disease, a laboratory false positive (also called phantom hCG), or a physiologic artifact (pituitary hCG). For example, a false positive hCG test result or pituitary hCG is commonly found in women who also have a history of gestational trophoblastic disease.

Unless a tumor is evident, it is essential to exclude these possibilities before initiating chemotherapy for assumed persistence of disease. Persistent low-level positive hCG results can be defined as hCG levels varying by no more than two-fold over at least a three-month period in the absence of a tumor on imaging studies.

13.15 FALSE POSITIVE HCG

The capture and tracer antibodies used for hCG testing may be goat, sheep, or rabbit polyclonal antibodies, or mouse, goat, or sheep monoclonal antibodies. Humans extensively exposed to animals or certain animal by-products can develop human antibodies against animal antibodies that are collectively called heterophilic antibodies. Human antimouse antibody (HAMA) is a common example of a heterophilic antibody. Individuals with recent exposure to mononucleosis are prone to develop heterophilic antibodies. False positive hCG tests due to the presence of heterophilic antibody in the serum specimen have been well documented in the literature. Such false positive results in the absence of pregnancy have led to many men and women misdiagnosed with cancer, confusion and misunderstanding, and needless surgery and chemotherapy. Because heterophilic antibodies are found mainly in serum, plasma, or whole blood (but not in urine), such interference is absent in analysis of urine specimens for the same analyte. This gives an excellent way to detect the interference for analytes that may be present in both matrices. Although among tumor markers serum hCG assay is mostly affected by the presence of heterophilic antibody, false positive test results can occur with other tumor markers, including PSA, CA-125, CA-19-9, CEA, alpha-fetoprotein, and even β_2-microglobulin. In fact, interference of heterophilic antibody is the major problem in the assay of various cancer markers. Even the IgM lambda antibody to *Escherichia coli* can produce false positive test results with the determination of various tumor markers (as well as troponin I). The prevalence of heterophilic antibody in the general population is difficult to estimate as published literature reports indicate prevalence of heterophilic antibody from 1 to 11.7% [11]. There are two main methods for identifying false positive hCG:

- The most readily available approach is to show the absence of hCG in the patient's urine.
- A second useful way of identifying a false positive serum hCG result is to send the serum to two laboratories using different commercial assays. If the assay results vary greatly or are negative in one or both alternative tests, then a false positive hCG can be presumed.

CASE REPORT

A 44-year-old HIV-positive man presented with a painless swelling of his left testicle. He underwent left radical orchiectomy for a pathological stage T1 non-seminomatous germ cell tumor (NSGCT). However, after the procedure, his serum hCG was persistently elevated and he went through four cycles of chemotherapy with etoposide and cisplatin. Despite chemotherapy, his serum hCG values did not return to normal. However, further investigation revealed that the patient was cancer-free. Suspecting false positive serum hCG levels, the authors re-analyzed the sample after adding a heterophilic antibody blocking agent. Serum hCG levels became undetectable, which indicated the presence of a heterophilic antibody in the serum specimen that had caused a false positive serum hCG. Other tumor markers, including alpha-fetoprotein, were not elevated [12].

KEY POINTS

- Tumor markers can be used for one of five purposes, including (1) screening a healthy population or a high-risk population for the probable presence of cancer, (2) diagnosis of cancer or of a specific type of cancer, (3) evaluating prognosis in a patient, (4) predicting potential response of a patient to therapy, and (5) monitoring recovery of a patient during surgery, radiation, or chemotherapy.
- Commonly measured tumor markers in clinical laboratories include: prostate-specific antigen (PSA) for prostate cancer, human chorionic gonadotropin (hCG) for gestational trophoblastic tumors and some germ cell tumors, alpha-fetoprotein (AFP) for certain germ cell tumors and HCC, carbohydrate antigen or cancer antigen-125 (CA-125) for ovarian cancer, carbohydrate antigen-19-9 (CA-19-9) for pancreatic and gastro-intestinal cancers, and carcinoembryonic antigen (CEA) for colon and rectal cancers.
- PSA is expressed by both normal and neoplastic prostate tissue. Under normal conditions, PSA is produced as a proantigen (proPSA) by the secretory cells that line the prostate glands and secrete into the lumen (where the propeptide moiety is removed to generate active PSA). The active PSA can then undergo proteolysis to generate inactive PSA, of which a small portion then enters the bloodstream and circulates in an unbound state (free PSA). Alternatively, active PSA can diffuse directly into the circulation where it is rapidly bound to protease inhibitors, including alpha-1-antichymotrypsin (ACT) and alpha-2-macroglobulin.
- In men with a normal prostate, the majority of free PSA in the serum reflects the mature protein that has been inactivated by internal proteolytic cleavage. In contrast, free PSA is relatively decreased in patients with prostate cancer. Thus, the percentage of free or unbound PSA is lower in the serum of men with prostate cancer. This finding has been used in the use of the ratio of free-to-total PSA and complexed PSA (cPSA) as a means of distinguishing between prostate cancer and BPH as a cause of an elevated PSA. Causes of elevated PSA include benign prostatic hyperplasia (BPH), prostate cancer, prostatic inflammation/infection, and perineal trauma.

- Emerging concepts regarding PSA testing that may help refine the interpretation of an elevated concentration include PSA density, PSA velocity, and free versus complexed or bound PSA.
- These modifications would presumably be most useful for prostate cancer screening when the total PSA is between 2.5 and 10.0 ng/mL, the range in which decisions regarding further diagnostic testing are most difficult.
- The major cause of false elevation of PSA is the presence of heterophilic antibody in the serum.
- CA-125 concentrations may be elevated in the blood of some patients with specific types of cancers (such as ovarian cancer). CA-125 is also increased in a variety of benign and malignant conditions, including endometriosis, uterine leiomyoma, cirrhosis with or without ascites, pelvic inflammatory disease, cancers of the endometrium, breast, lung, and pancreas, and pleural or peritoneal fluid inflammation due to any cause.
- Serum AFP is the most commonly used marker for diagnosis of hepatocellular carcinoma (HCC). Serum levels of AFP do not correlate well with other clinical features of HCC, such as size, stage, or prognosis. Elevated serum AFP may also be seen in patients without HCC, such as acute or chronic viral hepatitis. AFP may be slightly elevated in patients with liver cirrhosis due to chronic hepatitis C infection. A significant rise in serum AFP in a patient with cirrhosis should raise concerns that HCC may have developed. It is generally accepted that serum levels greater than 500 µg/L (the upper limit of normal in most laboratories is between 10 and 20 µg/L) in a high-risk patient are diagnostic of HCC. However, HCC is often diagnosed at a lower AFP level in patients undergoing screening.
- Carcinoembryonic antigen (CEA) is found in the sera of patients with colorectal carcinoma (CRC), gastric carcinoma, pancreatic carcinoma, lung carcinoma, and breast carcinoma. Patients with medullary thyroid carcinoma also have higher levels of CEA compared to healthy individuals (>2.5 ng/mL). However, a CEA blood test is not reliable for diagnosing cancer or as a screening test for early detection of cancer. Serum levels of CEA should be routinely measured preoperatively in patients undergoing potentially curative resections for CRC for two reasons:
  - Elevated preoperative CEA levels that do not normalize following surgical resection imply the presence of persistent disease and the need for further evaluation.
  - Preoperative CEA values are of prognostic significance. CEA levels ≥ 5.0 ng/mL are associated with an adverse impact on survival that is independent of tumor stage
- CA-19-9, also called cancer antigen-19-9 or sialylated Lewis (a) antigen, is used primarily in the management of pancreatic cancer. Guidelines from the American Society of Clinical Oncology discourage the use of CA-19-9 as a screening test for cancer, particularly pancreatic cancer. However, in individuals with pancreatic

masses, CA-19-9 can be useful in distinguishing between cancer and other pathologies of the gland.

- Laboratory tests for hCG are essentially very sensitive and specific for diagnosis of trophoblast-related conditions, including pregnancy and the gestational trophoblastic diseases.

- At hCG levels above 500,000 mIU/mL, a "hook effect" can occur that results in an artifactually low value for hCG (i.e. 1 to 100 mIU/mL). This is because the sensitivity of most hCG tests is set to the pregnancy hCG range (i.e. 27,300 to 233,000 mIU/mL at 8 to 11 weeks of gestation). Therefore, when an extremely high hCG concentration is present, both the capture and tracer antibodies used in assays become saturated, thus preventing the binding of the two to create a sandwich. For this reason, a suspected diagnosis of gestational trophoblastic disease must be communicated to the laboratory so that the hCG assay can also be performed at a 1:1,000 dilution in order to get a true hCG value.

REFERENCES

[1] Lilja H, Christensson A, Dahlén U, Matikainen MT, et al. T. Prostate-specific antigen in serum occurs predominantly in complex with alpha 1-antichymotrypsin. Clin Chem 1991;37:1618−25.

[2] Hudson MA, Bahnson RR, Catalona WJ. Clinical use of prostate specific antigen in patients with prostate cancer. J Urol 1989;142(4):1011−7.

[3] Tanguay S, Bégin LR, Elhilali MM, Behlouli H, et al. Comparative evaluation of total PSA, free/total PSA, and complexed PSA in prostate cancer detection. Urology 2002;59:261−5.

[4] Sokoll LJ, Sanda MG, Feng Z, Kagan J, et al. A prospective, multicenter, National Cancer Institute Early Detection Research Network study of [-2]proPSA: improving prostate cancer detection and correlating with cancer aggressiveness. Cancer Epidemiol Biomarkers Prev 2010;19:1193−200.

[5] Kilpelainen TP, Tammela TL, Roobol M, Hugosson J, et al. False positive screening results in the European randomized study of screening for prostate cancer. Eur J Cancer 2011;47:2698−705.

[6] Morgan BR, Tarter TH. Serum heterophilic antibodies interfere with prostate specific antigen test and result in over treatment in a patient with prostate cancer. J Urol 2001;166:2311−2.

[7] Henry N, Sebe P, Cussenot O. Inappropriate treatment of prostate cancer caused by heterophilic antibody interference. Nat Clin Pract Urol 2009;6:164−7.

[8] Hu KQ, Kyulo NL, Lim N, Elhazin B, et al. Clinical significance of elevated alphafetoprotein (AFP) in patients with chronic hepatitis C, but not hepatocellular carcinoma. Am J Gastroenterol 2004;99(5):860−4.

[9] Howaizi M, Abboura M, Krespine C, Sbai-Idrissi MS, et al. A new case of CA-19-9 elevation: heavy tea consumption. Gut 2003;52:913−4.

[10] Pang YP, Rajesh H, Tan LK. Molar pregnancy with false negative urine hCG; The hook effect. Singapore Med J 2010;51:e58−61.

[11] Koshida S, Asanuma K, Kuribayashi K, Goto M, et al. Prevalence of human anti-mouse antibodies (HAMAs) in routine examination. Clin Chim Acta 2010;411:391–4.

[12] Gallagher DJ, Riches J, Bajorin DF. False positive elevation of human chorionic gonadotropin in a patient with testicular cancer. Nat Rev Urol 2010;7:230–3.

CHAPTER 14

Therapeutic Drug Monitoring

14.1 WHAT IS THERAPEUTIC DRUG MONITORING?

There are over 6,000 prescription and non-prescription (over-the-counter) drugs available for clinical use in the United States. Most drugs have a wide therapeutic index (the difference in the therapeutic and toxic drug levels) and do not require therapeutic drug monitoring. For example, acetaminophen has a therapeutic range between 5 and 20 µg/mL, and toxicity is encountered at a concentration of 150 µg/mL and above. Therefore, acetaminophen therapy does not require therapeutic drug monitoring. In contrast, digoxin has a therapeutic range of 0.8–1.8 ng/mL, but toxicity can be observed at a level of 2.0 ng/mL, and even sometimes at a digoxin concentration within the therapeutic range. Therefore, digoxin therapy requires routine monitoring. The highlights of the current state of therapeutic drug monitoring practice include:

- Approximately 20–26 prescription drugs are frequently monitored in the majority of hospital-based laboratories because these drugs have narrow therapeutic indexes.
- In addition, 25–30 drugs are subjected to therapeutic drug monitoring less frequently. Usually monitoring of these drugs is offered in large academic medical centers and national reference laboratories.
- The goal of therapeutic drug monitoring is to optimize pharmacological responses of a drug while avoiding adverse effects.
- For most drugs, serum or plasma is used for monitoring and measuring the trough level of the drug (15–30 min prior to next dose), except immunosuppressants (cyclosporine, tacrolimus, sirolimus, and everolimus), where trough drug levels are measured in whole blood. Interestingly, another immunosuppressant drug, mycophenolic acid, is monitored in serum or plasma.
- For aminoglycosides and vancomycin, both peak and trough concentrations in serum or plasma can be measured.

CONTENTS

249

A. Dasgupta and A. Wahed: Clinical Chemistry, Immunology and Laboratory Quality Control
DOI: http://dx.doi.org/10.1016/B978-0-12-407821-5.00014-0

14.2 DRUGS THAT REQUIRE THERAPEUTIC DRUG MONITORING

As mentioned earlier, only a small fraction of prescription drugs require therapeutic drug monitoring because for most prescription drugs there is a wider difference between therapeutic and toxic concentrations. The characteristics of a drug where therapeutic drug monitoring is beneficial include:

- Difficulty in interpreting therapeutic range or a low toxicity of a drug based on clinical evidence alone.
- Narrow therapeutic range.
- Toxicity of a drug may lead to hospitalization, irreversible organ damage, and even death, but an adverse drug reaction can be avoided by therapeutic drug monitoring.
- There is a correlation between serum or whole blood concentration of the drug and its therapeutic response or toxicity.

14.3 FREE VERSUS TOTAL DRUG MONITORING

A drug may be bound to a serum protein, and only the unbound (free) drug is pharmacologically active. Protein binding of drugs may vary from 0% (not bound to protein) to over 99%. Usually when a drug concentration is measured in serum/plasma, it is the total drug concentration (free drug + protein-bound drug). However, if a drug is <80% protein bound, total concentration can adequately predict free drug concentration and direct measurement of free drug may not be necessary. However, for a strongly protein-bound drug (>80%), direct measurement of free drug may be clinically beneficial in patients with hypoalbuminemia, uremia, or liver disease. Important issues regarding free drug monitoring include:

- Usually free phenytoin is the most commonly measured drug in clinical laboratories.
- Measurement of other strongly protein-bound anticonvulsants such as valproic acid and carbamazepine may be useful.
- Free mycophenolic acid concentration (an immunosuppressant) may be useful in a uremic patient.

14.4 THERAPEUTIC DRUG MONITORING BENEFITS

In general, many drugs are used as a prophylactic to prevent clinical symptoms, and non-compliance has serious clinical consequences. Therapeutic drug monitoring is very helpful in identifying such non-compliant patients. Mattson et al. noted that zero, sub-therapeutic levels, or variable drug levels, are indicators of non-compliance to a medication where therapeutic drug monitoring is available [1]. Chandra et al. noted that poor patient compliance is one of the major causes of non-responsiveness to anti-epileptic drug therapy [2]. Therapeutic drug monitoring can greatly reduce the chances of treatment failure by personalization of drug dosage based on drug levels in serum/plasma or whole blood. The benefits of therapeutic drug monitoring include:

- Identification of non-compliance.
- Personalization of drug dosage for maximum therapeutic benefit and avoidance of drug toxicity.
- Identification of clinically significant drug–drug, drug–food, and drug–herb interactions.
- Identification of a non-responder to a drug.

If a patient does not respond to a drug despite the drug level being in the therapeutic range, it may be an indication that the patient is a non-responder to that drug. A different drug choice must be made to treat the patient.

14.5 BASIC PHARMACOKINETICS

When a drug is administered orally, it undergoes several steps in the body that determine the concentration of that drug in serum/plasma or whole blood. These steps include:

- Liberation: The release of a drug from the dosage form (tablet, capsule, extended release formulation).
- Absorption: Movement of drug from the site of administration (for drugs taken orally) to blood circulation. Many factors affect this stage, including gastric pH, presence of food particles, as well as the efflux mechanism (if present) in the gut. First-pass metabolism plays an important role in determining bioavailability of a drug given orally.
- Distribution and Protein Binding: Movement of a drug from the blood circulation to tissues/target organs. Drugs may also be bound to serum proteins ranging from zero protein binding to 99% protein binding.
- Metabolism: Chemical transformation of a drug to the active and inactive metabolites. The liver is responsible for metabolism of many drugs, although drugs may also be metabolized by a non-hepatic path or are subjected to minimal metabolism.
- Excretion: Elimination of the drug from the body via renal, biliary, or pulmonary mechanism.

Liberation of a drug after oral administration depends on the formulation of the drug. Immediate release formulation releases the drugs at once from the dosage form when administered, while the same drug may also be available in sustained release formulation. Absorption of a drug depends on the route of administration. Generally, an oral administration is the route of choice, but under certain circumstances (nausea, vomiting, and convulsion) rectal administration may present a practical alternative for delivering anticonvulsants, non-narcotic and narcotic analgesics, theophylline, antibacterial, and antiemetic agents. This route can also be used for inducing anesthesia in children. Although the rate of drug absorption is usually lower after rectal administration compared to oral administration, for certain drugs, rectal absorption is higher than oral absorption due to avoidance of the hepatic first-pass metabolism These drugs include lidocaine, morphine, metoclopramide, ergotamine, and propranolol. Local irritation is a possible complication of rectal drug delivery [3]. When a drug is administered by direct injection, it enters the blood circulation immediately.

When a drug enters the blood circulation, it is distributed throughout the body into various tissues, and the pharmacokinetic parameter is called the volume of distribution (V_d). This is the hypothetical volume to account for all drugs in the body and is also called the apparent volume of distribution, where V_d is the Dose/Plasma concentration of drug.

The amount of a drug that interacts with the receptor or target site is usually a small fraction of the total drug administered. Muscle and fat tissues may serve as a reservoir for lipophilic drugs. For neurotherapeutics, penetration of blood–brain barrier is essential. Drugs usually undergo chemical transformation (metabolism) before elimination. Drug metabolism can occur in any tissue, including the blood. For example, plasma butylcholinesterase metabolizes drugs such as succinylcholine. The role of metabolism is to convert lipophilic non-polar molecules to water-soluble polar compounds for excretion in urine. Many drugs are metabolized in the liver in two phases by various enzymes. Major steps in drug metabolism involve:

- Phase I: The step that involves manipulation of a functional group of a drug molecule in order to make the molecule more polar by oxidation or reduction of a functional group in the drug molecule or hydrolysis. The cytochrome P-450 mixed-function family of enzymes (CYP) plays a major role in Phase I reactions, although other enzymes may also be involved.
- Phase II: This step may involve acetylation (adding acetate group to the drug molecule), sulfation (adding inorganic sulfate), methylation (adding methyl group), amino acid conjugation, or glucuronidation (adding sugar such as glucuronic acid) in order to increase the polarity of the drug metabolite.

The cytochrome P-450 mixed-function oxidase family of enzymes (CYP) plays a major role in the Phase I metabolism of many drugs. These enzymes are found in abundance in the liver, but also may be found in other organs such as the gut. Nicotinamide adenine dinucleotide phosphate (NADPH) is a required cofactor for CYP-mediated biotransformation; oxygen serves as a substrate. At present, 57 human genes are known to encode CYP isoforms. Of these, at least 15 are associated with xenobiotic metabolism. CYP isoenzymes are named according to sequence homology: amino acid sequence similarity >40% assigns the numeric family (e.g. CYP1, CYP2); >55% similarity determines the subfamily letter (e.g. CYP2C, CYP2D); isoforms with >97% similarities are given an additional number (e.g. CYP2C9, CYP2C19) to distinguish them. The major CYP isoforms responsible for metabolism of drugs include CYP1A2, CYP2B6, CYP2C9, CYP2C19, CYP2D6, CYP2E1, and CYP3A4/CYP3A5. However, CYP3A4 is the predominant isoform of the CYP family (almost 30%) and is usually responsible for metabolism of approximately 37% of drugs. In addition to CYP, other enzymes are also involved in Phase I metabolism.

One enzyme that plays a vital role in Phase II metabolism is uridine-5-phosphate glucuronyl transferase (UDP-glucuronyl transferase). This enzyme is responsible for conjugation of glucuronic acid with the drug molecule in Phase II metabolism, thus inactivating the drug. This enzyme is mostly found in liver, but may also be present in other organs. Major enzymes involved in drug metabolism are summarized in Table 14.1. Genetic predisposition can determine the activities of such enzymes that ultimately affect drug metabolism. This is an important topic known as pharmacogenomics. Please see Chapter 20 for an introduction to this important subject. The half-life of a

Table 14.1 Major Enzymes Involved in Drug Metabolism

Reaction Type	Phase	Name of Enzyme
Oxidation	Phase I	Cytochrome P-450
		Alcohol dehydrogenase
		Aldehyde dehydrogenase
		Monoamine oxidase
Reduction	Phase I	Various reductase
Hydrolysis	Phase I	Butylcholinesterase, epoxide hydrolase, amidases
Glucuronidation	Phase II	Glucuronosyltransferase
Acetylation	Phase II	N-Acetyltransferase
Methylation	Phase II	Methyltransferase
Amino acid conjugation	Phase II	Glutathione transferase
Sulfation	Phase II	Sulfotransferase

drug is the time required for the serum concentration to be reduced by 50%. Key points regarding steady state concentration of a drug include:

- With repeat administered doses, the steady state is reached after 5 half-lives.
- Therapeutic drug monitoring is recommended when a drug reaches a steady state.

Half-life of a drug can be calculated from the elimination rate constant (K) of a drug: 0.693/K.

The elimination rate constant can be easily calculated from the serum concentrations of a drug at two different time points using the formula where Ct_1 is the concentration of drug at a time point t_1, and Ct_2 is the concentration of the same drug at a later time point t_2 (Equation 14.1):

$$K = \frac{\ln Ct_1 - \ln Ct_2}{t_2 - t_1} \tag{14.1}$$

A drug may also undergo extensive metabolism before fully entering the circulation. This process is called first-pass metabolism. If a drug undergoes significant first-pass metabolism, then the drug should not be delivered orally (such as lidocaine). Renal excretion is a major pathway for the elimination of drugs and their metabolites. Drugs may also be excreted via other routes, such as biliary excretion. A drug excreted in bile may also be reabsorbed from the gastrointestinal tract. A drug conjugate can also be hydrolyzed by gut bacteria, thus liberating the original drug (which can then return to the blood circulation). Enterohepatic circulation can prolong the effects of a drug. Cholestatic disease, where normal bile flow is reduced, may reduce bile clearance of the drug, thus causing drug toxicity.

14.6 EFFECT OF GENDER AND PREGNANCY ON DRUG METABOLISM AND DISPOSITION

Men and women can show differences in response to certain drugs. In addition, pregnancy can also significantly alter the metabolism and disposition of certain drugs. Gender differences affect bioavailability, distribution, metabolism, and elimination of drugs due to variations between men and women in body weight, blood volume, gastric emptying time, drug–protein binding, activities of drug metabolizing enzymes, drug transporter function, and excretion activity. Other important gender differences in drug metabolism include:

- Hepatic metabolism of drugs by Phase I (via CYP1A2 and CYP2E1) and Phase II (by glucuronyl transferase, methyltransferases, and

dehydrogenases) reactions appears to be faster in males than females, although metabolism of drugs by other enzymes appears to be the same.
- Women may have higher activity of CYP3A4.

In general, women are also more susceptible to adverse effects of drugs than men. Women are at increased risk of QT prolongation with many antiarrhythmic drugs, which can lead to critical conditions such as torsade de pointes; this is compared to men even at the same levels of serum drug concentrations.

Drug therapy in pregnant women usually focuses on potential teratogenic effects of a drug, and therapeutic drug monitoring during pregnancy aims to improve individual dosage improvement, taking into account pregnancy-related changes in drug disposition. Gastrointestinal absorption and bioavailability of many drugs vary in pregnancy due to changes in gastric secretion and small intestine motility. Elevated concentrations of various hormones in pregnancy, such as estrogen, progesterone, placental growth hormone, and prolactin, could be related to altered drug metabolism observed in pregnant women. The renal excretion of unchanged drugs is increased in pregnancy. In general, dosage adjustments are required for anticonvulsants, lithium, digoxin, certain beta-blockers, ampicillin, cefuroxime, and certain antidepressants in pregnant women [4].

14.7 EFFECT OF AGE ON DRUG METABOLISM AND DISPOSITION

In the fetus, CYP3A7 is the major hepatic enzyme, but CYP3A5 may also be present in significant levels in half of all children. However, in adults CYP3A4 is the major functional cytochrome P-450 enzyme responsible for metabolism of many drugs. CYP1A1 is also present during organogenesis while CYP2E1 may be present in some second-trimester fetuses. After birth, hepatic CYP2D6, CYP2C8/9, and CYP2C18/19 are activated. CYP1A2 becomes active during the fourth to fifth months after birth [5]. Neonates and infants have increased total body water-to-body fat ratio compared to adults, whereas the reverse is observed in elderly people. These factors may affect volume of distribution of drugs depending on their lipophilic character. Moreover, altered plasma binding of drugs may be observed in both neonates and some elderly people due to low albumin, thus increasing the fraction of pharmacologically active free drug. General features of drug metabolizing capacity as a function of age include:

- Neonates and infants (0−4 months) may metabolize drugs slower than adults because of the lack of mature (fully functioning) drug metabolizing enzymes.

- Renal function at the time of birth is reduced by more than 50% of the adult value, but then increases rapidly in the first two to three years of life. However, activities of drug metabolizing enzymes are higher in children than adults. Therefore, children may need a higher per-kilogram dosage of a drug than an adult.
- Activities of drug metabolizing enzymes decrease with advanced age. In addition, renal clearance of drugs may also start to decline with age. Therefore, careful adjustment of dosage is needed to treat elderly patients.
- Elderly patients (>70 years of age) may also have lower albumin. Therefore, protein binding of strongly protein-bound drugs may be impaired.

14.8 DRUG METABOLISM AND DISPOSITION IN UREMIA

Renal disease causes impairment in the clearance of many drugs by the kidney. Correlations have been established between creatinine clearance and the clearance of digoxin, lithium, procainamide, aminoglycoside, and several other drugs, but creatinine clearance does not always predict renal excretion of all drugs. Moreover, elderly patients can have unrecognized renal impairment, and caution should be exercised when medications are prescribed to elderly patients. Serum creatinine remains normal until glomerular filtration rate (GFR) has fallen by at least 50%. Nearly half of the older patients have normal serum creatinine but reduced renal function. Dose adjustments based on renal function are recommended for many medications in elderly patients, even for medications that exhibit large therapeutic windows [6]. Other characteristics of drug metabolism and disposition in elderly patients include:

- Patients with chronic renal failure show reduced activities of cytochrome P-450 enzymes. Therefore, metabolism of many drugs mediated by these enzymes is significantly reduced.
- Chronic renal failure can also significantly reduce non-renal clearance of many drugs, including drugs that are metabolized by Phase II reactions and drug transporter proteins, such as P-glycoprotein and organic anion transporting polypeptide.
- Renal disease also causes impairment of drug–protein binding because uremic toxins compete with drugs for binding to albumin. In addition, uremic patients may also have hypoalbuminemia. Such interaction leads to increases in the pharmacologically active free drug concentration, especially for classical anticonvulsants such as phenytoin, carbamazepine, and valproic acid. Therefore, monitoring free phenytoin, free valproic acid, and, to some extent, free carbamazepine, is recommended in uremic patients in order to avoid drug toxicity.

14.9 DRUG METABOLISM AND DISPOSITION IN LIVER DISEASE

Liver dysfunction not only reduces the clearance of a drug metabolized through hepatic enzymes or biliary mechanisms, but also affects plasma protein binding due to reduced synthesis of albumin and other drug-binding proteins. Even mild-to-moderate hepatic disease may cause an unpredictable effect on drug metabolism. Portal-systemic shunting present in patients with advanced liver cirrhosis can cause a significant reduction in first-pass metabolism of high-extraction drugs, thus increasing bioavailability as well as the risk of drug overdose and toxicity. Important points to remember regarding drug metabolism and disposition in patients with liver disease include:

- Although the Phase I reaction involving cytochrome P-450 enzymes may be significantly impaired in liver disease, the Phase II reaction (glucuronidation) seems to be unaffected.
- Patients with liver disease often suffer from hypoalbuminemia. Therefore, protein bindings of strongly protein-bound drugs are impaired and cause elevation of pharmacologically active free fraction of the drug.
- Non-alcoholic fatty liver disease is the most common chronic liver disease. This type of liver disease also affects the activity of drug-metabolizing enzymes in the liver with the potential to produce adverse drug reactions from the standard dosage.
- Mild to moderate hepatitis infection may also decrease drug clearance.

14.10 EFFECT OF CARDIOVASCULAR DISEASE ON DRUG METABOLISM AND DISPOSITION

Cardiac failure is often associated with disturbances in cardiac output, and can influence the extent and pattern of tissue perfusion, sodium and water metabolism, and gastrointestinal motility that eventually may affect absorption and disposition of many drugs. Hepatic elimination of drugs via oxidative Phase I metabolism is impaired in patients with congestive heart failure due to decreased blood supply in the liver. Theophylline metabolism is reduced in patients with severe cardiac failure, and dose reduction is strongly recommended. Digoxin clearance is also decreased. Quinidine plasma levels may also be high in these patients due to a lower volume of distribution [7]. Therefore, therapeutic drug monitoring is crucial in avoiding drug toxicity in these patients.

14.11 THYROID DYSFUNCTION AND DRUG METABOLISM

Patients with thyroid disease may have an altered drug disposition because thyroxine is a potent activator of the cytochrome P-450 enzyme system. The key points on the effect of thyroid dysfunction on drug metabolisms are listed below:

- Drug metabolism is increased in patients with hyperthyroidism due to excessive levels of thyroxine.
- Drug metabolism is decreased in patients with hypothyroidism due to lower levels of thyroxine.
- Amiodarone is an antiarrhythmic drug associated with thyroid dysfunction because, due to high iodine content, it inhibits 5-deiodinase activity. Screening of thyroid disease before amiodarone therapy and periodic monitoring of thyroid function are recommended for patients treated with amiodarone.

CASE REPORT

A 48-year-old woman presented to the emergency department with a one-day history of dysarthria, visual disturbances, inco-ordination, and difficulty in mobilizing. She had been suffering from epilepsy for the past 27 years and was taking phenytoin (400 mg/day) and carbamazepine 500 mg twice a day. She was also diagnosed with hypothyroidism and was stable with 250 micrograms of thyroxine a day for the past 4 years. Her blood level of phenytoin was 42.9 µg/mL on admission (therapeutic: 10−20 µg/ml), but her carbamazepine level was within normal range at 6.9 µg/mL (therapeutic: 4−12 µg/mL). Her TSH was highly elevated at 139.72 mIU/L (normal: 0.4−4.0 mIU/L), indicating severe hypothyroidism. Her free thyroxine level was <0.4 ng/dL (normal: 9.8−2.4 ng/dL). A diagnosis of phenytoin toxicity due to hypothyroidism was made. She was treated with thyroxine and was discharged after 12 days. At that time her phenytoin level was therapeutic (11.6 µg/mL) and she was back to a euthyroid stage. At a three-week follow-up, her TSH (1.42 mIU/L) and FT4 (1.31 ng/dL) were both normal, and her phenytoin level was also within therapeutic range (17.9 µg/mL) [8].

14.12 EFFECT OF FOOD, ALCOHOL CONSUMPTION, AND SMOKING ON DRUG DISPOSITION

Drug−food interactions can be pharmacokinetic or pharmacodynamic in nature. Certain foods can affect absorption of certain drugs and can also alter the activity of enzymes that metabolize drugs (especially CYP3A4). It has been documented that the intake of charcoal-broiled food or cruciferous vegetables induces the metabolism of multiple drugs. The most important interaction between fruit juice and a drug involves consumption of grapefruit juice. It was reported in 1991 that a single glass of grapefruit juice caused a two- to three-fold increase in the plasma concentration of felodipine, a cal-cium channel blocker, after oral intake of a 5-mg tablet, but a similar amount of orange juice showed no effect [9]. Subsequent investigations demonstrated that the pharmacokinetics of approximately 40 other drugs are also affected by intake of grapefruit juice [10]. The main mechanism for enhanced bioavailability of drugs after intake of grapefruit juice is as follows:

- Furanocoumarins found in grapefruit juice inhibit CYP3A4 in the small intestine, thus inhibiting the metabolism of drugs in the small intestine and increasing the concentration of available drugs. Grapefruit juice does not inhibit liver CYP3A4. Therefore, if the drug is injected, no change in pharmacokinetics is observed.
- Grapefruit juice also inhibits P-glycoprotein, thus inhibiting its drug efflux metabolism, which indirectly increases the bioavailability of a drug that is a substrate for P-glycoprotein.
- Common drugs that interact with grapefruit juice include alprazolam, carbamazepine, cyclosporine, erythromycin, methadone, quinidine, simvastatin, and tacrolimus.

There are two types of interactions between alcohol and a drug: pharmacokinetic and pharmacodynamic. Pharmacokinetic interactions occur when alcohol interferes with the hepatic metabolism of a drug. Pharmacodynamic interactions occur when alcohol enhances the effect of a drug, particularly in the central nervous system. In this type of interaction, alcohol alters the effect of a drug without changing its concentration in the blood. The package insert of many antibiotics and other drugs states that the medication should not be taken with alcohol due to drug–alcohol interactions. Fatal toxicity can occur from alcohol and drug overdoses due to pharmacodynamic interactions. In a Finnish study, it was found that median amitriptyline and propoxyphene concentrations were lower in alcohol-related fatal cases compared to cases where no alcohol was involved. The authors concluded that when alcohol was present, a relatively small overdose of a drug could cause fatality [11].

Approximately 4,800 compounds are found in tobacco smoke, including nicotine and carcinogenic compounds such as polycyclic aromatic hydrocarbons (PAHs) and N-nitroso amines. PAHs induce CYP1A1, CYP1A2, and possibly CYP2E1, and may also induce Phase II metabolism. Key points regarding the effect of smoking on drug metabolism include:

- Cigarette smoke (not nicotine) is responsible for the alteration of drug metabolism.
- Increased theophylline metabolism in smokers due to induction of CYP1A2 is well documented.

In one study, the half-life of theophylline was reduced by almost two-fold in smokers compared to non-smokers [12]. Significant reductions in drug concentrations with smoking have been reported for caffeine, chlorpromazine, clozapine, flecainide, fluvoxamine, haloperidol, mexiletine, olanzapine, propranolol, and tacrine due to increased metabolism of these drugs. Smokers may therefore require higher doses than non-smokers in order to achieve pharmacological responses [13]. Warfarin disposition in smokers is also different than in non-smokers. One case report described an increase in International Normalization Ratio (INR) to 3.7 from a baseline of 2.7 to 2.8 in an 80-year-old man when he stopped smoking. Subsequently, his warfarin dose was reduced by 14% [14].

14.13 MONITORING OF VARIOUS DRUG CLASSES: GENERAL CONSIDERATIONS

For a meaningful interpretation of a serum drug concentration, the time of specimen collection should be noted along with the time and date of the last dose and route of administration of the drug. This is particularly important for aminoglycosides because, without knowing the time of specimen collection, the serum drug concentration cannot be interpreted. The information needed for proper interpretation of drug levels for the purpose of therapeutic drug monitoring is listed in Table 14.2. Reference ranges of various therapeutic drugs are provided with the result. Therapeutic ranges of common drugs are given in Table 14.3. However, therapeutic ranges may vary slightly between different laboratories due to variations in the patient population.

Usually therapeutic drug monitoring should be ordered after a drug reaches its steady state. It typically takes at least five half-lives after initiation of a drug therapy to reach steady state. For example, the half-life of digoxin is 1.6 days, and the steady state of digoxin is reached after 7 days of therapy. However, for a drug with a shorter half-life than digoxin, for example, valproic acid (half-life 11–17 hours), it takes only three days to reach the steady state. Pre-analytical errors can contribute significantly to an erroneous result for therapeutic drug monitoring. For example, collecting a specimen in a serum separator tube can affect the concentrations of a few therapeutic drugs (phenytoin, valproic acid, and lidocaine).

14.14 MONITORING OF ANTICONVULSANTS

Phenytoin, phenobarbital, primidone, ethosuximide, valproic acid, and carbamazepine are considered as conventional anticonvulsant drugs. All of these antiepileptic drugs have a narrow therapeutic range requiring therapeutic drug monitoring. Phenytoin, carbamazepine, and valproic acid are also strongly bound to serum proteins. Therefore, for a selected patient population, monitoring free phenytoin, free valproic acid, and, to a lesser

Table 14.2 Information Required for Interpretation of Therapeutic Drug Monitoring Results

Patient-related Information Required on the Request

Name of the patient
Hospital identification number
Age
Gender (pregnant female?)
Race

Other Essential Information

Time of last dosage
Type of and number of specimen (serum, whole blood urine, saliva, other body fluid)
Identification of peak versus trough specimen (for aminoglycosides and vancomycin only)
Special request (such as free phenytoin)

Essential Information Needed for Interpretation of Result

Dosage regimen
Other drugs the patient is receiving
Concentration of the drug
Pharmacokinetic parameters of the drug
Is the patient critically ill or suffering from hepatic, cardiovascular or renal disease?
Albumin level, creatinine clearance

extent, free carbamazepine, is clinically useful. However, free phenobarbital monitoring is not required because this drug is only moderately bound to serum protein. Free phenytoin is the most commonly ordered free drug monitoring request in the hospital where this author works. Monitoring free phenytoin (and also free valproic acid) is recommended in the following patients:

- Uremic patients.
- Patients with liver disease.
- Pediatric population (small children often show impaired protein binding or altered disposition).
- Pregnant women.
- Critically ill patients, elderly patients, and patients with hypoalbuminemia.

In addition, monitoring free phenytoin concentration is useful in patients where a drug–drug interaction is suspected. Several strongly protein-bound drugs such as valproic acid, non-steroidal anti-inflammatory drugs (aspirin,

Table 14.3 Therapeutic Level of Commonly Monitored Drugs

Drug Class/Drug	Recommended Therapeutic Range (Trough)
Anticonvulsants	
Phenytoin	10–20 µg/mL
Carbamazepine	4–12 µg/mL
Phenobarbital	15–40 µg/mL
Primidone	5–12 µg/mL
Valproic acid	50–100 µg/mL
Clonazepam	10–75 ng/mL
Lamotrigine	3–14 µg/mL
Cardioactive Drugs	
Digoxin	0.8–1.8 ng/mL
Procainamide	4–10 µg/mL
N-Acetyl Procainamide	4–8 µg/mL
Quinidine	2–5 µg/mL
Lidocaine	1.5–5 µg/mL
Antiasthmatics	
Theophylline	10–20 µg/mL
Caffeine	5–15 µg/mL
Antidepressants	
Amitriptyline + nortriptyline	120–250 ng/mL
Nortriptyline	50–150 ng/mL
Doxepin + nordoxepin	150–250 ng/mL
Imipramine + desipramine	150–250 ng/mL
Lithium	0.8–1.2 mEq/L
Immunosuppressants	
Cyclosporine*	100–400 ng/mL
Tacrolimus*	5–15 ng/mL
Sirolimus*	4–20 ng/mL
Everolimus*	3–8 ng/mL
Mycophenolic acid	1–3.5 µg/mL
Antineoplastic	
Methotrexate	Varies with therapy type
Antibiotics	
Amikacin	20–35 µg/mL, Peak
	4–8 µg/mL, Trough
Gentamicin	5–10 µg/mL, Peak
	<2 µg/mL, Trough

Continued...

Table 14.3 Therapeutic Level of Commonly Monitored Drugs
Continued

Drug Class/Drug	Recommended Therapeutic Range (Trough)
Tobramycin	5–10 µg/mL, Peak
	<2 µg/mL, Trough
Vancomycin	20–40 µg/mL, Peak
	5–15 µg/mL, Trough

Therapeutic ranges are based on published literature, including books and adaptations from reputed national reference laboratories (e.g. Mayo Medical Laboratories and ARUP laboratories). Please note that therapeutic ranges can vary widely among different patient populations and that each institute should establish its own guidelines. These values are provided as examples only.
*Monitored in whole blood instead of serum or plasma.

ibuprofen, naproxen, tolmetin, etc.), and certain antibiotics (ceftriaxone, nafcillin, oxacillin, etc.) can displace phenytoin from the protein-binding site, thus causing an elevated free phenytoin level.

CASE REPORT

A 72-year-old man with coronary artery disease was hospitalized for coronary revascularization. On the 5th day after surgery, new onsets of focal seizure led to initiation of phenytoin therapy. Because the seizures were not controlled completely by phenytoin, phenobarbital was also introduced. His biochemical tests were normal. The patient suffered a respiratory arrest on the tenth hospital day, was intubated, and his condition improved. On the 25th hospital day (9 days after cessation of seizure activity, and 2 days after discontinuation of phenobarbital) the patient was still receiving phenytoin and showed lethargy accompanied by nystagmus (indicating phenytoin toxicity), although his total phenytoin was within therapeutic range (19.6 µg/mL; therapeutic range: 10–20 µg/mL). At that time, his free phenytoin level was determined and was found to be toxic (4.4 µg/mL; therapeutic: 1–2 µg/mL). Although free phenytoin represents 10% of total phenytoin level, this patient showed a free fraction of 22.4% due to severe hypoalbuminemia (2.4 g/dL). As a result of this finding, phenytoin was withheld for 12 h and subsequently dosage of phenytoin was reduced to 400 mg from an initial dosing of 800 mg per day. The patient continued to improve and was eventually discharged from the hospital in a stable condition [15].

Carbamazepine is metabolized to carbamazepine 10,11-epoxide, which is an active metabolite. Although in the normal population epoxide concentrations may be 10–14% of total carbamazepine concentration, patients with renal failure may show an over 40% epoxide concentration relative to the carbamazepine concentrations. Monitoring active metabolite concentration using chromatographic methods may be useful in these patients as there is no immunoassay available for monitoring epoxide levels. Certain drug therapies, such as treatment with both valproic acid and carbamazepine, tend to increase the epoxide concentration. Monitoring the epoxide level may be helpful if a patient experiences drug toxicity from an elevated epoxide level.

Primidone is an anticonvulsant that is metabolized to phenobarbital, another anticonvulsant. Although pharmacological activities of primidone are partly due to phenobarbital, primidone itself has anticonvulsant activity.

For routine therapeutic drug monitoring of classical anticonvulsants, immunoassays are commercially available and can be easily adopted on various automated analyzers. Since 1993, fourteen new antiepileptic drugs have been approved: eslicarbazepine acetate, felbamate, gabapentin, lacosamide, lamotrigine, levetiracetam, oxcarbazepine, pregabalin, rufinamide, stiripentol, tiagabine, topiramate, vigabatrin, and zonisamide. In general, these antiepileptic drugs have better pharmacokinetic profiles, improved tolerability in patients, and are less involved in drug interactions compared to traditional anticonvulsants. However, felbamate is a very toxic drug with a risk of fatal aplastic anemia, and the use of this drug is reserved for a few patients where the benefits may override the risks. Therapeutic drug monitoring of some of these new anticonvulsants is not needed, although a few drugs may benefit from therapeutic drug monitoring. Therapeutic drug monitoring of levetiracetam and pregabalin is justified in patients with renal impairment. Monitoring active metabolites of oxcarbazepine (10-hydroxycarbazepine) has some justification. In addition, therapy with lamotrigine, zonisamide, and topiramate may also benefit from therapeutic drug monitoring. Usually chromatographic techniques are employed for therapeutic drug monitoring of these newer anticonvulsants. These methods are usually free from interferences. However, there are commercially available immunoassays for lamotrigine, zonisamide, and topiramate.

14.15 MONITORING OF CARDIOACTIVE DRUGS

Therapeutic drug monitoring of several cardioactive drugs, including digoxin, procainamide, lidocaine, and quinidine, is routinely performed in clinical laboratories due to the established correlation between serum drug concentrations and pharmacological response of these drugs. Moreover, drug toxicity can be mostly avoided by therapeutic drug monitoring. Digoxin monitoring is challenging for following reasons:

- Digoxin has a very narrow therapeutic window, and there are overlaps between the therapeutic and toxic ranges. A classical therapeutic window of 0.8−1.8 ng/mL is problematic. Although digoxin toxicity is common with a digoxin level >2 ng/mL, some patients may experience digoxin toxicity at a level of 1.5 ng/mL or higher.

- Digoxin immunoassays are affected by both endogenous and exogenous factors (see Chapter 15).
- Digoxin overdose can be treated with Digibind or DigiFab. For these patients, progress of therapy must be monitored by measuring the free digoxin level because the total digoxin level may be misleading due to interference of Digibind/DigiFab with digoxin immunoassays.

Procainamide is metabolized to an active metabolite, N-acetyl procainamide (NAPA). During therapeutic drug monitoring of procainamide, NAPA should also be monitored because it contributes to the toxicity of procainamide. In patients with renal insufficiency, NAPA concentration increases in blood due to impaired renal clearance.

CASE REPORT

An 83-year-old patient with a history of adenocarcinoma of prostate (stage D) and placement of a pacemaker 5 years prior to his recent admission to the hospital was admitted for a non-healing ulcer in his left foot (which required amputation). Two days prior to amputation, the patient showed decreased urine output, atrial tachycardia, and shortness of breath. He was admitted to the intensive care unit and was treated with digoxin and furosemide. The next day he developed ventricular tachycardia and was managed with intravenous lidocaine. Later he was switched to intramuscular procainamide. Three days later he developed renal insufficiency with a creatinine level of 5.5 mg/dL and BUN of 42 mg/dL. The patient showed a procainamide level of 14.1 µg/mL, but the N-acetyl procainamide (NAPA) level was at the toxic level of 60.5 µg/mL. It was postulated that accumulation of NAPA in his blood was due to renal failure. The combined procainamide and NAPA toxicity was treated with hemodialysis, hemoperfusion, and combined hemodialysis–hemoperfusion. He was eventually discharged from the hospital when his creatinine was reduced to 3.3 mg/dL [16].

Key points to remember regarding therapeutic drug monitoring of lidocaine and quinidine include:

- Lidocaine cannot be given orally due to high first-pass metabolism. However, tocainide, an analog of lidocaine, can be administered orally.
- Lidocaine is strongly bound to α-acid glycoprotein, but free lidocaine is not usually monitored.
- Lidocaine after topical application may be absorbed significantly in some patients, causing toxicity. Therapeutic drug monitoring of lidocaine is needed for these patients.
- Lidocaine is metabolized into monoethylglycinexylidide, and this conversion can be used as a liver function test.
- Quinidine is infrequently used today. Although this drug is strongly bound to α-acid glycoprotein, free quinidine is usually not monitored.

Less frequently monitored cardioactive drugs include tocainide, flecainide, mexiletine, verapamil, propranolol, and amiodarone. Tocainide was developed as an oral analog of lidocaine, because lidocaine cannot be administered orally due to high first-pass metabolism; tocainide and lidocaine have similar electrophysiological properties.

14.16 MONITORING OF ANTI-ASTHMATIC DRUGS

Theophylline and caffeine are two anti-asthmatic drugs that require therapeutic drug monitoring. Theophylline is a bronchodilator and a respiratory stimulant effective in the treatment of acute and chronic asthma. The drug is readily absorbed after oral absorption, but peak concentration may be observed much later with sustained-release tablets. Theophylline is metabolized by hepatic cytochrome P-450; altered pharmacokinetics of theophylline in disease states have been reported. In infants, theophylline is partly metabolized to caffeine, but in adults this metabolite is not formed. Theophylline is metabolized to 3-methylxanthine and other metabolites in adults.

Apnea with or without bradycardia is a common medical problem in premature infants. Caffeine is effective in treating apnea in neonates. Because the effectiveness of caffeine therapy can be readily observed clinically, therapeutic drug monitoring of caffeine is only indicated when caffeine toxicity is apparent from clinical symptoms, including tachycardia, gastrointestinal intolerance, and jitteriness. In addition, therapeutic drug monitoring of caffeine is also indicated if a neonate is unresponsive to caffeine therapy despite a high dose.

14.17 MONITORING OF ANTIDEPRESSANTS

Tricyclic antidepressants (TCAs), including amitriptyline, doxepin, nortriptyline, imipramine, desipramine, protriptyline, trimipramine, and clomipramine were introduced in the 1950s and 1960s. These drugs have a narrow therapeutic window, and therapeutic drug monitoring is essential for efficacy of these drugs as well as to avoid drug toxicity. The efficacy of lithium in acute mania and for prophylaxis against recurrent episodes of mania has been well established. Therapeutic drug monitoring of lithium is essential for efficacy as well as to avoid lithium toxicity. Key points to remember in therapeutic drug monitoring of antidepressants:

- Although immunoassays are available for determination of tricyclic antidepressants, such assays should only be used for diagnosis of tricyclic overdose (usually total tricyclic antidepressant concentrations >500 ng/mL are considered critical).

- For routine therapeutic drug monitoring of tricyclic antidepressants, chromatographic techniques must be used (high-performance liquid chromatography or gas chromatography) because only such methods can differentiate between different tricyclics, for example, amitriptyline from its metabolite nortriptyline. Immunoassays can provide a total concentration only, because both amitriptyline and nortriptyline have almost 100% cross-reactivity. This is true for other tricyclic antidepressants.
- A common mistake for therapeutic drug monitoring of lithium is to collect specimens in a lithium heparin tube (which will falsely elevate the true lithium concentration). This must be avoided and either a sodium heparin tube or serum specimen (with no anticoagulant, such as a red-top tube) must be collected for lithium analysis.

CASE REPORT

A 37-year-old woman delivered a female infant at full term (birth weight: 3.1 kg). The mother was on 900 mg/day of lithium throughout the pregnancy, and her serum lithium level was 0.9 mmol/L at the time of delivery and 0.7 mmol/L (0.7 mEq/L) 11 days later. The infant was breastfed, and her serum lithium level was undetectable at 3 days after birth; the infant's blood lithium level was increased to 0.7 mmol/L at Day 6 and 1.1 mmol/L at Day 10. The authors questioned the validity of the serum lithium level in the infant because the infant showed no sign of lithium toxicity nor did she have any renal insufficiency. Later it was found that the infant's blood was wrongly collected in a tube containing lithium heparin as an anticoagulant. A blood specimen correctly collected later from the infant showed an undetectable level of lithium [17].

More recently introduced antidepressants are selective serotonin reuptake inhibitors (SSRIs), for example, citalopram, fluoxetine, fluvoxamine, paroxetine, and sertraline. This class of drugs has a wide therapeutic index. Usually most of these drugs do not require routine therapeutic drug monitoring, but some drugs may benefit from infrequent monitoring, especially in certain patient populations like children, the elderly, pregnant women, and individuals with intelligence disabilities.

14.18 MONITORING OF IMMUNOSUPPRESSANTS

Therapeutic drug monitoring of all immunosuppressants is important. Key points are the following:

- Immunosuppressant drugs cyclosporine and tacrolimus are calcineurin inhibitors, but sirolimus and everolimus (the most recently approved drug that is a 2-hydroxyethyl derivative of sirolimus) are m-TOR (mammalian target of rapamycin) inhibitors. All these drugs are monitored in whole blood.

- Everolimus was developed to improve pharmacokinetic parameters of sirolimus. The half-life of sirolimus is 60 hours, but the half-life of everolimus is 18−35 hours.
- Mycophenolic acid is a potent non-competitive inhibitor of inosine monophosphate dehydrogenase enzymatic activity and thus selectively inhibits lymphocyte proliferation. This is the only immunosuppressant that is monitored in serum or plasma.
- In patients with uremia and hypoalbuminemia, monitoring free mycophenolic acid can be clinically useful.
- Although immunoassays are available for monitoring immunosuppressants, metabolite interferences in the immunoassays are a significant problem. Chromatographic methods, especially liquid chromatography combined with tandem mass spectrometry, is a gold standard for therapeutic drug monitoring of immunosuppressants.
- Although cyclosporine, sirolimus, everolimus and mycophenolic acid can be determined by high-performance liquid chromatography combined with ultraviolet detector (HPLC-UV), tacrolimus cannot be monitored by HPLC-UV due to lack of an absorption peak in the ultraviolet region.

14.19 MONITORING OF SELECTED ANTIBIOTICS

The most commonly monitored antibiotics in clinical laboratories are aminoglycosides and vancomycin. The aminoglycoside antibiotics consist of two or more amino-sugars joined by a glycosidic linkage to a hexose or aminocyclitol. These drugs are used in the treatment of serious and often life-threatening systemic infections. However, aminoglycosides can produce serious nephrotoxicity and ototoxicity. Aminoglycosides are poorly absorbed from the gastrointestinal tract, and these drugs are administered intravenously or intramuscularly. Children have a higher clearance of aminoglycosides. Patients with cystic fibrosis usually exhibit altered pharmacokinetics of the antibiotics. After a conventional dose of an aminoglycoside, a patient with cystic fibrosis shows a lower serum concentration compared to a patient not suffering from cystic fibrosis. Although there are several aminoglycosides used in the United States, the most commonly monitored aminoglycosides are tobramycin, amikacin, and gentamicin. Aminoglycosides are administered either in traditional dosing (2−3 times a day) or once daily. Other less common types of dosing such as synergy dosing are also practiced. Key points in therapeutic drug monitoring of aminoglycosides are as follows:

- If aminoglycoside therapy is needed for three days or less, therapeutic drug monitoring may not be needed.
- During traditional dosing, a peak concentration blood level should be drawn 30−60 minutes after each dose, and a trough concentration specimen must be drawn 30 minutes prior to the next dose.
- In once-daily dosing, a larger dose of aminoglycoside is administered compared to traditional dosing. There is no firm established guideline for therapeutic drug monitoring after once-daily dosing. Peak and trough concentration can be monitored. Alternatively, a specimen can be drawn 6−14 hours after the first dose to calculate dosing intervals using various nomograms.
- During aminoglycoside therapy, serum creatinine must be monitored at least twice a week to ensure there is no significant renal insufficiency.

Therapeutic drug monitoring is also frequently employed during vancomycin therapy (vancomycin is not an aminoglycoside). The drug is excreted in the urine with no metabolism. Vancomycin therapy warrants therapeutic drug monitoring if the patient receives vancomycin for five or more days, receives a higher dosage of vancomycin, or is receiving both vancomycin and aminoglycosides. Both peak and trough concentrations should be monitored. Ranges for peak concentrations of 20−40 µg/mL have been widely quoted, and a trough concentration range of 5−15 µg/mL has reasonable literature support. A trough concentration of 5−15 µg/mL is recommended because nephrotoxicity and other complications are observed at vancomycin concentrations higher than this level.

Rapid infusion of vancomycin may be associated with pruritus, a rash involving the upper torso, head and neck, and occasionally hypotension. Known as "red man" or "red neck" syndrome, this phenomenon is caused by non-immunologically mediated release of histamine and can be avoided by slower administration of vancomycin over at least 60 minutes. Heparin and vancomycin are incompatible if mixed in intravenous solution or infused one after another through a common intravenous line. Aminophylline, amobarbital, aztreonam, chloramphenicol, dexamethasone, and sodium bicarbonate are also incompatible with vancomycin if mixed in the same container. However, there are other antibiotics which are monitored infrequently. Examples of less frequently monitored antibiotics include ciprofloxacin, chloramphenicol, isoniazid, rifampin, and rifabutin.

14.20 MONITORING OF ANTINEOPLASTIC DRUGS

Methotrexate is a competitive inhibitor of dihydrofolate reductase, a key enzyme for the biosynthesis of nucleic acids. The cytotoxic activity of this drug was discovered in 1955. The use of leucovorin to rescue normal host cells has permitted the higher doses of methotrexate therapy in clinical practice. Methotrexate is used in the treatment of acute lymphoblastic leukemia (ALL), brain tumors, carcinomas of the lung, and other cancers. Most of the toxicities of this drug are related to serum concentrations and pharmacokinetic parameters. Methotrexate is also approved for the treatment of refractory rheumatoid arthritis. Usually low doses of methotrexate are used for treating rheumatoid arthritis (5 to 25 mg once weekly). Although therapeutic drug monitoring of methotrexate is not indicated in patients receiving low-dose methotrexate for rheumatoid arthritis, therapeutic drug monitoring is essential during high-dose treatment with methotrexate because of frequent adverse reactions leading to leucopenia and thrombocytopenia.

Pharmacokinetic studies have shown that the clinical response and toxicity of 5-fluorouracil are related to the area under the curve (AUC). Monitoring 5-fluorouracil may have clinical benefits. In addition, therapeutic drug monitoring of busulfan may also be valuable.

14.21 MONITORING OF ANTIRETROVIRALS

Human immunodeficiency virus (HIV) is the virus that causes AIDS (acquired immunodeficiency syndrome). Six classes of drugs are used today to treat people with AIDS, including: (1) nucleoside reverse transcriptase inhibitors (NRTIs) such as zidovudine; (2) non-nucleoside reverse transcriptase inhibitors (NNRTIs), which include nevirapine, delavirdine, and efavirenz; (3) protease inhibitors (PIs) such as saquinavir, ritonavir, indinavir, nelfinavir, amprenavir, lopinavir, and atazanavir; (4) entry inhibitors such as maraviroc; (5) a fusion inhibitor such as enfuvirtide; (6) and integrase inhibitors such as raltegravir. Although some antiretroviral agents do not require therapeutic drug monitoring, patients receiving protease inhibitors may benefit from therapeutic drug monitoring. Currently there is no commercially available immunoassay for any antiretroviral agent. Therefore, therapeutic drug monitoring for these drugs is only available in major academic medical centers and reference laboratories; most hospital laboratories do not provide service for therapeutic drug monitoring of antiretroviral drugs.

KEY POINTS

- Therapeutic drug monitoring is only required for drugs with a narrow therapeutic window where there is a better correlation between serum or plasma (or whole blood) drug level and therapeutic efficacy (as well as toxicity).

- Therapeutic drug monitoring should be ordered after a drug reaches its steady state. It usually takes at least five half-lives after initiation of a drug therapy to reach steady state.
- Usually a trough specimen (15–30 min before next dose) is used for therapeutic drug monitoring, except for vancomycin and aminoglycosides where both peak and trough levels are monitored.
- Serum or plasma is the preferred specimen for therapeutic drug monitoring, except for certain immunosuppressants (cyclosporine, tacrolimus, sirolimus, and everolimus, which are monitored only using whole blood). However, another immunosuppressant drug, mycophenolic acid, is usually monitored using a serum or plasma specimen.
- Usually in therapeutic drug monitoring, total drug concentration (free drug–drug bound to protein) is measured. Only free drug is pharmacologically active. If a drug is <80% bound to serum protein, direct monitoring of unbound drug (free drug) is not necessary. However, for strongly protein-bound drugs, free drug monitoring may be useful for certain patient populations where protein binding may be impaired.
- Free phenytoin is the most commonly monitored free drug. It is important to do free phenytoin monitoring in patients with uremia, liver disease, and any condition that may cause hypoalbuminemia, such as the elderly, critically ill patients, pregnant women, etc.
- The major CYP isoforms responsible for metabolism of drugs include CYP1A2, CYP2B6, CYP2C9, CYP2C19, CYP2D6, CYP2E1, and CYP3A4/CYP3A5. However, CYP3A4 is the predominant isoform of the CYP family (almost 30%) responsible for metabolism of approximately 37% of all drugs.
- Neonates and infants (0–4 months) may metabolize drugs slower than adults because of the lack of mature (fully functioning) drug-metabolizing enzymes, but activities of drug-metabolizing enzymes are higher in children than adults. Therefore, children may need a higher per kilogram dosage of a drug than an adult. However, the drug metabolism rate is reduced in the elderly and an appropriate dosage adjustment is needed. In addition, elderly patients (>70 years of age) may also have lower albumin, and protein binding of strongly protein-bound drugs such as phenytoin may be impaired, causing an elevated concentration of the pharmacologically active free drug level.
- Uremia may impair clearance of a drug/metabolite by the kidney while liver disease may impair metabolism of certain drugs. Dosage adjustment may be needed for these patients.
- Drug metabolism is increased in patients with hyperthyroidism due to excessive levels of thyroxine, but decreased in patients with hypothyroidism due to lower levels of thyroxine. Amiodarone has high iodine content and may cause thyroid dysfunction. Screening for thyroid disease before amiodarone therapy and periodic monitoring of thyroid function are recommended for patients treated with amiodarone.

- Grapefruit juice inhibits intestinal cytochrome P-450 and P-glycoprotein, thus increasing the bioavailability of many drugs. However, if a drug is given intramuscularly or intravenously, no such interaction is observed because grapefruit juice does not inhibit liver cytochrome P-450 enzymes. Common drugs that interact with grapefruit juice if taken orally include alprazolam, carbamazepine, cyclosporine, erythromycin, methadone, quinidine, simvastatin, and tacrolimus.
- Increased theophylline metabolism in smokers due to induction of CYP1A2 is well documented.
- Carbamazepine is metabolized to carbamazepine 10,11-epoxide, which is an active metabolite. Although in the normal population epoxide concentrations may be 10–14% of total carbamazepine concentration, patients with renal failure can show an over 40% epoxide concentration relative to the carbamazepine concentrations. Monitoring active metabolite concentrations using chromatographic methods can be useful in these patients as there is no immunoassay available for monitoring epoxide levels.
- Procainamide is metabolized to N-acetylprocainamide (NAPA); monitoring both procainamide and NAPA is essential. NAPA accumulates in patients with renal failure, causing toxicity where the procainamide level may be within a therapeutic range. Immunoassays are available for therapeutic drug monitoring of both procainamide and NAPA.
- Lidocaine cannot be given orally due to high first-pass metabolism. However, tocainide, an analog of lidocaine, can be administered orally.
- Immunoassays for tricyclic antidepressants (TCA) should only be used in a case of suspected overdose; such assays should not be used for therapeutic drug monitoring because almost all tricyclic antidepressants show significant cross-reactivity. In a suspected overdose, total TCA concentration (drug + metabolite) matters, but in drug monitoring individual concentration is important.
- Although immunoassays are available for monitoring immunosuppressants, metabolite interferences in the immunoassay are a significant. Chromatographic methods, especially liquid chromatography combined with tandem mass spectrometry, are a gold standard for therapeutic drug monitoring of immunosuppressants.
- If aminoglycoside therapy is needed for three days or less, therapeutic drug monitoring may not be needed; but it may be needed if continued for more than three days. During traditional dosing, a peak concentration blood level should be drawn 30–60 minutes after each dose, and a trough concentration specimen must be drawn 30 minutes prior to the next dose.

REFERENCES

[1] Mattson RH, Cramer JA, Collins JF. Aspects of compliance: taking drugs and keeping clinic appointments. Epilepsy Res Suppl 1988;1:111−7.

[2] Chandra RS, Dalvi SS, Karnad PD, Kshirsagar NA, et al. Compliance monitoring in epileptic patients. J Assoc Physicians India 1993;41:431−2.

[3] Babalik A, Babalik A, Mannix S, Francis D, et al. Therapeutic drug monitoring in the treatment of active tuberculosis. Can Respir J 2011;18:225−9.

[4] Jeong H. Altered drug metabolism during pregnancy: hormonal regulation of drug metabolizing enzymes. Expert Opin Drug Metab Toxicol 2010;6:689−99.

[5] Anderson GD. Pregnancy induced changes in pharmacokinetics: a mechanistic based approach. Clin Pharmacokinet 2005;44:989−1008.

[6] Terrell KM, Heard K, Miller DK. Prescribing to older ED patients. Am J Emerg Med 2006;24:468−78.

[7] Benowitz NL, Meister W. Pharmacokinetics in patients with cardiac failure. Clin Pharmacokinet 1976;1:389−405.

[8] Betteridge T, Fink J. Phenytoin toxicity and thyroid dysfunction. NZ Med J 2009;122:102−4.

[9] Bailey DG, Spence JD, Munoz C, Arnold JM. Interaction of citrus juices with felodipine and nifedipine. Lancet 1991;337:268−9.

[10] Saito M, Hirata-Koizumi M, Matsumoto M, Urano T, Hasegawa R. Undesirable effects of citrus juice on the pharmacokinetics of drugs: focus on recent studies. Drug Saf 2005;28:677−94.

[11] Koski A, Vuori E, Ojanpera I. Relation of postmortem blood alcohol and drug concentrations in fetal poisonings involving amitriptyline, propoxyphene and promazine. Hum Exp Toxicol 2005;24:389−96.

[12] Zevin S, Benowitz NL. Drug interactions with tobacco smoking: an update. Clin Pharmacokinetic 1999;36:425−38.

[13] Kroon LA. Drug interactions and smoking: raising awareness for acute and critical care provider. Crit Care Nurs Clin N Am 2006;18:53−62.

[14] Colucci VJ, Knapp JE. Increase in International normalization ratio associated with smoking cessation. Ann Pharmacother 2001;35:385−6.

[15] Discoll DE, McMahon M, Blackburn GL, Bistrian BR. Phenytoin toxicity in a critically ill hypoalbuminemic patient with normal serum drug concentration. Crit Care Med 1998;16:1248−9.

[16] Rosansky SJ, Bradt ME. Procainamide toxicity in a patient with acute renal failure. Am J Kidney Dis 1986;7:502−6.

[17] Tanaka T, Moretti ME, Verjee ZH, Shupak M, et al. A pitfall of measuring lithium levels in neonates. Ther Drug Monit 2008;30:752−4.

CHAPTER 15

Interferences in Therapeutic Drug Monitoring

15.1 METHODOLOGIES USED IN THERAPEUTIC DRUG MONITORING AND ISSUES OF INTERFERENCES

Methods used for therapeutic drug monitoring include:

- Immunoassays: Most commonly used method but suffers from interference.
- Gas Chromatography (GC) with Flame Ionization or Nitrogen Detection: Can be used for selective drugs which are relatively volatile (e.g. pentobarbital).
- Gas Chromatography Combined with Mass Spectrometry (GC/MS): Most commonly used for analysis of abused drugs, but volatile drugs can also be analyzed by this method. Very specific and relatively free from interferences.
- High-Performance Liquid Chromatography (HPLC) combined with Ultraviolet or Fluorescence Detection: Widely used for monitoring drugs where immunoassays are not commercially available. Both polar (less or non-volatile) and non-polar (relatively volatile) drugs can be analyzed by this method.
- High-Performance Liquid Chromatography combined with Mass Spectrometry (LC/MS) or Tandem Mass Spectrometry (LC/MS/MS): Very specific method. Considered the gold standard for therapeutic drug monitoring because this method is virtually free from interferences.

Immunoassays are most commonly used for measuring concentrations of various drugs in serum, plasma, and, less commonly, in whole blood. Specimens can in most cases be analyzed directly without any pretreatment (except for certain immunosuppressants). Moreover, specimens can be batched, and results can be obtained within 20–40 minutes from the beginning of the run. However, the antibody used in an immunoassay can

CONTENTS

A. Dasgupta and A. Wahed: Clinical Chemistry, Immunology and Laboratory Quality Control
DOI: http://dx.doi.org/10.1016/B978-0-12-407821-5.00015-2

cross-react with another molecule with a similar structure to the analyte drug molecule, most commonly the drug metabolite. In addition, other structurally related drugs (and even endogenous compounds such as high bilirubin), hemolysis, and elevated lipids, can interfere with immunoassays. In contrast, chromatographic techniques are more labor intensive, require highly experienced medical technologists, and are expensive; however, such methods are relatively free from interferences. Therefore, LC/MS or LC/MS/MS methods are used mostly for therapeutic monitoring of drugs where immunoassays are not commercially available, such as in therapeutic drug monitoring of antiretrovirals. In addition, chromatographic methods (especially LC/MS/MS) are also used where immunoassays may suffer from significant interferences, e.g. immunoassays for various immunosuppressants that suffer from significant metabolite cross-reactivity.

15.2 EFFECT OF ENDOGENOUS FACTORS ON THERAPEUTIC DRUG MONITORING

Endogenous factors such as bilirubin, hemolysis, and high lipids, if present in a specimen, can interfere with therapeutic drug monitoring of various drugs. Bilirubin is derived from the hemoglobin of aged or damaged red blood cells. Some part of serum bilirubin is conjugated as glucuronides ("direct" bilirubin); the unconjugated bilirubin is also referred to as indirect bilirubin. In normal adults, bilirubin concentrations in serum range from 0.3 to 1.2 mg/dL (total) and <0.2 mg/dL (conjugated). A total bilirubin concentration of up to 20 mg/dL is not usually a problem for most assays, but a bilirubin level over 20 mg/dL, which is uncommon, can cause some interference in certain immunoassays. Currently, the interference of elevated bilirubin in the colorimetric assay for acetaminophen and salicylate (Trinder salicylate assay) is of major concern. However, immunoassays for acetaminophen and salicylate are free from such interferences [1,2].

Hemolysis can occur in vivo, during venipuncture and blood collection, or during processing of the specimen. Hemoglobin interference depends on its concentration in the specimen. Serum appears hemolyzed when the hemoglobin concentration exceeds 20 mg/dL. Hemoglobin interference is caused not only by the spectrophotometric properties of hemoglobin, but also by its participation in chemical reaction with sample or reactant components as well. The absorbance maxima of the heme moiety in hemoglobin are at 540 to 580 nm. However, hemoglobin begins to absorb around 340 nm, with absorbance increasing at 400−430 nm as well. Methods that use the absorbance properties of NAD or NADH (340 nm) can therefore be affected by hemolysis. Lipids in serum or plasma exist as complexed with proteins called lipoproteins. Lipoproteins consist of various proportions of lipids, and range from 10 nm to 1,000 nm in size (the higher the percentage of the lipid, the lower the density and the larger the particle size of the resulting lipoprotein). The lipoprotein particles with high lipid contents (such as chylomicrons and VLDL) are micellar and are the main source of assay interference, especially in turbidimetric assays.

15.3 EFFECT OF COLLECTING SPECIMEN IN GEL-SEPARATOR TUBE ON THERAPEUTIC DRUG MONITORING RESULTS

Serum separator gel tubes (SSTs) are widely used for blood collection for clinical laboratory tests, including therapeutic drug monitoring. Different barrier materials are available among tube manufacturers, but all are thixotropic materials that facilitate separation of serum or plasma from the cells and prevent hemolysis upon prolonged storage. The base material for preparing the gel is acrylic, silicone, or a polyester polymer. The stability of various analytes when blood is collected in gel tubes has been studied extensively. Important points to remember are the following:

- Various companies manufacture serum separator gel tubes, and different gel tubes absorb different drugs. However, in general, drugs affected by collection in gel separation tubes include phenytoin, carbamazepine, phenobarbital, lidocaine, quinidine, and tricyclic antidepressants. If analysis is made within two hours of collection, serum separator gel tubes can be used for all drugs because absorption of affected drugs should be minimal.
- Usually absorption of these few drugs depends on specimen volume, and absorption is much less if the tube is completely filled to capacity rather than only half-filled.

15.4 DIGOXIN IMMUNOASSAYS: SO MUCH INTERFERENCE

Digoxin immunoassays are subjected to the most interference compared to other immunoassays for therapeutic drugs [3]. Sources of interferences in digoxin immunoassays include:

- Endogenous factors such as digoxin-like immunoreactive substances (DLIS).
- Endogenous heterophilic antibody (very rare interference).
- Digibind and DigiFab interferences.
- Digoxin metabolites.

- Interferences from spironolactone, potassium canrenoate, and their common metabolite canrenone.
- Chinese medicines such as Chan Su, Lu-Shen-Wan, and oleander-containing herbs.

Interestingly, some of these interferences can be eliminated by monitoring free digoxin concentrations in the protein-free ultrafilter (Table 15.1).

The presence of endogenous digoxin-like immunoreactive substances (DLIS) was first described in a volume expanded dog in 1980 [4]. After publication of that initial report, many investigators confirmed the presence of endogenous DLIS in serum and other biological fluids in volume expanded patients, not limited to patients with uremia, liver disease, essential hypertension, transplant recipients, eclampsia, pregnant women, or pre-term babies. Usually high amounts of DLIS are encountered in premature babies. DLIS, like digoxin, can inhibit Na,K-ATPase. Although DLIS interference with digoxin immunoassays was a significant problem in the past (due to use of polyclonal antibodies in the assay design), more recently introduced digoxin immunoassays utilizing highly specific monoclonal antibodies against digoxin are relatively free from DLIS interference. However, taking advantage of strong protein binding of DLIS and only 25% protein binding of digoxin, interference of DLIS in digoxin immunoassays can be eliminated by monitoring free digoxin concentration in protein free ultrafiltrate [5].

The presence of human anti-animal antibodies (especially those directed against mice) in serum may cause interference with certain immunoassays.

Table 15.1 Interferences in Digoxin Immunoassays and Elimination of Interference by Monitoring Free Digoxin

Interfering Substance	Elimination by Free Digoxin Monitoring
Endogenous Factors	
Digoxin-like immunoreactive substances (DLIS)	Yes
Heterophilic antibody	Yes
Endogenous Factors	
Digoxin metabolites	No
Spironolactone	Yes
Potassium canrenoate	Partial elimination of interference
Digibind/DigiFab	Yes
Chan Su, Lu-Shen-Wan	Depending on dosage, with low dosage interference can be eliminated by monitoring free digoxin.

The clinical use of mouse monoclonal antibody for radioimaging and treatment of certain cancers may cause accumulation of human anti-mouse antibody (HAMA). Anti-animal antibodies are also found among veterinarians, farm workers, or pet owners due to exposure to animals; these antibodies are broadly classified as heterophilic antibodies. Usually the presence of heterophilic antibodies in serum will interfere with sandwich assays designed for measuring relatively large molecules such as beta-hCG (human chorionic gonadotropin). Nevertheless, Liendo et al. described a case report of a patient with cirrhotic liver disease and atrial fibrillation who had been treated with spironolactone and digoxin and showed an elevated digoxin concentration of 4.2 ng/mL. Despite a toxic digoxin level the patient was asymptomatic and, after discontinuation of both drugs, digoxin values over 3.0 ng/mL were observed for approximately five weeks in the patient's serum. Because such interference was eliminated by measuring digoxin in protein free ultrafiltrate (heterophilic antibodies due to large molecular weights are absent in protein free ultrafiltrate), the authors concluded that the falsely observed digoxin level was due to interference from heterophilic antibodies [6].

The major metabolites of digoxin are digoxigenin, digoxigenin monodigitoxoside, digoxigenin bisdigitoxoside, and dihydrodigoxin. These metabolites exhibit significantly different cross-reactivities against various antidigoxin antibodies. However, due to relatively low levels of digoxin metabolites in serum in comparison to digoxin, the effects of metabolite cross-reactivities are minimal on serum digoxin measurements by immunoassays in patients with normal renal function. In contrast, for patients with renal disease, liver disease, or diabetes, the digoxin immunoassays can significantly overestimate digoxin values compared to chromatographic methods [7].

Digibind and DigiFab are Fab fragments of the antidigoxin antibody used for treating life-threatening acute digoxin overdose. Digibind was marketed in 1986 while DigiFab was approved for use in 2001. Both products are Fab fragments of antidigoxin antibody. The molecular weight of DigiFab (46,000 Daltons) is similar to the molecular weight of Digibind (46,200 Daltons), and both compounds can be excreted in urine. Both Digibind and DigiFab interfere with serum digoxin measurement using immunoassays. In patients overdosed with digoxin and being treated with Digibind or DigiFab, only the unbound digoxin (free digoxin) measurement is clinically useful, because unbound (free) digoxin is responsible for digoxin toxicity. In addition, neither Digibind nor DigiFab is present in the protein-free ultrafiltrate (the molecular weight of the filter in the ultrafiltration device is usually 30,000 Daltons) due to their high molecular weight. Therefore, free digoxin measurements by immunoassays are not affected by Digibind or DigiFab.

CASE REPORT

A 35-year-old woman intentionally swallowed 100 Lanitop tablets (0.1 mg methyldigoxin per tablet, 10 mg of methyldigoxin) in a suicide attempt. Methyldigoxin is a semisynthetic cardiac glycoside which is rapidly converted into digoxin after oral administration. On admission, approximately 19 hours after ingestion, her serum digoxin level was 7.4 ng/mL (therapeutic 0.8–1.8 ng/mL), and the patient was treated immediately with 80 mg of Digibind. A total of 395 mg of Digibind was administered to the patient. Her total serum digoxin level peaked at 125 ng/mL 23 hours after ingestion, but the free serum digoxin level immediately decreased to a non-toxic level, indicating that Digibind therapy was effective in treating her overdose. Her symptoms of digoxin toxicity (electrocardiogram) as well as nausea and vomiting resolved within three hours of initiation of therapy, and the patient was discharged from the hospital after three days [8].

Although various diuretics interact with digoxin and can increase serum digoxin levels, potassium-sparing diuretics such as spironolactone, potassium canrenoate, and eplerenone, not only pharmacokinetically interact with digoxin, but also can interfere with serum digoxin measurements using digoxin immunoassays. After oral administration, spironolactone is rapidly and extensively metabolized to several metabolites, including canrenone, which is an active metabolite. Potassium canrenoate is also metabolized to canrenone, but this drug is not approved for clinical use in the United States; it is, however, still used clinically in Europe and other countries throughout the world. Because of structural similarities between spironolactone and related compounds with digoxin, these substances interfere with serum digoxin assays (especially assays utilizing polyclonal antibody against digoxin). Although spironolactone and related compounds falsely elevate serum digoxin levels, negative interference was observed with the microparticle enzyme immunoassay (MEIA) digoxin assay [9]. Monitoring free digoxin may eliminate some interference. However, relatively new digoxin assays utilizing specific monoclonal antibodies are free from such interferences.

The Chinese medicines Chan Su and Lu-Shen-Wan contain bufalin, which has structural similarity to digoxin. Therefore, if a patient takes such Chinese medicines, a false digoxin level may be observed in serum using an immunoassay even though the patient is not taking digoxin. Oleander-containing herbal supplements also interfere with serum digoxin measurement because oleandrin, the active component of oleander, has structural similarity to digoxin.

15.5 INTERFERENCES IN ANALYSIS OF ANTIEPILEPTICS

Monitoring classical anticonvulsants, such as phenytoin, carbamazepine, phenobarbital, and valproic acid, are essential for proper patient management, and immunoassays are commercially available for determination of serum plasma concentrations of these drugs. Usually phenobarbital and valproic

Table 15.2 Interferences in Carbamazepine and Phenytoin Immunoassays

Anticonvulsant	Interfering Substance	Comment
Carbamazepine	Carbamazepine 10,11-epoxide	Cross-reactivity varies:* PETINIA: 96% EMIT: 0.4% CEDIA: 10.5% Beckmann SYNCHRON: 7.6%
	Hydroxyzine, Cetirizine	Affect PETINIA assay only
Phenytoin	Fosphenytoin	Fosphenytoin metabolite in uremic patients may falsely elevate phenytoin level.

Abbreviations: EMIT, Enzyme multiplied immunoassay technique (Syva); PETINIA, Particle enhanced turbidimetric inhibition immunoassay (Siemens Diagnostics); CEDIA, Cloned enzyme donor immunoassay (Roche Diagnostics).
*Examples of cross-reactivities of carbamazepine 10,11-epoxide with various commercially available carbamazepine immunoassays. These are representative examples only as there are more commercially available immunoassays for carbamazepine.

acid immunoassays are robust, and interference is observed only rarely. Overdose with amobarbital or secobarbital can cause a falsely elevated phenobarbital level, but this is rarely observed. However, carbamazepine and phenytoin immunoassays are subjected to interferences (Table 15.2). Cross-reactivity of carbamazepine 10,11-epoxide with carbamazepine immunoassays can vary from 0% to 96%. Therefore, true carbamazepine concentration may be overestimated if a carbamazepine immunoassay with high cross-reactivity towards its active epoxide metabolite is used for therapeutic drug monitoring of carbamazepine; for example, PETINIA assay (PETINIA: particle enhanced turbidimetric inhibition immunoassay, Siemens Diagnostics; 96% cross-reactivity with epoxide). Hydroxyzine and cetirizine are antihistamine drugs that interfere only with the PETINIA carbamazepine immunoassay.

CASE REPORT

A 32-year-old male epileptic patient with co-morbid psychotic disorders and multiple substance abuses was admitted to the emergency department for the complaint of feeling faint. This patient had been admitted before to the emergency department for overdose with antiepileptics (valproic acid and carbamazepine). Examination revealed only drowsiness and slurred speech with a Glasgow coma scale score of 10. Laboratory investigation showed a toxic carbamazepine concentration of 40.6 µg/mL using the PETINIA assay (therapeutic: 4–12 µg/mL), but no valproic acid was detected. However, a subsequent drug screening analysis by liquid chromatography–diode array detection (LC-DAD) showed very low carbamazepine and epoxide levels (<0.5 µg/mL), but showed a toxic hydroxyzine concentration (0.55 µg/mL; therapeutic range <0.1 µg/mL). No other drugs or alcohol were detected. On questioning, the patient admitted regularly taking Atarax (hydroxyzine HCl) to feel "high." A diagnosis of hydroxyzine toxicity was made, and it was determined that the false positive carbamazepine level measured by the PETINIA assay was due to interference of hydroxyzine. The authors did not observe any interference of hydroxyzine when serum carbamazepine levels were measured by another immunoassay (EMIT: Syva) [10].

Fosphenytoin, a pro-drug of phenytoin, is rapidly converted into phenytoin after administration. Fosphenytoin, unlike phenytoin, is readily water-soluble and can be administered intravenously (IV) or via intramuscular (IM) routes. Unlike phenytoin, fosphenytoin does not crystallize at the injection site, and no discomfort is experienced by the patient. Fosphenytoin is not typically monitored clinically because of its short half-life and lack of pharmacological activity. However, phenytoin is monitored in a patient after administration of fosphenytoin, but in this case monitoring of phenytoin must be initiated after complete conversion of fosphenytoin into phenytoin. In uremic patients after fosphenytoin administration, phenytoin concentrations measured by various immunoassays may be significantly higher than true phenytoin levels as determined by high-performance liquid chromatography. Annesley et al. identified a unique oxymethylglucuronide metabolite derived from fosphenytoin in sera of uremic patients and demonstrated that this unusual metabolite was responsible for the cross-reactivity [11].

15.6 INTERFERENCES IN ANALYSIS OF TRICYCLIC ANTIDEPRESSANTS

The major interference in immunoassays for determining tricyclic antidepressant concentrations in serum or plasma is the interference from their metabolites. Other drugs that may interfere with immunoassays for tricyclic antidepressants are listed in Table 15.3. In general, a tertiary amine tricyclic antidepressant (TCA) is metabolized to a secondary amine, and this metabolite usually has almost 100% cross-reactivity with an antibody used for immunoassay of TCA. Therefore, for monitoring tertiary amines, immunoassays in general indicate total concentration of the parent drug along with the active metabolite; TCA immunoassays should be used only for diagnosis of overdose (see also Chapter 14).

Table 15.3 Common Interferences In Immunoassays For Tricyclic Antidepressants (TCA)

Interfering Substance
Metabolites of TCAs
Phenothiazines and metabolites
Carbamazepine and its active metabolite carbamazepine 10,11-epoxide
Quetiapine
Cyproheptadine, if present in toxic concentrations
Diphenhydramine, if present in toxic concentrations

Phenothiazines and their metabolites can interfere with TCA immunoassays, and even a therapeutic concentration of such drugs can cause a falsely elevated serum TCA level. Carbamazepine is metabolized to carbamazepine 10,11-epoxide, an active metabolite. Both the parent drug and the epoxide metabolite interfere with immunoassays for TCA due to structural similarities. Another structurally related drug, oxcarbazepine, also interferes with immunoassays for TCA. Quetiapine can also interfere with TCA immunoassays. However, interference of diphenhydramine (Benadryl) and cyproheptadine with TCA immunoassays occurs only in overdosed patients.

15.7 INTERFERENCES IN ANALYSIS OF IMMUNOSUPPRESSANTS

Immunosuppressant drugs cyclosporine, tacrolimus, sirolimus, and everolimus must be monitored in whole blood; mycophenolic acid is the only immunosuppressant that is monitored in serum or plasma. Although immunoassays are commercially available for therapeutic drug monitoring of all immunosuppressants, they suffer from metabolite cross-reactivities. Therefore, positive bias is commonly observed in therapeutic drug monitoring of immunosuppressants using various immunoassays. The positive bias can vary 10–40% depending on the immunosuppressant and type of immunoassay compared to the corresponding values obtained by the more specific chromatographic method. Important points regarding monitoring of immunosuppressants include:

- Chromatographic methods, especially liquid chromatography combined with mass spectrometry (LC/MS) or tandem mass spectrometry (LC/MS/MS), are the gold standard for therapeutic drug monitoring of immunosuppressants.
- Most immunoassays for cyclosporine, tacrolimus, sirolimus, and everolimus require specimen pretreatment to extract the whole blood, except for antibody-conjugated magnetic immunoassays (ACMIA, Siemens Diagnostics) for cyclosporine and tacrolimus, which utilize online mixing and ultrasonic lysis of whole blood. However, ACMIA cyclosporine and tacrolimus immunoassays may show falsely elevated levels due to interference of heterophilic antibody.
- Immunoassays for mycophenolic acid do not require serum pretreatment because the assays can be run using serum or plasma. Usually acyl glucuronide (minor active metabolite) is responsible for interferences in some immunoassays.
- Low hematocrit can cause interference (false positive results) with the microparticle enzyme immunoassay (MEIA) for tacrolimus (Abbott

Laboratories), but newer tacrolimus assays from the same manufacturer on the Abbott Architect analyzer are free from such interferences.

CASE REPORT

A 59-year-old man underwent a kidney transplant and was managed with tacrolimus and corticosteroids. For the first three weeks after transplant the patient's tacrolimus whole blood concentrations were consistent with dosage and were below 12 ng/mL. Twenty-five days after transplant, his tacrolimus level measured by the ACMIA tacrolimus assay was found to be highly elevated to 21.5 ng/mL. Tacrolimus was discontinued, but the tacrolimus level was still elevated. Suspecting interference, tacrolimus was analyzed using microparticle enzyme immunoassay (MEIA assay, Abbott Laboratories), and the observed value was below 2 ng/mL, indicating interference in the tacrolimus measurement using the ACMIA assay. The authors suggested that if the tacrolimus value measured by the ACMIA assay did not match the clinical picture, tacrolimus must be measured by an alternative method before any clinical intervention. They concluded that the interference was probably due to the presence of heterophilic antibody [12].

15.8 INTERFERENCES IN ANALYSIS OF ANTIBIOTICS

Usually immunoassays used for monitoring various aminoglycosides and vancomycin are robust and relatively free from interferences. There are only a few case reports of paraprotein interference in vancomycin immunoassay causing falsely lower values.

CASE REPORT

A 68-year-old woman with a history of lymphoplasmacytic lymphoma with an IgM kappa monoclonal component of 42.8 g/L was started on vancomycin, and on Day three her trough vancomycin concentration was <0.1 µg/mL as measured by Beckman Coulter SYNCHRON competitive turbidimetric immunoassay. This was inconsistent with the vancomycin therapy. When the specimen was sent to another laboratory and analyzed using a competitive enzyme-linked immunoassay (Olympus analyzer), a value of 9.8 µg/mL was obtained, indicating that the vancomycin level was falsely lowered (negative interference) due to the presence of paraprotein in the specimen [13].

Gentamicin is not a single molecule, but a complex of three major (C_1, C_{1a}, and C_2) and several minor components. In addition, the C_2 component is a mixture of stereoisomers. Most immunoassay methods can measure a total gentamicin concentration in serum or plasma but are incapable of measuring individual components.

KEY POINTS

- Immunoassays are commonly used for therapeutic drug monitoring. Chromatographic methods used for therapeutic drug monitoring include gas chromatography (GC) with flame ionization or nitrogen detection, gas chromatography combined with mass spectrometry (GC/MS, commonly used for confirming abused drugs), high-performance liquid chromatography (HPLC) combined with ultraviolet or fluorescence detection and high-performance liquid chromatography combined with mass spectrometry (LC/MS) or tandem mass spectrometry (LC/MS/MS). LC/MS/MS is a very specific method and is considered the gold standard for therapeutic drug monitoring because it is virtually free from interferences.

- If a specimen for therapeutic drug monitoring is collected in a gel separation tube, the concentration of drugs such as phenytoin, carbamazepine, phenobarbital, lidocaine, quinidine, and tricyclic antidepressants may be reduced unless analysis is performed within two hours of collection (insignificant absorption). Usually absorption of these few drugs depends on specimen volume, and absorption is much less if the tube is completely filled to capacity rather than half-filled.

- Interferences in digoxin immunoassays may be observed due to the presence of endogenous factors such as digoxin-like immunoreactive substances (DLIS), heterophilic antibody (very rare interference), Digibind or DigiFab, or may be seen in patients receiving spironolactone. However, currently marketed digoxin immunoassays that use very specific monoclonal antibodies are minimally affected by these interfering substances (except Digibind and DigiFab).

- For a patient overdosed with digoxin and being treated with Digibind or DigiFab, only free digoxin should be monitored because pharmacologically active free digoxin correlates better with therapeutic success compared to the total digoxin level. In addition, Digibind and DigiFab are absent in protein-free ultrafiltrates (molecular weight around 46,000, but the filter used for making ultrafiltrate has a molecular weight cutoff of 30,000 Daltons), and monitoring free digoxin eliminates interference from Digibind and DigiFab in digoxin immunoassays.

- Chinese medicines such as Chan Su, Lu-Shen-Wan, and oleander-containing herbs interfere with digoxin immunoassays even with more recently introduced monoclonal antibody-based assays. This is because the bufalin found in Chan Su and Lu-Shen-Wan and the oleandrin in the oleander extract have structural similarities to digoxin.

- Cross-reactivity of carbamazepine 10,11-epoxide with carbamazepine immunoassays can vary from 0 to 96%. Therefore, true carbamazepine concentration may be overestimated if a carbamazepine immunoassay with high cross-reactivity towards its active epoxide metabolite is used for therapeutic drug monitoring of carbamazepine; for example, PETINIA assay (PETINIA: particle enhanced turbidimetric inhibition immunoassay, Siemens Diagnostics; 96% cross-reactivity with epoxide). Hydroxyzine and cetirizine are antihistamine drugs that interfere only with the PETINIA carbamazepine immunoassay.

■ Fosphenytoin is a prodrug of phenytoin. In uremic patients, after fosphenytoin administration, phenytoin concentrations measured by various immunoassays may be significantly higher than true phenytoin levels as determined by high-performance liquid chromatography. This is due to the presence of a unique oxymethylglucuronide metabolite derived from fosphenytoin in the sera of uremic patients that cross-reacts with phenytoin immunoassays.

■ Phenothiazine and its metabolites, carbamazepine, oxcarbazepine, and quetiapine, interfere with immunoassays for tricyclic antidepressants. However, interferences of diphenhydramine (Benadryl) and cyproheptadine with TCA immunoassays occur only in overdosed patients.

■ Chromatographic methods, especially liquid chromatography combined with mass spectrometry (LC/MS) or tandem mass spectrometry (LC/MS/MS), are the gold standard for therapeutic drug monitoring of immunosuppressants because all immunoassays for various immunosuppressants suffer significantly from metabolite cross-reactivity and show significant positive bias when compared to values obtained by LC/MS or LC/MS/MS methods.

■ There are only a few case reports of paraprotein interference in vancomycin immunoassay causing falsely lower values.

REFERENCES

[1] Polson J, Wians FH, Orsulak P, Fuller D, et al. False positive acetaminophen concentrations in patients with liver injury. Clin Chim Acta 2008;391:24−30.

[2] Dasgupta A, Zaldi S, Johnson M, Chow L, et al. Use of fluorescence polarization immunoassay for salicylate to avoid positive/negative interference by bilirubin in the Trinder salicylate assay. Ann Clin Biochem 2003;40:684−8.

[3] Dasgupta A. Therapeutic drug monitoring of digoxin: impact of endogenous and exogenous digoxin-like immunoreactive substances. Toxicol Rev 2006;25:273−81.

[4] Gruber KA, Whitaker JM, Buckalew VM. Endogenous digitalis-like substances in plasma of volume expanded dogs. Nature 1980;287:743−5.

[5] Dasgupta A, Trejo O. Suppression of total digoxin concentration by digoxin-like immunoreactive substances in the MEIA digoxin assay: elimination of interference by monitoring free digoxin concentrations. Am J Clin Pathol 1999;111:406−10.

[6] Liendo C, Ghali JK, Graves SW. A new interference in some digoxin assays: anti-murine heterophilic antibodies. Clin Pharmacol Ther 1996;60:593−8.

[7] Tzou MC, Reuning RH, Sams RA. Quantitation of interference in digoxin immunoassay in renal, hepatic and diabetic disease. Clin Pharm Ther 1997;61:429−41.

[8] Fyer F, Steimer W, Muller C, Zilker T. Free and total digoxin in serum during treatment of acute digoxin poisoning with Fab fragments; Case study. Am J Crit Care 2010;19:387−91.

[9] Steimer W, Muller C, Eber B. Digoxin assays: frequent, substantial and potentially dangerous interference by spironolactone, canrenone and other steroids. Clin Chem 2002;48:507−16.

[10] Parant F, Moulsma M, Gagnieu MV, Lardet G. Hydroxyzine and metabolite as a source of interference in carbamazepine particle-enhanced turbidimetric inhibition immunoassay (PETINIA). Ther Drug Monit 2005;27:457−62.

[11] Annesley T, Kurzyniec S, Nordblom G, Buchanan N, et al. Glucuronidation of prodrug reactive site: isolation and characterization of oxymethylglucuronide metabolite of fosphenytoin. Clin Chem 2001;46:910−8.

[12] D'Alessandro M, Mariani P, Mennini G, Severi D, et al. Falsely elevated tacrolimus concentrations measures using the ACMIA method due to circulating endogenous antibodies in a kidney transplant recipient. Clin Chim Acta 2011;412:245−8.

[13] Simons SA, Molinelli AR, Sobhani K, Rainey PM, Hoofnagle A. Two cases with unusual vancomycin measurements. Clin Chem 2009;55:578−82.

CHAPTER 16

Drugs of Abuse Testing

16.1 COMMONLY ABUSED DRUGS IN THE UNITED STATES

The National Survey on Drug Use and Health conducted by the Substance Abuse and Mental Health Services Administration (SAMHSA) reported that an estimated 22.6 million Americans (8.9% of the population) aged 12 or older were current illicit drug users. Illicit drugs abused by Americans include marijuana/hashish, cocaine (including crack cocaine), amphetamines, heroin, opioids, benzodiazepines, and barbiturates [1]. Abuse of prescription drugs (including those obtained illegally) is also becoming an epidemic in the U.S. The most commonly abused prescription drugs are benzodiazepines, opioids, and some psychoactive drugs. In general, drugs that are commonly abused can be classified under different categories (Table 16.1). Key points regarding drug abuse include:

- Marijuana is the most popular abused drug in the U.S. according to the SAMHSA survey.
- Abuse of barbiturates is declining, but abuse of various benzodiazepines, especially alprazolam, lorazepam, and temazepam, is on the rise.
- Abuse of narcotic analgesics, particularly oxycodone, is also increasing.

Recent reports also indicate an increase in the rates of hospitalization due to drug abuse in the U.S. Excessive consumption of alcohol combined with drug abuse is prevalent in young adults ages 18 to 24. The authors commented that strong efforts are needed to educate the public regarding the risk of drug overdose, especially if drugs are consumed with alcohol [2]. Currently, drug and alcohol testing is conducted in almost all hospital laboratories. Drugs of abuse testing is usually conducted using a urine specimen and commercially available immunoassays. Immunoassays are the first step in drugs of abuse analysis, primarily because they can be easily automated using automated analyzers (especially chemistry analyzers). Alternatively, immunoassay

CONTENTS

A. Dasgupta and A. Wahed: Clinical Chemistry, Immunology and Laboratory Quality Control
DOI: http://dx.doi.org/10.1016/B978-0-12-407821-5.00016-4

16.11 Issues of
Adulterated Urine
Specimens in
Workplace Drug
Testing 302

16.12 Miscellaneous
Issues in Drugs of
Abuse Testing 303

Key Points 304

References 305

Table 16.1 Commonly Abused Drugs in the U.S.

Stimulants: amphetamine, methamphetamine, MDMA, MDA, cocaine.

Narcotic Analgesics: heroin, codeine, morphine, oxycodone, hydrocodone, hydromorphone, oxymorphone, buprenorphine, meperidine, methadone, fentanyl, and other synthetic or designer drugs.

Sedatives/Hypnotics: barbiturates (e.g. amobarbital, pentobarbital, secobarbital), benzodiazepines (e.g. diazepam, alprazolam, clonazepam, lorazepam, temazepam, zolpidem, flunitrazepam), and others (e.g. GHB).

Hallucinogens: LSD, marijuana.

Anesthetics: ketamine, phencyclidine.

Date Rape Drugs: GHB, Rohypnol (flunitrazepam).

Abbreviations: GHB, Gamma-Hydroxybutyric Acid; LSD, Lysergic Acid Diethylamide; MDMA, 3,4-Methylenedioxymethamphetamine; MDA, 3,4-Methylenedioxyamphetamine.

principles can also be adopted in a point of care device. However, immunoassays may produce false positive test results due to interferences.

CASE REPORT

A 38-year-old gravida 2, para 0 woman was transferred to the hospital at 28-2/7 weeks of gestation. At a routine appointment two days before admission, she had severe hypertension, pitting edema, hyperreflexia, and proteinuria. On admission, her systolic blood pressure was more than 170 mmHg, and her 24-hour urine sample showed a total protein level of 1,744 mg. A diagnosis of superimposed preeclampsia on chronic hypertension was made. She received betamethasone for fetal lung maturity, magnesium for seizure prophylaxis, and labetalol for blood pressure control. Her urine drug screen tested positive for amphetamine, but she was in the hospital over 48 h with no access to any illicit drugs. The medical team determined that her positive amphetamine in the immunoassay screen was due to labetalol interference because the detection window of amphetamine in urine is only 48 hours [3].

16.2 MEDICAL VS. WORKPLACE DRUG TESTING

The majority of drug testing conducted by a hospital laboratory is medical drug testing. Emergency room physicians routinely order drug testing if a patient appears overdosed based on clinical presentation. In addition, medical drug testing may be ordered by a physician in an outpatient clinic if drug abuse is suspected. On the other hand, workplace drug testing is considered as legal drug testing. A chain of custody must be maintained for legal drug testing where there is a record of the personnel who had the possession of the specimen from the time of collection until the time of analysis and reporting of the result. Moreover, after collection, the specimen must be sealed in front of the donor to ensure specimen integrity. This is important because, based on a positive report, a person may be denied employment or face a disciplinary action, including termination. Therefore, it is also mandatory that all immunoassay positive results be confirmed by an alternative method, preferably gas chromatography/mass spectrometry (GC/MS). If the immunoassay test result is positive, but GC/MS confirmation is negative, the drug testing must be reported as negative. A Medical Review Officer (MRO) who is not affiliated with the testing laboratory must also certify a positive drug test result as a true positive in a legal drug testing setting. This adds another level of safety for the donor because the MRO ensures that there is no alternative explanation for the positive drug testing result. The criteria for medical drug testing are much less stringent, and the privacy of the test result is guaranteed by law. The three most important points regarding legal drug testing include:

- Initial immunoassay test results must be confirmed by a second method, preferably GC/MS.
- The MRO must review and certify the result, indicating there is no alternative explanation for a positive drug test.
- Special certification, such as College of American Pathology (CAP) Forensic Drug Testing certification or SAMHSA certification, is needed for operation of a laboratory that performs workplace drug testing.

CASE REPORT

A 31-year-old female was undergoing pre-employment drug testing, and she stated that she was taking Vicks® DayQuil Cough medicine and oral contraceptives. During pre-employment drug testing, a specimen tested positive for phencyclidine (PCP). However, during the GC/MS confirmation step, no phencyclidine was detected. Therefore, the certifying scientist reported the drug test as "negative" to the MRO. The most likely cause of this false positive immunoassay test result was the presence of dextromethorphan in the specimen, which is known to cause false positive test results. Vicks DayQuil Cough medicine contains dextromethorphan.

16.3 SAMHSA VS. NON-SAMHSA DRUGS

The workplace drug testing program in its present form was initiated in 1986 when President Reagan issued Executive Order 12564, requiring federal agencies to conduct drug testing for federal workers employed in sensitive positions. Initially the National Institute on Drug Abuse (NIDA) was given the responsibility of developing the program. Currently, SAMHSA is the lead agency for a substance abuse program. SAMHSA also certifies laboratories that perform drugs of abuse testing and publishes new guidelines. SAMHSA guidelines mandate testing for five abused drugs, including:

- Amphetamine, including 3,4-methylenedioxymethamphetamine (MDMA, ecstasy).

- Cocaine, tested as benzoylecgonine.
- Opiates.
- Marijuana, tested as a marijuana metabolite.
- Phencyclidine.

These five drugs are usually called the five SAMHSA drugs. Private employers, in addition to these five drugs, may also test for the presence of benzodiazepines, barbiturates, methadone, methaqualone, propoxyphene, oxycodone, and methadone. Although propoxyphene, methadone, and oxycodone are opioids, these drugs cannot be detected by using opiate immunoassays because they show poor cross-reactivity with the morphine antibody used in opiate immunoassays.

16.4 DETECTION WINDOW OF VARIOUS DRUGS IN URINE

Although drug testing can be conducted by using hair, saliva (oral fluid), blood, and sweat, the urine drug test is the most common, and consists of over 90% of the drugs of abuse testings. The advantage of the urine specimen is that it can be collected non-invasively (although hair, oral fluid, and sweat specimens can also be collected non-invasively), and drug metabolites can be detected for a longer time period (window of detection) than in blood. However, an abused drug may be detected for up to six months in a hair specimen. Relative windows of detection for various drugs using urine specimens are summarized in Table 16.2.

Table 16.2 Window of Detection of Abused Drugs

Drug/Drug Class	Detection Window in Urine
SAMHSA-Mandated Drugs	
Amphetamines	2 days
3,4-Methylenedioxymethamphetamine	2 days
Cannabinoids	2–3 days single use/30 days after chronic abuse
Cocaine metabolites	2 days single use/4 days after repeated use
Opiates	2–3 days
Phencyclidine	14 days
Non-SAMHSA Drugs	
Barbiturates	Short-acting (secobarbital, etc.), 1 day long-acting (phenobarbital), 21 days
Benzodiazepines	Short-acting (alprazolam, etc.), 3 days long-acting (diazepam, etc.), 30 days
Methadone	3 days
Methaqualone	3 days
Propoxyphene	1–2 days
Oxycodone	2–4 days

Although a parent drug is present in blood, oral fluid, or hair, many drugs are detected through their respective metabolites in urine specimens. For example, cocaine abuse is detected in urine by confirming the presence of benzoylecgonine, the major inactive metabolite of cocaine. Immunoassays for detecting various drugs in urine often target a major metabolite of the parent drug, although some drugs are also detected as the parent drug in the urine specimen.

16.5 METABOLISM OF ABUSED DRUGS/TARGET OF IMMUNOASSAY ANTIBODIES

The liver metabolizes most drugs of abuse, and these metabolites are excreted in the urine. Most drugs are metabolized in the liver by cytochrome P-450 mixed-function oxidase (CYP450), while CYP3A4 and CYP2D6, the two isoforms of cytochrome P-450, are responsible for the metabolism of the majority of drugs in the liver. However, there are some drugs that are not extensively metabolized by the liver (e.g. amphetamine). For such drugs the antibody used in the immunoassay can be designed to detect the parent drug. Some abused drugs are rapidly metabolized, such as cocaine, which is broken down by plasma butyrylcholinesterase into ecgonine methyl ester. Another major metabolite of cocaine, benzoylecgonine, probably arises spontaneously in plasma by hydrolysis of cocaine in vivo. Benzoylecgonine and ecgonine methyl ester represent the major urinary excretions of cocaine. A small amount of unchanged cocaine can also be recovered in urine. Liver enzymes can also metabolize cocaine into nor-cocaine. The elimination half-life of cocaine is approximately 45 minutes. Abusing cocaine and alcohol simultaneously is dangerous because benzoylecgonine undergoes transesterification in the presence of alcohol (ethyl alcohol) to form cocaethylene; this process is facilitated by the liver enzyme carboxylesterase. Two important points regarding cocaine metabolism are:

- The antibody used in immunoassays for cocaine usually targets benzoylecgonine.
- Cocaethylene is an active metabolite responsible for life-threatening toxicity if cocaine and alcohol are abused simultaneously.

Heroin (diacetyl morphine) is first metabolized to 6-acetylmorphine (also called 6-monoacetyl morphine), and then to morphine by hydrolysis of the ester linkage by pseudocholinesterase in serum and by human carboxylesterase-1 and carboxylesterase-2 in the liver.

$$\text{Heroin (diacetyl morphine)} \rightarrow \text{6-Acetylmorphine}$$
$$\text{6-Acetylmorphine} \rightarrow \text{Morphine}$$

6-Acetylmorphine, which can be confirmed by gas chromatography/mass spectrometric analysis (GC/MS), is considered as the marker compound for heroin abuse. The majority of morphine is excreted in urine as morphine-3-glucuronide. This metabolite is formed by conjugation in the liver by the action of the liver enzyme uridine diphosphate glucuronosyltransferase. Codeine is metabolized to morphine in liver, mostly by CYP2D6. Although morphine is a stronger narcotic analgesic than codeine, morphine is poorly absorbed from the stomach while codeine is easily absorbed after oral administration. However, the conversion of codeine to morphine is essential for pain relief. Almost 10% of the Caucasian population has a genetic polymorphism of CYP2D6 that makes the enzyme less active; these patients may not get adequate pain relief during codeine therapy [4]. Hydromorphone is also excreted in the urine, mostly in conjugated form, but a small part of free hydromorphone can also be recovered in urine. Oxycodone is metabolized to oxymorphone, which is then conjugated in liver. However, the antibody used in opiate immunoassays only targets morphine. Therefore, the presence of oxycodone in urine cannot usually be detected by opiate immunoassays due to low cross-reactivity with the antibody.

Δ^9-Tetrahydrocannabinol (marijuana, THC) is the most active component (out of over 60 related compounds identified) of marijuana. THC is found in various parts of the cannabis plant (*Cannabis sativa*), including flowers, stems, and leaves. THC is rapidly oxidized by cytochrome P-450 enzymes to the active metabolite, 11-hydroxy Δ^9-tetrahydrocannabinol (11-OH-THC), and the inactive metabolite, 11-nor-9-carboxy Δ^9-tetrahydrocannabinol (THC-COOH). In immunoassays for THC, the antibody recognizes the inactive metabolite THC-COOH.

Phencyclidine (PCP) undergoes extensive metabolism by liver cytochrome P-450 enzymes (especially CYP3A4) into several hydroxy metabolites, including cis-1-(1-phenyl-4-hydroxycyclohexyl) piperidine, trans-1-(1-phenyl-4-hydroxycyclohexyl) piperidine, 1-(1-phenylcyclohexyl)-4-hydroxypiperidine, and 5-(1-phenylcyclohexylamino) pentanoic acid. The elimination half-life of PCP varies significantly in humans (7 to 57 h; average 17 h). Despite extensive metabolism of PCP, a significant amount of the parent drug is also present in urine.

Barbiturates can be ultra-short-acting, short-acting, or long-acting (phenobarbital). Barbiturates are extensively metabolized. However, many barbiturate immunoassays use antibodies that recognize secobarbital. Although there are more than 50 benzodiazepines, only about 15 are used in the U.S. Benzodiazepines are also extensively metabolized. The antibody in many benzodiazepine immunoassays targets oxazepam because it is a common metabolite of diazepam and temazepam. Antibody targets of various

Table 16.3 Target Analytes in Immunoassays Used for Drugs of Abuse Testing

Drugs of Abuse	Antibody Target
Amphetamine/methamphetamine	Methamphetamine or amphetamine
3,4-Methylenedioxymethamphetamine	3,4-Methylenedioxymethamphetamine
Cocaine	Benzoylecgonine (metabolite)
Opiate	Morphine
Oxycodone	Oxycodone
Heroin	6-Acetylmorphine (metabolite and marker of abuse)
Marijuana	11-Nor-9-carboxy Δ^9-tetrahydrocannabinol (THC-COOH, metabolite)
Phencyclidine (PCP)	Phencyclidine
Benzodiazepine	Commonly oxazepam or nordiazepam
Barbiturates	Commonly secobarbital

immunoassays used for drugs of abuse testing are summarized in Table 16.3.

16.6 IMMUNOASSAYS VS. GC/MS CUT-OFF CONCENTRATIONS

Immunoassays for various drugs have different cut-off concentrations as mandated by SAMHSA guidelines. If the concentration of the drug is below the cut-off concentration, the immunoassay test result is considered negative. In workplace drug testing, a positive immunoassay test result must be confirmed by using a different analytical technique, preferably GC/MS, but the confirmation concentration could be lower than the immunoassay screening cut-off concentration. Immunoassay and GC/MS cut-off concentrations of various drugs are listed in Table 16.4. Important points regarding the GC/MS confirmation step for amphetamines as recommended by SAMHSA include:

- If methamphetamine is confirmed by GC/MS, its metabolite (amphetamine) must also be confirmed by GC/MS to at least a 100 ng/mL concentration.
- This protocol ensures that no pseudoephedrine or ephedrine present in the urine is converted into methamphetamine during the GC/MS procedure as an artifact.
- If the injector temperature in the GC is high, ephedrine/pseudoephedrine may be thermally converted into methamphetamine.

Table 16.4 Immunoassay and GC/MS Confirmation Cut-Off Concentrations for Various SAMHSA and Non-SAMHSA Drugs

Drug/Drug Class	Immunoassay (ng/mL)	GC-MS (ng/mL)	
SAMHSA Drugs			
Amphetamines/ methamphetamine	500	Amphetamine	250
		Methamphetamine	250
Cannabinoids	50	THC-COOH	15
Cocaine metabolites	150	Benzoylecgonine	100
Opiates	2,000	Morphine	2,000
		Codeine	2,000
		6-Acetylmorphine	10
Phencyclidine	25	Phencyclidine	25
Non-SAMHSA Drugs			
Barbiturates	200	Usually 200 for individual drugs	
Benzodiazepines	200	Usually 200 for individual drugs	
Methadone	300	Methadone	300
Propoxyphene	300	Propoxyphene	300
Oxycodone	100 or 300	Oxycodone	100

16.7 FALSE POSITIVE IMMUNOASSAY TEST RESULTS WITH VARIOUS ABUSED DRUGS

Immunoassays are subject to false positive test results. Amphetamine immunoassays suffer from more interferences than any other immunoassays used for screening of abused drugs in urine. Many over-the-counter cold and cough medications containing ephedrine or pseudoephedrine can produce false positive test results with amphetamine/methamphetamine immunoassays. Many sympathomimetic amines found in over-the-counter medications can also cause false positive test results. In addition, other drugs such as bupropion (mostly metabolites), trazodone, labetalol, and doxepin can also cause false positive test results (Table 16.5). Amphetamine and methamphetamine have optical isomers designated as d (+) for dextrorotatory and l (−) for levorotatory. The d isomer (which is the abused isomer) is the intended target of immunoassays. Ingestion of medications containing the l isomer can cause false positive results. For example, Vicks® inhaler contains the active ingredient l-methamphetamine[5], and extensive use of this product can cause false positive results for immunoassay screening.

Table 16.5 Most Commonly Encountered Drugs in False Positive Test Results Using Immunoassays for Various Abused Drugs

Immunoassay	Interfering Drugs*
Amphetamine	Ephedrine, pseudoephedrine, phentermine, tyramine, methylphenidate, perazine, doxepin, labetalol
Benzodiazepines	Oxaprozin, sertraline
Opiates	Diphenhydramine, rifampin, dextromethorphan, verapamil, fluoroquinolones
Phencyclidine	Dextromethorphan, diphenhydramine, ibuprofen, imipramine, ketamine, meperidine, thioridazine, tramadol
Tetrahydrocannabinol	Non-steroidal antiinflammatory drugs, pantoprazole, efavirenz
Methadone	Diphenhydramine

These drugs are the most commonly encountered clinically, but this list does not provide names of all drugs that are known to interfere with various immunoassays (as based on published literature). Please see any toxicology reference book for in-depth information on this topic.

CASE REPORT

An emergency medical team responded at the residence of a 77-year-old man who had difficulty breathing. The patient was transported to the emergency department and was admitted. Approximately 12 h after admission he was unresponsive and died. His medical history included heart failure, atrial flutter, and bronchial asthma. His urine screen was positive for amphetamine, and GC/MS confirmation was positive for acetaminophen, nicotine, cotinine, caffeine, diltiazem, doxylamine, and methamphetamine. When chiral analysis was performed, the methamphetamine was identified as the *l*-isomer, not the *d*-isomer. Further investigation revealed that the decedent frequently used a Vicks inhaler for his bronchial asthma, which led to the positive amphetamine screen as well as the GC/MS confirmation [6].

Certain fluoroquinolone antibiotics can cause false positive test results with opiate immunoassay screening. PCP is rarely abused today, but the dextromethorphan present in many over-the-counter cough mixtures can cause false positive test results with PCP immunoassays. However, GC/MS confirmation is negative, indicating the absence of PCP in the urine specimen. Studies indicate that non-steroidal antiinflammatory drugs can produce false positive results in immunoassay screening [7]. Some of the drugs that can cause false positive test results with immunoassays used for detecting various drugs of abuse in urine are summarized in Table 16.5.

CASE REPORT

A 3-year-old girl was hospitalized due to behavioral disturbances of unknown origin. She was treated with niflumic acid (including suppositories) five days prior to hospitalization. The urine toxicology screen was positive for cannabinoid (marijuana), but her parents strongly denied any exposure to marijuana. Another specimen from the same patient was analyzed using chromatographic techniques and no cannabinoid was detected. The authors suspected that niflumic acid interfered with the immunoassay for cannabinoid and caused the false positive test result [8].

Important points regarding false positive immunoassay test results include:

- Amphetamine/methamphetamine immunoassay screening tests are most commonly subject to interferences from over-the-counter cold medications containing ephedrine/pseudoephedrine.
- Although Vicks® inhaler contains *l*-methamphetamine, which has less cross-reactivity with amphetamine/methamphetamine immunoassays that target *d*-methamphetamine, excessive use of Vicks® inhaler can cause false positive test results with both immunoassays and GC/MS confirmations. Chiral derivatization is essential to resolve this issue (it distinguishes the *d*-from the *l*-isomer).
- Positive opiate test results may be observed due to therapy with fluoroquinolone antibiotics, but GC/MS confirmation should be negative.
- A common cause of positive PCP immunoassay screening results is the presence of dextromethorphan in the specimen. GC/MS confirmation should be negative.

16.8 FALSE NEGATIVE TEST RESULTS

Although false positive test results with immunoassays are more common, false negative test results (where the drug was present in the specimen but was not detected) can also be encountered during immunoassay screening steps. Obviously, if the concentration of a particular drug is below the cut-off concentration of the immunoassay, the test result should be negative. Major reasons for false negative test results during immunoassay screenings include:

- Drug concentration may be below the cut-off concentration. This is a problem when detecting various benzodiazepines.
- Cross-reactivity of a drug is poor with the antibody of the assay. The best example of this is poor cross-reactivity of oxycodone with various opiate immunoassays.
- The drug may not have been metabolized and appears in the urine. A common example is a negative urine test for benzoylecgonine (cocaine metabolite) in a patient with acute cocaine overdose.

Various benzodiazepines are common prescription drugs in the U.S. They are widely abused. Physicians sometimes order urine drug tests in their patients to ensure compliance. However, when using a 200 ng/mL cut-off concentration for benzodiazepine immunoassays, certain drugs may not be detected after taking the recommended dosage due to low concentrations of the drug in urine (in particular alprazolam, clonazepam, and lorazepam). Clonazepam is metabolized to 7-aminoclonazepam. West et al. reported that, when urine specimens collected from subjects taking clonazepam were tested using the DRI benzodiazepine immunoassay at a 200 ng/mL cut-off, only 38 specimens out of 180 tested positive by the immunoassay (21% positive). However, using liquid chromatography combined with tandem mass spectrometry (LC/MS/MS), 126 specimens out of 180 specimens tested positive (70% positive) when the detection limit of the LC/MS/MS assay was set at 200 ng/mL, the same cut-off used by the DRI benzodiazepine assay. This indicated poor detection capability of the benzodiazepine immunoassay for clonazepam and its metabolite in urine. The authors concluded that the 200 ng/mL cut-off may not be adequate for monitoring patient compliance with clonazepam therapy [9].

Opiate immunoassays usually utilize a morphine-specific antibody, and certain opioids cannot be detected by opiate immunoassays due to low cross-reactivities. These opioids include:

- Oxycodone and oxymorphone (keto-opioids).
- Methadone.
- Propoxyphene.
- Fentanyl and its analogs.

CASE REPORT

A 40-year-old man suffering from severe chronic migraine attack was treated with oxycodone (20 mg dosage twice a day). Although his pain was under control, he routinely called the clinic stating that he had finished his medication faster and needed a refill. Suspecting that the patient was selling oxycodone on the street, a urine toxicology test was ordered (it was negative for opiates) and the patient was dismissed from the pain clinic. On his behalf, a family member contacted a toxicologist who informed that oxycodone can cause a false negative test result in a urine opiate drug screen due to poor cross-reactivity of oxycodone with the antibody used in certain opiate assays. An aliquot of the original urine specimen was re-tested using GC/MS and the presence of oxycodone at 1,124 ng/mL was confirmed. This value confirmed that the patient was compliant with his oxycodone usage [10].

Cocaine overdose can be fatal. Sometimes the victim dies from cocaine overdose, but a urine toxicology screen for the cocaine metabolite benzoyl-ecgonine can be negative because death can occur quickly from cocaine overdose and sufficient time has not passed for the metabolite to be excreted in the urine in enough concentration to trigger a positive response.

CASE REPORT

A 44-year-old Caucasian man was found dead in his apartment and police transported the body to the coroner's office. The urine toxicology screen using immunoassays (EMIT assays) was negative for benzoylecgonine (cocaine metabolite) using a 300 ng/mL cut-off concentration. However, GC/MS analysis of a post-mortem heart blood specimen showed a very high cocaine concentration of 18,330 ng/mL, and it was concluded that the cause of death was cocaine overdose [11].

16.9 DERIVATIZATION IN GC/MS CONFIRMATION TESTING

Although urine specimens can be directly used for immunoassay analysis for various abused drugs, a GC/MS confirmation test requires extraction of drugs from the urine specimen using organic solvents or solid-phase extraction (using BondElute or another solid-phase extraction column). In general, hexane, acetone, chloroform, methanol, ethyl acetate, 1-chlorobutane, or dichloromethane are typically used for extraction. Some drugs, such as amphetamine and methamphetamine, can be extracted directly from the urine specimen using an organic solvent. However, a drug such as morphine is present in urine as a glucuronide conjugate, which is water-soluble. Therefore, prior to extraction, it is important to break this conjugate with acid or alkali hydrolysis. It is also possible to cleave the conjugate by using beta-glucuronidase enzyme. Although acid or alkali hydrolysis can be conducted by heating specimens for 30 min, usually several hours or more are needed for enzymatic hydrolysis; enzymatic hydrolysis is a gentler method.

Derivatization is a chemical reaction where a polar group in a molecule (e.g. a carboxyl or hydroxyl group) is chemically converted to a non-polar group in order to make the molecule volatile so that it can be analyzed by GC/MS. In general, molecules containing a carboxylic acid (−COOH, e.g. THC-COOH and benzoylecgonine), hydroxyl (−OH, e.g. morphine), or primary or secondary amine (such as amphetamine and methamphetamine) require derivatization. Usually amphetamine and methamphetamine are analyzed as trifluoroacetyl or pentafluoropropyl derivatives, although many other derivatization methods have been described in the literature. Benzoylecgonine is usually analyzed as a trimethylsilyl or acetyl derivative. Codeine and morphine can also be analyzed as a trimethylsilyl derivative, while phencyclidine can be analyzed without derivatization.

16.10 ANALYTICAL TRUE POSITIVE DUE TO USE OF PRESCRIPTION DRUGS AND OTHER FACTORS

As expected, if a person is taking codeine for pain control, the opiate test should be positive. Similarly, a positive benzodiazepine test result should be expected if a person is taking a prescription medication of the benzodiazepine class. Amphetamine and methamphetamine, as well as their analogs, are used in treating attention deficit disorders, and, as expected, use of such drugs could lead to a positive amphetamine/methamphetamine drug test result. Although not employed frequently, cocaine is still used as a local anesthetic in ENT surgery, and a patient may test positive for cocaine metabolites 1−2 days after the procedure. Poppy seeds contain both codeine and morphine. Therefore, eating foods containing poppy seeds can cause false positive opiate test results. In order to circumvent this problem, the cut-off concentration of opiate was be increased to 2,000 ng/mL from the original cut-off level of 300 ng/mL in federal drug testing programs. However, some private employers still use the 300 ng/mL cut-off concentration for opiate testing in their workplace drug testing protocols. Eating poppy seed-containing foods prior to drug testing (1−2 days) can cause a positive opiate test result if the 300 ng/mL cut-off concentration is used. Hemp oil is prepared from hemp seed, which may contain trace amounts of cannabinoid (marijuana). However, ingesting hemp oil should not cause a positive marijuana test result at the 50 ng/mL cut-off level. Analytical true positive test results due to the use of prescription medications or other factors are summarized in Table 16.6.

CASE REPORT

A 54-year-old Caucasian male was involved in a physical altercation and suffered a broken nose. Three days later a surgical procedure was performed to repair his nose and he was released from the hospital later that day with a prescription of Percocet (oxycodone, 5 mg; and acetaminophen, 325 mg) for pain relief. He took one tablet at 4 PM and went to bed, but at 8 PM when his wife tried to wake him, he was unconscious. She called the emergency medical team, and, despite performing CPR, he was pronounced dead at the scene at 8:40 PM. Toxicological analysis revealed the presence of cocaine (48 ng/mL) and benzoylecgonine (482 ng/mL) in his urine specimen, but no cocaine or its metabolite was detected in the subclavian blood specimen. In addition, oxycodone (73 ng/mL) was present in his blood as expected. Upon detection of cocaine and its major metabolite in his urine specimen, it was initially suspected that the subject had abused illicit cocaine, but his family denied any drug abuse by the deceased. It was later revealed that cocaine was used during his nasal surgery as a local anesthetic, which explained the presence of both cocaine and its metabolite (benzoylecgonine) in his urine specimen: it was from medical use of cocaine [12].

Table 16.6 Analytical Positive Drug Test Results (GC/MS Confirmation) Due to Use of Prescription Medication and Other Factors

Positive Drug Test	Prescription Medication
Amphetamine	Adderall (contains amphetamine), lisdexamfetamine, desoxyn (contains methamphetamine), clobenzorex, ethyl amphetamine, mefenorex, benzphetamine, famprofazone.
Cocaine	Cocaine use (as topical anesthetic) during ENT procedure. Drinking Inca tea or Coca de Mate tea (South American teas) prepared from coca leaves.
Barbiturates	Short-and long-acting barbiturates.
Benzodiazepines	Short-and long-acting benzodiazepines.
Opiates	Codeine, morphine, oxymorphone, hydrocodone, hydromorphone, oxycodone. Eating poppy-seed-containing food.
Phencyclidine	None.
Tetrahydrocannabinol	Marinol (synthetic marijuana); drinking hemp oil should not cause a positive test at 50 ng/mL cut-off.
Methadone	None.

Important points regarding analytical true positives include:

- Prescription use of various drugs can cause analytical true positive test results.
- Positive opiate test results can occur in a person who eats poppy seed-containing food prior to drug testing, especially if the private employer still uses a 300 ng/mL cut-off concentration for opiates.
- Drinking coca tea (Health Inca tea, Inca tea, or Mate de Coca) can cause a positive cocaine test result because these teas originate from South America and are prepared from coca leaves.
- Use of cocaine as a local anesthetic during ENT surgery can cause positive cocaine test results 1−2 days after surgery.
- Ingestion of hemp oil or passive inhalation of marijuana should not cause positive marijuana test results.

16.11 ISSUES OF ADULTERATED URINE SPECIMENS IN WORKPLACE DRUG TESTING

People try to beat workplace drug tests. Usually people drink detoxifying agents purchased off the Internet. Contrary to the claim that these agents can purge drugs out of circulation, these products contain only diuretics such as caffeine or hydrochlorothiazide, and if a person taking such products also drinks plenty of water (as suggested in the package insert), it produces diluted urine. Because specimen integrity checks (pH, creatinine, specific gravity, and temperature) are routinely used in all urine specimens collected for workplace drug testing, low creatinine (below 20 mg/dL) indicates that urine is diluted and no further analysis is conducted. Specimen adulteration is considered as a "refuse to test" and the person is denied employment. Various household chemicals such as table salt (sodium chloride), vinegar, liquid soap, liquid laundry bleach, sodium bicarbonate (baking soda), and lemon juice are added to urine after collection in order to beat drug tests. Although some of these products can invalidate the immunoassay screening step, all of them can be detected indirectly by specimen integrity testing, and thus the specimen will be rejected for further testing due to adulteration. Although Visine® Eye Drops significantly interfere with enzyme multiplied immunoassay technique (EMIT) assays and other immunoassays, the presence of Visine Eye Drops in an adulterated specimen cannot be detected by routine specimen integrity testing.

More recently, several products have become available through the Internet for use as in vitro adulterants. These products usually contain strong oxidizing agents such as potassium nitrite, pyridinium chlorochromate, glutaraldehyde, and Stealth (hydrogen peroxide and peroxidase). These adulterants cannot be detected by routine specimen integrity testing, and such products can mask tests of marijuana metabolites not only in the immunoassay screening step, but also in the GC/MS confirmation step because THC-COOH is slowly oxidized by these products. However, spot tests and special urine dipstick analysis (AdultaCheck 6 or 10, Intect 7, etc.) are available for detecting these adulterants [13].

Sometimes pain management patients sell their prescribed narcotic analgesics on the street for more money. In addition, they may go to various doctors (doctor shopping) to obtain extra prescriptions for narcotics. Drug testing in urine is often conducted to ensure they are compliant with therapy. These people want their urine drug test to be positive for the prescription drug and often grind up the tablet to produce a fine powder that easily dissolves in the urine. However, during analysis, high quantities of the parent drug are recovered in the urine as expected, but no metabolite is detected, indicating that these patients are attempting to cheat on their drug test. Important points regarding urine adulteration:

- If household chemicals are used as urinary adulterants, specimen integrity testing (pH, creatinine, specific gravity, and temperature) should be able to identify such adulterants.
- The presence of Visine® Eye Drops in adulterated urine cannot be determined by specimen integrity testing.
- The presence of adulterants purchased off the Internet, such as potassium nitrite, glutaraldehyde, pyridinium chlorochromate, and Stealth, cannot be detected by specimen integrity testing. However, spot tests and special urine dipsticks (AdultaCheck 6 or 10, Intect 7) can be used for detection.
- The presence of an adulterant in a urine specimen subject to workplace drug testing is considered as a "refusal to test" and the person may be denied the job (no further testing is usually conducted).

16.12 MISCELLANEOUS ISSUES IN DRUGS OF ABUSE TESTING

Lysergic acid diethylamide (LSD) was widely abused in the 1960s and 1970s, and in some U.S. cities this drug is reappearing at rave parties. There is a commercially available immunoassay for LSD testing. The rave party drug 3,4-methylenedioxymethamphetamine (MDMA) can be detected by some amphetamine immunoassays and also by using special immunoassays that detect both MDMA and amphetamine. However, other rave party drugs like ketamine and gamma-hydroxybutyric acid cannot be detected by routine toxicology testing, although reference toxicology laboratories offer testing for such drugs. Many designer drugs cannot be detected by routine toxicological analysis, including the new designer drugs spice, K2 (synthetic marijuana), and bath salts (synthetic amphetamine analogs).

As mentioned earlier, over 90% of drug testing is conducted using urine specimens. Hair testing provides a larger window of detection (up to six months), while saliva (oral fluid) testing can identify impairment due to recent abuse of a drug. In European countries, oral fluid testing is gaining popularity for identifying impaired drivers. In addition, sweat testing can be used for continuous monitoring of a drug in a person undergoing drug rehabilitation.

KEY POINTS

- Urine drug testing represents approximately 90% of all drug testing. Although initial immunoassay screening results using urine specimens may not be confirmed by gas chromatography/mass spectrometry (GC/MS) in medical drug testing, GC/MS confirmation (or confirmation by another method) is mandatory in legal drug testing. In addition, in legal drug testing the medical review officer (MRO) must review and certify the positive drug test result, thus confirming that there is no alternative explanation for a positive drug test. Moreover, a chain of custody must be maintained.
- Currently, the Substance Abuse and Mental Health Services Administration (SAMHSA) have guidelines for federal workplace drug testing. SAMHSA guidelines require testing for five abused drugs, including amphetamines (amphetamine, methamphetamine, and 3,4-methylenedioxymethamphetamine; MDMA, ecstasy), cocaine (tested as benzoylecgonine), opiates, marijuana (tested as marijuana metabolites), and phencyclidine (PCP).
- In urine, most drugs can be detected for only 2–3 days after abuse, except for PCP (14 days) and marijuana (up to 30 days, from chronic abuse).
- Amphetamine/methamphetamine immunoassay screening tests are subject to interferences, most commonly from over-the-counter cold medications containing ephedrine/pseudoephedrine.
- Although Vicks Inhaler contains *l*-methamphetamine, which has less cross-reactivity with amphetamine/methamphetamine immunoassays (*d*-methamphetamine is abused and measured by amphetamine/methamphetamine immunoassays), excessive use of Vicks Inhaler can cause false positive test results with both immunoassay and GC/MS confirmations. Chiral derivatization is essential to resolve this issue.
- Cocaine abuse is detected by the presence of benzoylecgonine (inactive metabolite of cocaine) in urine.
- Abuse of cocaine and ethanol (alcohol) is dangerous because cocaethylene is formed due to the interaction between ethanol and benzoylecgonine; cocaethylene is an active metabolite with a long half-life. Deaths have been

reported in individuals who abuse both cocaine and alcohol, although pure cocaine abuse can also be lethal.

- Cocaine may be used in ENT surgery as a local anesthetic. Such use can cause a positive urine specimen analysis for up to two days.
- Certain herbal teas (Mate de Coca and Health Inca tea) may contain cocaine, and drinking such teas can cause a positive cocaine drug test.
- Heroin is metabolized to 6-monoacetylmorphine (also called 6-acetylmorphine) and then into morphine. Detection of 6-monoacetylmorphine (the marker compound of heroin abuse) in urine is only possible if a person abuses heroin.
- Eating poppy seed-containing food may cause positive opiate test results, but in a person who has eaten poppy seeds, both codeine and morphine must be present, they can be confirmed by GC/MS.
- Opiate immunoassays typically utilize a morphine-specific antibody, so certain opioids cannot be detected by opiate immunoassays due to low cross-reactivities. These opioids include: oxycodone and oxymorphone (keto-opioids), methadone, propoxyphene, and fentanyl and its analogs.
- A common cause of positive phencyclidine immunoassay test results is the presence of dextromethorphan in the specimen. GC/MS confirmation should be negative.
- Marijuana is analyzed by detecting carboxylic acid derivatives in urine. Use of synthetic marijuana (Marinol) should cause positive marijuana test results, but passive inhalation of marijuana should not. This is because urine concentrations of the metabolite should be well below the 50 ng/mL cut-off concentration.
- People try to beat drug tests by adulterating their urine. Visine Eye Drops can interfere with various immunoassays, but the presence of Visine Eye Drops in urine cannot be determined by specimen integrity testing (pH, creatinine, specific gravity, and temperature). Adulterants purchased off the Internet, such as potassium nitrite, pyridinium chlorochromate, Stealth (peroxidase enzyme and hydrogen peroxide), and glutaraldehyde, can invalidate drug tests, but their presence cannot be detected by specimen integrity tests. However, special urine dipstick tests (AdultaCheck 6 or 10, Intect 7, etc.), spot tests, and other laboratory-based tests, are available for their detection. Such specimen integrity tests are essential in workplace drug testing.

REFERENCES

[1] Results from the 2010 National Survey on Drug Use and Health: National Findings: U.S. DEPARTMENT OF HEALTH AND HUMAN SERVICES Substance Abuse and Mental Health Services Administration (SAMHSA) Office of Applied Studies.

[2] White AM, Hingson RW, Pan IJ, Yi HY. Hospitalization for alcohol and drug overdoses in young adults ages 18-24 in the United States, 1999-2008: results from the nationwide inpatient sample. J Stud Alcohol Drugs 2011;72:774−86.

[3] Yee LM, Wu D. False positive amphetamine toxicology screen results in three pregnant women using labetalol. Obstet Gynecol 2011;117:503–6.

[4] Kreek MJ, Bart G, Lilly C, LaForge KS, et al. Pharmacogenetics and human molecular genetics of opiate and cocaine addictions and their treatments. Pharmacol Rev 2005;57:1–26.

[5] Solomon MD, Wright JA. False-positive for (+)-methamphetamine. Clin Chem 1977;23:1504.

[6] Wyman JF, Cody JT. Determination of l-methamphetamine: a case history. J Anal Toxicol 2005;29:759–61.

[7] Rollins DE, Jennison TA, Jones G. Investigation of interference by non-steroidal antiinflammatory drugs in urine tests for abused drugs. Clin Chem 1990;36:602–6.

[8] Boucher A, Vilette P, Crassard N, Bernard N, et al. Urinary toxicology screening: analytical interference between niflumic acid and cannabis. Arch Pediatr 2009;16:1457–60 [article in French].

[9] West R, Pesce A, West C, Crews B, et al. Comparison of clonazepam compliance by measurement of urinary concentration by immunoassay and LC-MS/MS in patient management. Pain Physician 2010;13:71–8.

[10] Von Seggern RL, Fitzgerald CP, Adelman LC, Adelman JU. Laboratory monitoring of OxyContin (oxycodone): clinical pitfalls. Headache 2004;44:44–7.

[11] Baker JE, Jenkins AJ. Screening for cocaine metabolite fails to detect an intoxication. Am J Forensic Med Pathol 2008;29:141–2.

[12] Bailey KM, Clay DJ, Gebhardt MA, Schmidt MJ, et al. Cocaine detection in postmortem samples following therapeutic administration. J Anal Toxicol 2009;33:550–2.

[13] Dasgupta A, Chughtai O, Hannah C, Davis B, Wells A. Comparison of spot tests with AdultaCheck 6 and Intect 7 urine test strips for detecting the presence of adulterants in urine specimens. Clin Chem Acta 2004;34:19–25.

CHAPTER 17

Challenges in Drugs of Abuse Testing: Magic Mushrooms, Peyote Cactus, and Designer Drugs

17.1 NEGATIVE TOXICOLOGY REPORT

A negative toxicology report does not mean that a patient does not have any abused drugs in his or her circulation. Although abused less frequently, magic mushroom and peyote cactus abuse cannot be detected by routine toxicological analysis. In addition, designer drugs are not usually detected by routine toxicology analysis, except for 3,4-methylenedioxymethamphetamine (MDMA, ecstasy), and 3,4-methylenedioxyamphetamine (MDA), which can have sufficient cross-reactivity with amphetamine immunoassays. Designer drugs were initially synthesized by clandestine laboratories to avoid the legal consequences of manufacturing and selling illicit drugs. Then in 1986 the United States Controlled Substances Act was amended in order to make manufacturing and selling of designer drugs illegal. The common types of designer drugs include amphetamine analogs such as MDMA and MDA, opiate analogs (including fentanyl derivatives), piperazine analogs, tryptamine-based hallucinogens, phencyclidine analogs, and gamma-hydroxybutyric acid (GHB) analogs. GHB and its analogs are used in date rape crimes (also called drug-facilitated rape). These drugs cannot be detected by routine toxicology analysis. Common designer drugs and their street names are listed in Table 17.1.

17.2 MAGIC MUSHROOM ABUSE

Magic mushrooms (psychoactive fungi) that grow in the United States, Mexico, South America, and many other parts of the world, contain psilocybin and psilocin, which are hallucinogens and are Class I controlled substances. Magic mushrooms can be eaten raw, cooked with food, or dried, and then consumed. These mushrooms can be mistaken for other non-hallucinogenic mushrooms or even poisonous mushrooms such as the

CONTENTS

A. Dasgupta and A. Wahed: Clinical Chemistry, Immunology and Laboratory Quality Control
DOI: http://dx.doi.org/10.1016/B978-0-12-407821-5.00017-6

Table 17.1 Common Designer Drugs and Their Street Names

Designer Drug	Common Street Name
Amphetamine Analogs	
3,4-Methylenedioxymethamphetamine (MDMA)	Ecstasy, XTC, adam, lover's speed
3,4-Methylenedioxyamphetamine (MDA)	Adam
3,4-Methylenedioxyethylamphetamine (MDEA)	Eve
4-Bromo-2,5-dimethoxyphenylethylamine (2-CB)	Venus, bromo, erox, nexu
Para-methoxyamphetamine (PMMA)	Killer, Dr. Killer
4-Methylthioamphetamine (4-MTA)	Golden eagle
Mephedrone, 3,4-methylenedioxypyrovalerone	Bath salt
Fentanyl Analogs	
α-Methylfentanyl	China white
Gamma-hydroxybutyrate (GHB)*	Liquid ecstasy, G, nitro
GHB Analogs	
Gamma-hydroxybutyrolactone (GBL)	Blue nitro, GH
1,4-Butanediol	Weight belt
Flunitrazepam*	Roofies, roche, Mexican valium
Synthetic Cannabinoids	
JWH-018, JWH-073, JW-250, etc.	Spice, spice gold, legal high, K3
Ketamine*	Special K, vitamin K, new ecstasy, psychedelic heroin

*Not a designer drug.

Amanita Class. After ingestion, psilocybin, often the major component of mushrooms, is rapidly converted by dephosphorylation into psilocin, which has psychoactive effects similar to lysergic acid diethylamide (LSD). In general, the duration of the trip from abusing magic mushrooms can last between 2 and 6 h, and the effects range from the intended feeling of relaxation, uncontrollable laughter, joy, euphoria, visual enhancement of colors, hallucination, and altered perception to a negative experience such as depression or paranoia. Intoxication from use of magic mushrooms is common. Some species of magic mushrooms contain phenylethylamine, which can cause cardiac toxicity. Fatality from abusing magic mushrooms alone is rare [1]. Currently there is no immunoassay for the determination of psilocybin and psilocin in body fluids. Therefore, chromatographic methods must be employed for their analysis, especially during a forensic investigation.

17.3 PEYOTE CACTUS ABUSE

Peyote cactus (*Lophophora williamsii*) is a small spineless cactus that grows in the Southwestern part of the United States and Mexico. The top of the Peyote

Table 17.2 Active Ingredients of Magic Mushrooms and Peyote Cactus	
Abused Substance	**Active Ingredient**
Magic Mushroom	Psilocybin and psilocin (psilocybin is also converted in vivo into psilocin)
Peyote Cactus	Mescaline

cactus, known as the "crown," contains the psychoactive compound mescaline. Mescaline is classified as a Class I controlled substance, but approximately 300,000 members of the Native Americans Church can ingest peyote cactus legally as a religious sacrament during all night services [2]. The mescaline content of peyote cactus is usually 0.4% in fresh cactus and 3−6% in dried cactus. The highest psychedelic effect can be achieved within two hours of ingestion, but the effect may last up to 8 h [3]. The psychoactive effects of mescaline are similar to LSD, including deeply mystical feelings. Abuse of peyote cactus can cause serious toxicity that requires medical attention, and may even cause fatality. Currently, there is no commercially available immunoassay for analysis of mescaline in body fluids, and only chromatographic methods are available for its analysis. Although mescaline is metabolized into several different metabolites, a large amount of mescaline can be recovered unchanged in urine. Therefore, detection of mescaline in serum or urine can be used for establishing diagnosis of peyote cactus abuse. For death investigations, confirmation of the presence of mescaline in body fluid is essential in establishing the cause of death. Severe toxicity and even death can occur from mescaline overdose. The active ingredients of magic mushrooms and peyote cactus are also listed in Table 17.2.

17.4 RAVE PARTY DRUGS AND DATE RAPE DRUGS (INCLUDING DESIGNER DRUGS)

At rave parties, commonly abused drugs are methamphetamine, 3,4-methylenedioxymethamphetamine (MDMA, ecstasy), 3,4-methylenedioxyamphetamine (MDA), ketamine, and the date rape drug gamma-hydroxybutyric acid (GHB) and its analogs. Both MDMA and MDA are designer drugs that are synthesized and have structural similarity with amphetamine and methamphetamine. Rave party drugs, including date rape drugs, are listed in Table 17.3. Rave party drugs are also known as club drugs.

Abuse of methamphetamine can be easily detected in urine using amphetamine immunoassays or immunoassays specifically designed to detect MDMA. However, detection of other rave party drugs such as GHB and its

Table 17.3 Rave Party and Date Rape Drugs

Type	Examples
Rave Party (Club) Drugs	3,4-methylenedioxyamphetamine (MDA) 3,4-Methylenedioxymethamphetamine (MDMA)
	Ketamine, methamphetamine, LSD[#]
	Gamma-hydroxybutyric acid (GHB)
	Rohypnol (tlunitrazepam)
Date Rape Drugs*	Gamma-hydroxybutyric acid (GHB)
	Rohypnol (flunitrazepam)

Rave party drugs most commonly encountered in date rape crimes.
[#]*Lysergic acid diethylamide (LSD): Less commonly encountered at rave parties compared to other drugs.*

analogs, ketamine, Rohypnol (flunitrazepam), and LSD is difficult. Although immunoassays are commercially available for screening of LSD in urine, many clinical laboratories do not routinely perform LSD screen. Flunitrazepam, a benzodiazepine, is banned in the United States. However, this drug can be obtained illegally, and usually a benzodiazepine screening assay of urine at the usual cut-off concentration of 200 ng/mL does not detect the presence of flunitrazepam due to its low concentration in urine. Forsman et al., using the CEDIA benzodiazepine assay at a cut-off of 300 ng/mL, failed to obtain a positive result in the urine of volunteers after they received a single dose of 0.5 mg flunitrazepam. In addition, only 22 out of 102 urine specimens collected from volunteers after receiving the highest dose of flunitrazepam (2 mg) showed a positive screening test result using the CEDIA benzodiazepine assay [4]. However, methods such as gas chromatography/mass spectrometry (GC/MS) or liquid chromatography combined with tandem mass spectrometry (LC/MS/MS) at a cut-off concentration of 40 ng/mL or lower should detect both flunitrazepam and its metabolite, 7-aminoflunitrazepam.

CASE REPORT

The victim was a female sales representative who had been acquainted with a male assailant through her work. During a meeting in a café she went to the bathroom and on return consumed a soft drink. Approximately 40 min after she consumed half of the drink she felt feverish and had a progressive loss of sensation. She had no recollection of any event approximately 4.5 h following consumption of the drink, and only partial recollection of events for the next 2.5 h, indicating a period of amnesia that lasted almost 7 h. She was still confused and drowsy when she returned home. The next day she went to a hospital and a urine specimen was collected. When the specimen was analyzed in a police crime laboratory, 7-aminoflunitrazepam, the major metabolite of flunitrazepam, was detected. It was discovered that the male assailant had dissolved a 1-mg flunitrazepam tablet in her drink when she went to the bathroom [5].

There is no commercially available immunoassay for ketamine or GHB and its analogs. Therefore, only chromatographic methods (GC/MS or LC/MS/MS) are available for their analysis, mostly in medical legal investigations. Ketamine is used through intravenous injection, but it is sometimes also added to a drink prior to a crime like drug-facilitated rape.

CASE REPORT

A 34-year-old woman lived with her husband. One day her husband found her unconscious after drinking a cup of tea shortly after bathing. She was transported to the hospital by an ambulance and pronounced dead. The case was investigated further and the investigators discovered that the victim was in good health but had episodes of unconsciousness starting in September 2002. However, no medical evidence was found regarding any malfunction of vital organs. Approximately a year before the first episode, her husband began preparing tea for her, which she consumed after taking a bath. Ketamine injection vials (100 mg) were also discovered in the refrigerator, and the husband's fingerprints were discovered on the vials and also on the coffee cup. Her husband, being a pediatric surgeon, had easy access to ketamine and knowledge of its toxicity. The police detained her husband and he confessed that he did not marry his wife for love and three years prior to the incident he had fallen in love with another girl. He also confessed that he poisoned his wife with ketamine because he did not want to lose any assets during the divorce settlement. At first he added 100 mg (one vial) of ketamine to her tea, but did not observe any toxicity. After five or six such attempts, he added two vials of ketamine to her coffee and the victim showed signs of toxicity. Each time he took his wife to the hospital so that the wife's family would be convinced that he was taking good care of her. During the murder, he added three vials (300 mg) of ketamine to her tea. Toxicology analysis of postmortem specimens indicated ketamine concentrations of 3.8 µg/mL in cardiac blood, 2.1 µg/mL in the stomach contents, and 1.2 µg/mL in urine, confirming the cause of death as ketamine poisoning [6].

Although sale of GHB and its analogs for human consumption is against the law in the United States, several GHB analogs are commercially available as industrial solvents and for the manufacture of plastics and other products (e.g. 1,4-butanediol). 1,4-Butanediol may be present in toys, especially toys manufactured outside the U.S. Because 1,4-butanediol is endogenously converted into GHB, licking toys containing 1,4-butanediol may cause serious toxicity in toddlers. Ortmann et al. reported coma in a 20-month-old child due to ingestion of a plastic toy containing 1,4-butanediol [7]. Gamma-valerolactone (GVL) is the most recent addition to the designer drug list; it is a GHB analog. GVL is quickly metabolized in vivo by lactonase enzymes into 4-methyl-gamma-hydroxybutyrate (gamma-hydroxyvaleric acid), which is responsible for its pharmacological effects similar to GHB (this metabolite can bind to the GHB receptor).

A 43-year-old man was found unconscious by two relatives at his home. Despite rapid emergency response, he was pronounced dead. He was a known drug abuser. His autopsy findings were unremarkable. There was no sign of violence and no needle mark was found. However, a very high concentration of GHB was detected in his cardiac blood (3,385 µg/mL), urine (33,727 µg/mL), bile (1,800 µg/mL), gastric contents (7.08 g in 100 mL), and vitreous humor (2,856 µg/mL), thus establishing the cause of death as GHB overdose. In addition, GHB was also detected in his pubic hair, indicating long-term abuse of GHB [8].

17.5 ABUSE OF AMPHETAMINE-LIKE DESIGNER DRUGS (INCLUDING BATH SALTS)

MDMA and MDA are the best examples of amphetamine-like designer drugs that are widely abused. More recently, the abuse of bath salts, which are synthetic compounds that resemble the structure of amphetamine, is increasing. Other common examples of amphetamine-like designer drugs include para-methoxyamphetamine (PMA), para-methoxymethamphetamine (PMMA), 3,4-methylenedioxy-N-ethylamphetamine (MDEA), 2,5-dimethoxy-4-methylamphetamine (DOM), and 2,5-dimethoxy-4-methylthioamphetamine (DOT).

Bath salts (synthetic cathinones) were placed in the Schedule 1 Drug Category in September 2011 due to high abuse potential. Cathinones are natural components of khat plants, which are abused in some parts of Africa where abusers chew leaves of the khat plant. Bath salts that are synthetic cathinones include methylenedioxypyrovalerone (MDPV), 4-methylmethcathinone (also known as mephedrone), and related synthetic drugs [9]. These compounds have stimulant-producing effects like amphetamine and cocaine, and abuse of bath salts may cause serious life-threatening toxicity and even death. Euphoria after bath salt use lasts for 2−4 h, and is more intense than the euphoria after abuse of MDMA. Key points regarding bath salt abuse are as follows:

- Bath salts cannot be detected by regular toxicological screening because these compounds do not cross-react with amphetamine immunoassays. However, Randox Corporation recently introduced an ELISA assay for bath salts.
- These compounds can be analyzed by GC/MS or LC/MS/MS.

CASE REPORT

A 36-year-old man had injured himself by smashing a window in a rage of fury. The police arrested him. He was brought to the hospital, and, despite resuscitation attempts, was pronounced dead. The autopsy showed multiple bruises and many superficial skin lacerations, but the cause of death could not be explained by these injuries. However, toxicological analysis showed a high level of mephedrone (5.1 µg/mL) in his femoral blood, explaining the cause of death as severe mephedrone (bath salt) overdose after oral ingestion of the drug. Traces of cocaine, MDMA, and oxazepam were also detected in his postmortem blood [10].

17.6 ABUSE OF SYNTHETIC MARIJUANA (SPICE AND K2)

Since 2008, synthetic marijuana compounds (sold as spice, K2, or herbal high) have been gaining popularity among drug abusers. Although less toxic than bath salts, these compounds are very addictive, and may cause severe overdose (including death). The first synthetic compound in this category, JWH-018, was synthesized by Dr. John W. Huffman at Clemson University to study the effect of this compound on cannabinoid receptors. It has been speculated that someone saw the paper and copied his method to produce JWH-018 illegally for abuse. Synthetic cannabinoids are called herbal high or legal high, and usually plant materials are sprayed with these compounds so that these active compounds are present at the surface of the herbal product for maximum effect. However, pure compounds (which are white powders) are also available for purchase on the black market. Currently more than 100 compounds are available which belong to this class, but the most common ones are JWH-018, JWH-073, JWH-250, JWH-015, JWH-081, HU-210, HU-211 (synthesized at Hebrew University), and CP 47,497 (synthesized at Pfizer).

When a synthetic cannabinoid compound is smoked, it produces an effect similar to (or greater than) smoking marijuana. By 2010, there had been 1,057 synthetic marijuana-related toxicities in 18 states and the District of Columbia. Synthetic marijuana compounds are abused due to their agonist activity at cannabinoid receptors 1 and 2 (CB1 and CB2). Therefore, the mechanism of causing euphoria and hallucination is similar to that produced by natural marijuana. However, users also report adverse effects such as rapid heartbeat, irritability, and paranoia. In addition, tremor, seizure, slurred speech, dilated pupils, hypokalemia, tachycardia, hypertension, and chest pain may also be present in a patient overdosed with spice. Long-term adverse effects are not known [11]. Key points regarding synthetic cannabinoids (marijuana) abuse include:

- Synthetic cannabinoids cannot be detected by immunoassays designed to detect marijuana metabolite in urine. However, American Screening Corporation and Randox Corporation have recently marketed assays for screening synthetic cannabinoids where these compounds can be detected in urine up to 3 days after abuse.
- Both GC/MS and LC/MS/MS can be used for confirmation of these compounds in both blood and urine.
- Detection of these compounds in urine is more challenging because many synthetic cannabinoids are extensively metabolized, thus requiring detection of the metabolite instead of the parent drug [12].

CASE REPORT

A 48-year-old healthy male experienced seizure 30 min after ingesting alcohol and a white powder purchased from the Internet to get high. On admission, serum chemistry was normal and a urine toxicology screen using an enzyme multiplied immunoassay technique (EMIT) assay was negative, including for marijuana. His seizure responded to lorazepam and the patient was discharged on Day 10 of his hospital stay. Using chromatography combined with tandem mass spectrometry, the white powder was identified as JW-018, and a monohydroxy metabolite of JW-018 was detected in the urine specimen [13].

17.7 DESIGNER DRUGS THAT ARE OPIOID ANALOGS

Heroin (diacetylmorphine), which was synthesized from morphine in 1874, can be considered as the first designer drug. Heroin is a Class I Scheduled Drug in the United States. Heroin is metabolized to 6-monoacetylmorphine, and then finally to morphine. Fentanyl is a widely used synthetic narcotic analgesic that is approximately 75 to 100 times more potent than morphine. Several analogs of fentanyl, such as sufentanil, alfentanil, lofentanil, and remifentanil, have been synthesized by the pharmaceutical industry and are in clinical use. Currently in the United States, fentanyl is a Schedule II Drug that is used as an anesthetic. Injection of 50 to 100 μg of fentanyl produces a rapid analgesic effect and unconsciousness. Fentanyl is also available as lozenges and transdermal patches (Duragesic®) for pain management. Oral transmucosal fentanyl citrate (Actiq®) is a relatively new formulation where fentanyl is incorporated into a sweetened matrix to produce fentanyl lozenges. The therapeutic range of fentanyl is 1−3 ng/mL in serum, and fentanyl toxicity is similar to opiate toxicity [14].

The fentanyl analog designer drug China White (α-methylfentanyl) appeared in the underground market of California in 1979 and caused over 100 deaths. In 1984 another illicit designer drug, 3-methylfentanyl, appeared as a street drug in California and was also related to fatal drug overdoses. Neither fentanyl nor its analogs can be detected in routine urine drug testing by regular opiate screening assays because these compounds do not cross-react with antibodies used in opiate assays that target morphine The concentrations of fentanyl and its analogs can be measured in serum, urine, and other biological matrices using GC/MS or liquid chromatography combined with tandem mass spectrometry.

KEY POINTS

- Magic mushrooms (psychoactive fungi) contain psilocybin and psilocin, which are hallucinogens and are Class I controlled substances. After ingestion, psilocybin, often the major component of magic mushrooms, is rapidly converted by dephosphorylation into psilocin, which has psychoactive effects similar to lysergic acid diethylamide (LSD). Magic mushroom abuse cannot be detected in routine immunoassay screening for abused drugs in urine.
- Peyote cactus (*Lophophora williamsii*) is a small spineless cactus that grows in the Southwestern part of the United States and Mexico and contains the psychoactive compound mescaline, a Class I controlled substance with effects similar to LSD. Peyote cactus abuse cannot be detected in routine toxicology screening.
- Rave party drugs 3,4-methylenedioxymethamphetamine (MDMA) and 3,4-methylenedioxyamphetamine (MDA) can be detected by amphetamine/ methamphetamine immunoassay or MDMA immunoassay. However, there is no immunoassay available for detection of ketamine or the date rape drug gamma-hydroxybutyric acid (GHB) and its analogs.
- Bath salts are synthetic derivatives of cathinones (a natural component of khat plants). Major bath salts are methylenedioxypyrovalerone (MDPV), and 4-methylmethcathinone (also known as mephedrone). These compounds have stimulant effects like amphetamine and cocaine. Abuse of bath salts can cause serious life-threatening toxicity and even death. Bath salts cannot be detected by regular toxicological screening because these compounds do not cross-react with amphetamine immunoassays.
- Since 2008, synthetic marijuana compounds (sold as spice, K2, or herbal high) are gaining popularity among drug abusers. The first synthetic compound in this category, JWH-018, was synthesized by Dr. John W. Huffman at Clemson University, and currently more than 100 compounds are available. The most common examples of spice are JWH-018, JWH-073, JWH-250, JWH-015, JWH-081, HU-210, HU-211 (synthesized at Hebrew University), and CP-47,497 (synthesized at Pfizer). Marijuana immunoassays cannot detect the presence of these illicit drugs if present in urine.
- The fentanyl analog designer drug China White (α-methylfentanyl) and 3-methylfentanyl are often abused, but neither fentanyl nor its analogs can be detected in routine urine drug testing by opiate immunoassays.

REFERENCES

[1] Gonomori K, Yoshioka N. The examination of mushroom poisonings at Akita University. Leg Med (Tokyo) 2003;5(Suppl. 1):S83−6.

[2] Halpern JH, Sherwood AR, Hudson JI, Yugerlum-Todd D, Pope Jr. HG. Psychological and cognitive effects of long term peyote use among Native Americans. Biol Psychiatry 2005;58:624−31.

[3] Nichols DE. Hallucinogens. Pharmacol Ther 2004;101:131−81.

[4] Forsman M, Nystrom I, Roman M, Berglund L, et al. Urinary detection times and excretion patterns of flunitrazepam and its metabolites after a single oral dose. J Anal Toxicol 2009;33:491−501.

[5] Ohshima T. A case of drug facilitated sexual assault by the use of flunitrazepam. J Clin Forensic Med 2006;13:44−5.

[6] Tao LY, Chen XP, Qin ZH. A fatal chronic ketamine poisoning. J Forensic Sci 2005;50:173−5.

[7] Ortmann LA, Jaeger MW, James LP, Schexnayder SM. Coma in a 20 month old child from ingestion of a toy containing 1,4-butanediol, a precursor of gamma-hydroxybutyrate. Pediate Emerg Care 2009;25:758−60.

[8] Kintz P, Villain M, Pelissier AL, Cirimele V, et al. Unusually high concentrations in a fatal GHB case. J Anal Toxicol 2005;29:582−5.

[9] Fass JA, Fass AD, Garcia A. Synthetic cathinones (bath salts): legal status and patterns of abuse. Annals Pharmacother 2012;46:436−41.

[10] Lusthof KJ, Oosting R, Maes A, Verschraagen M, et al. A case of extreme agitation and death after the use of mephedrone in the Netherlands. Forensic Sci Int 2011;206:e93−5.

[11] Wells D, Ott CA. The new marijuana. Annals Pharmacother 2011;45:414−7.

[12] ElSohly MA, Gul W, Elsohly KM, Murphy TP, et al. Liquid chromatography-tandem mass spectrometry analysis of urine specimens for K2 (JWH-018) metabolites. J Anal Toxicol 2011; 35:487−495.

[13] Lapoint J, James LP, Moran CL, Nelson LS, et al. Severe toxicity following synthetic cannabinoid ingestion. Clin Toxicol (Phil) 2011;49:760−4.

[14] Mystakidou K, Katsouda E, Parpa E, Vlahos L, et al. Oral transmucosal fentanyl citrate: overview of pharmacological and clinical characteristics. Drug Deliv 2006;13:269−76.

CHAPTER 18

Testing for Ethyl Alcohol (Alcohol) and Other Volatiles

18.1 ALCOHOL USE AND ABUSE

Man has used alcohol (ethanol) since prehistoric time (10,000 BC). Alcohol is produced by fermentation of sugar and starch. The normal fermentation process that uses yeast cannot produce alcoholic beverages with an alcohol content over 14%. Therefore, hard liquors and spirits are produced using fermentation followed by distillation. Alcoholic beverages are full of calories and can be classified under two broad categories: beer and wine (produced by direct fermentation; alcohol content <14%) and spirits (produced by fermentation followed by distillation; alcohol content as high as 40% or more). Alcohol content of various alcoholic beverages varies widely (Table 18.1), but, due to various amounts of fluid in a drink, one standard drink contains approximately 0.6 ounces or 14 g of pure alcohol [1]. Alcohol has beneficial effects if consumed in moderation. Moderate alcohol consumption is defined as follows[2]:

- Men: No more than two standard alcoholic drinks per day, not exceeding 14 drinks per week.
- Women: No more than one standard alcoholic drink per day, not exceeding 7 drinks per week.
- Adults over 65 (both male and female): No more than one drink per day.

Hazardous drinking is defined as the quantity or pattern of alcohol consumption that places individuals at high risk from alcohol-related disorders[3]:

- For men, 21 or more drinks per week or more than 7 drinks per occasion and at least three times a week.
- For women, more than 14 drinks per week or drinking more than 5 drinks on one occasion and at least three times a week.

Alcohol abuse is a leading cause of mortality and morbidity internationally, and is ranked by the World Health Organization (WHO) as one of the top

CONTENTS

A. Dasgupta and A. Wahed: Clinical Chemistry, Immunology and Laboratory Quality Control
DOI: http://dx.doi.org/10.1016/B978-0-12-407821-5.00018-8

Table 18.1 Alcohol Content of Various Alcoholic Beverages

Beverage	One Standard Drink	Alcohol Content
Standard American beer	12 ounce	4–6%
Table wine	5 ounce	7–14%
Sparkling wine	5 ounce	8–14%
Whiskey	1.5 ounce	40–75%
Vodka	1.5 ounce	40–50%
Gin	1.5 ounce	40–49%
Rum	1.5 ounce	40–80%
Tequila	1.5 ounce	45–50%

Table 18.2 Physiological Effect of Various Blood Alcohol Levels

Blood Alcohol	Physiological Effect
0.02–0.05% (20–50 mg/dL)	Relaxation and general positive mood-elevating effect of alcohol, including increased social interactions.
0.08% (80 mg/dL)	Legal limit of driving; minor impairment possible in a person who drinks rarely.
0.1%–0.15% (100–150 mg/dL)	Euphoria, but sensory impairment and decreased cognitive ability; difficulty driving a motor vehicle.
0.2% (200 mg/dL)	Worsening of sensory–motor impairment and inability to drive; decreased cognitive function and visual impairment.
0.3% (300 mg/dL)	Vomiting, incontinence, symptoms of alcohol intoxication.
0.4% (400 mg/dL)	Stupor, coma, respiratory depression, hypothermia.
0.5% and more (500 mg/dL)	Potentially lethal.

five risk factors for disease burden. Binge drinking means heavy consumption of alcohol within a short period of time with the intention to become intoxicated. Although there is no universally accepted definition for binge drinking, usually consumption of five or more drinks by males and four or more drinks by females is considered "binge drinking." Despite a legal drinking age of 21, binge drinking is very popular among college students. Physiological effects of various blood alcohol levels are listed in Table 18.2. Usually, moderate drinking (1–2 standard drinks) produces a blood alcohol level in the range of 0.05%, which can have a beneficial mood elevation effect (e.g. including increased social interaction). Binge drinking (drinking 5–8 drinks in one sitting) produces a blood alcohol level over 200 mg/dL which results in undesirable effects.

18.2 HEALTH BENEFITS OF MODERATE DRINKING

Currently, the best evidence of the health benefits of drinking in moderation is the reduced risk of cardiovascular disease. However, drinking in moderation has many other health benefits:

- Reduced risk of coronary heart diseases, including myocardial infarction and angina pectoris.
- Better survival chance after a heart attack.
- Reduced risk of stroke.
- Reduced risk of developing diabetes.
- Reduced risk of forming gallstones.
- Reduced risk of developing arthritis.
- Reduced risk of developing age-related dementia and Alzheimer's disease.
- Reduced risk of certain types of cancer.
- Increased longevity.
- Less chance of getting common cold.

The relationship between alcohol consumption and coronary heart disease was examined in the original Framingham Heart Study. Alcohol consumption showed a U-shaped curve with reduced risk of developing such diseases with moderate drinking, but high risk of developing such diseases with heavy drinking. For men, it is beneficial to drink one drink per day or at least six drinks per week to reduce the risk of coronary heart disease and heart attack, but women can get the health benefits of moderate drinking by consuming just one drink per week [4]. There are several hypotheses on how moderate drinking can reduce the risk of developing heart disease:

- Increases the concentration of high density lipoprotein (HDL) cholesterol.
- Decreases the concentration of low density lipoprotein (LDL) cholesterol.
- Reduces narrowing of coronary arteries by reducing plaque formation.
- Reduces risk of blood clotting.
- Reduces level of fibrinogen.

Studies have indicated that the level of increase in HDL cholesterol in blood may explain 50% of the protective effect of alcohol against cardiovascular disease, and the other 50% may be partly related to inhibition of platelet aggregation, which reduces blood clot formation in coronary arteries. It has been suggested that, although alcohol can increase HDL cholesterol levels and inhibit platelet aggregation, the polyphenolic antioxidant compound found in abundance in red wine can further reduce platelet activity via other mechanisms than alcohol. Therefore, it appears that red wine is more protective against cardiovascular disease than other alcoholic beverages [5].

18.3 HEALTH HAZARDS OF HEAVY DRINKING

All benefits of drinking alcohol are lost with heavy drinking. For example, drinking alcohol in moderation reduces the risk of cardiovascular disease, but heavy drinking increases the risk of cardiovascular disease. However, the most common alcohol-related organ damage is fatty liver, which may even lead to cirrhosis of liver, a potentially fatal disease. A pregnant woman or a woman planning to be pregnant should not consume any alcohol due to the risk of fetal alcohol syndrome in the newborn baby. According to a report by Dr. Ting-Kai Li, the Director of the National Institute of Alcohol Abuse and Alcoholism (NIAAA), alcohol-related problems cost the United States an estimated $185 billion annually. In the United States, over 18 million people age 18 and older suffer from alcohol abuse or dependency, and only 7% of these people receive any form of treatment. The highest prevalence of alcohol dependency in the United States is observed among young people between the ages of 18 and 24. The hazards of heavy drinking include:

- Alcoholic liver disease.
- Increased risk of cardiovascular disease.
- Brain damage.
- Increased risk of stroke.
- Damage to immune and endocrine systems.
- Anxiety and mood disorders.
- Increased risk of various cancers.
- Poor outcome of pregnancy.
- Reduced life span and increased mortality.

In addition, alcoholics are prone to deep depression and violent behavior. However, from a clinical point of view, alcoholic liver diseases and fetal alcohol syndrome have been studied in more detail than other alcohol-related problems. Alcohol-induced liver disease can be classified under three categories: fatty liver, alcoholic hepatitis, and liver cirrhosis. Heavy drinking for as little as a few days can produce fatty changes in the liver (steatosis); this can be reversed after abstinence. A person infected with hepatitis C should consult with his/her physician regarding safe consumption of alcohol because alcohol and hepatitis C act in synergy to cause liver damage. Alcohol is a small molecule, so it can easily pass through the placenta to the embryo and cause birth defects. These are collectively known as fetal alcohol spectrum disorders. If more severe signs of these birth defects are present in a newborn, the condition may be called "fetal alcohol syndrome." Drinking alcohol during pregnancy can even cause stillbirth. Poor outcomes associated with drinking alcohol during pregnancy include:

- Stillbirth or death of the newborn shortly after birth.
- Pre-term baby.

- Smaller birth weight and/or growth retardation of the baby.
- Neurological abnormality.
- Facial abnormalities.
- Intellectual impairment during development.

18.4 METABOLISM OF ETHYL ALCOHOL: EFFECT OF GENDER AND GENETIC FACTORS

After consumption, alcohol is absorbed from the stomach and metabolized by the liver. A small amount of alcohol is not absorbed and is found in the breath. This is the basis of breath analysis of alcohol in suspected drivers operating under impairment. Factors that affect how the body handles alcohol include:

- Age.
- Gender.
- Body weight.
- Amount of food consumed with alcohol.
- Race and ethnicity (genetic factors).

When alcohol is consumed, about 20% is absorbed from the stomach and the rest is absorbed from the small intestine. Food substantially slows down the absorption of alcohol, and sipping alcohol instead of drinking also slows the absorption. Peak blood alcohol concentration is also reduced if alcohol is consumed with food. A small amount of alcohol is metabolized by the enzyme present in gastric mucosa; another small amount is metabolized by the liver before it can enter the main bloodstream (first-pass metabolism). The rest of the alcohol enters the circulation. After drinking the same amount of alcohol, a man would have a lower peak blood alcohol level compared to a woman with the same body weight. This gender difference in blood alcohol level is related to the different body water contents between males and females. Other important points to remember regarding alcohol metabolism include:

- Alcohol follows zero-order kinetics during metabolism, which means no matter how high the blood alcohol level, only a certain amount is removed from the body per hour. In contrast, most drugs follow first-order kinetics, which means the higher the drug level in blood, the faster the metabolism.
- Women also metabolize alcohol slower than men. In women, alcohol concentration is reduced by 15−18 mg/dL (0.018%) per hour regardless of blood alcohol concentration. In men this rate is 18−20 mg/dL per hour.

- Hormonal changes also play a role in the metabolism of alcohol in women, although this finding has been disputed in the medical literature.

The human liver metabolizes alcohol using zero-order kinetics. Several enzyme systems are involved in the metabolism of ethanol, namely alcohol dehydrogenase (ADH), microsomal ethanol oxidizing system (MEOS), and catalase. These enzymes also metabolize other similar compounds such as methanol, isopropyl alcohol, and ethylene glycol. The most important enzyme for alcohol metabolism is alcohol dehydrogenase (ADH), which is found in hepatocytes. The enzyme catalyzes the following reaction (Scheme 18.1):

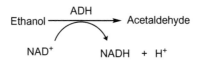

ADH activity is greatly influenced by the frequency of ethanol consumption. Adults who consume 2−3 alcoholic beverages per week metabolize ethanol at a rate much lower than alcoholics. For medium-sized male adults, the blood ethanol level declines at a rate of 18−20 mg/dL/h (0.018−0.020%/ hour). The average rate is slightly less in women than men. The major drug-metabolizing family of enzymes found in the liver is the cytochrome P-450 mixed function oxidase. Many members of this family of enzymes, most notably CYP3A4, CYP1A2, CYP2C19, and CYP2E1 isoenzymes, play vital roles in the metabolism of drugs. For non-alcoholics, this metabolic pathway is considered a minor, secondary route; but it becomes much more important in alcoholics, and the CYP2E1 isoenzyme plays a major role in metabolizing alcohol in addition to ADH. Because of the additional participation of CYP2E1, alcoholics can remove alcohol faster from their body compared to non-alcoholics (Scheme 18.2).

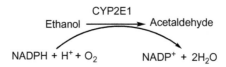

The acetaldehyde produced due to metabolism of alcohol (regardless of pathway) is subsequently converted to acetate as the result of the action of mitochondrial aldehyde dehydrogenase (ALDH2). Acetaldehyde is fairly toxic compared to ethanol and must be metabolized fast (Scheme 18.3).

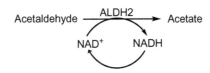

Acetate or acetic acid then enters the citric acid cycle (which is a normal metabolic cycle of living cells) and is converted into carbon dioxide and water. From a chemical point of view, the body oxidizes alcohol into carbon dioxide and water; this process generates calories. Therefore, alcoholic drinks are high in calories. Metabolism of alcohol changes with advancing age because the activity of the enzymes involved in alcohol metabolism diminishes with age. Water volume also reduces with advancing age. Therefore, an elderly person would have a higher blood alcohol level from consumption of the same amount of alcohol compared to a younger person of the same gender. Moreover, elderly persons consume more medications than younger people, and a medication may interact with the alcohol.

18.5 RELATION BETWEEN WHOLE BLOOD ALCOHOL AND SERUM ALCOHOL AND LEGAL LIMIT OF DRIVING

Usually the alcohol concentration in blood is measured in patients admitted to the emergency department who are suspected of drug and alcohol overdose. This is considered as medical blood alcohol determination because no chain of custody is maintained and alcohol concentration is confidential patient-related information that cannot be disclosed to a third party. Medical alcohol determination is usually conducted in serum using automated analyzers and enzymatic assays that can be easily automated. In addition, alcohol concentration in blood is measured in drivers suspected of driving with impairment (DWI). Legal alcohol testing is usually conducted using gas chromatography and whole blood:

- The legal limit of blood alcohol in all states in the U.S. is currently 0.08% whole blood alcohol (80 mg/dL).
- Serum alcohol concentration is higher than whole blood alcohol concentration due to higher amounts of water in serum (alcohol is freely water-soluble).
- In order to convert serum alcohol level to whole blood alcohol level, serum alcohol level must be divided by 1.15. Therefore, the serum alcohol concentration of 100 mg/dL (0.1%) is equivalent to 87 mg/dL (0.087%) whole blood concentration.

Rainey reported that the ratio between serum and whole blood alcohol ranged from 0.88 to 1.59, but the median was 1.15. Therefore, dividing serum alcohol value by 1.15 would calculate whole blood alcohol concentration. The serum-to-whole blood alcohol ratio was independent of serum alcohol concentration and hematocrit [6].

One popular defense of DWI is endogenous production of alcohol. Although substantial alcohol may be produced endogenously in a decomposed body by the action of various microorganisms, a living human body does not produce enough endogenous alcohol. In healthy individuals who do not drink, usually endogenous alcohol levels are significantly below the detection level. However, in a certain disease state known as "Auto-Brewery Syndrome," measurable blood alcohol may be detected in an individual who consumes no alcohol. Using a reliable gas chromatographic method, the concentration of alcohol in blood due to endogenous production of alcohol in patients suffering from various diseases (diabetes, hepatitis, cirrhosis, etc.) can reach up to 0.08 mg/dL, which is very low in comparison to the legal limit of driving (80 mg/dL). In rare cases, however, endogenous blood alcohol can reach or exceed the legal limit of driving due to "Auto-Brewery Syndrome." For example, a blood alcohol level over 80 mg/dL was reported in a Japanese subject with severe yeast infection. In these subjects endogenous alcohol is produced after the subject eats carbohydrate-rich foods [7].

CASE REPORT

A 3-year-old female patient with short bowel syndrome was first operated on 8 h after birth with closure of the abdomen and enterostomy (jejunum). The patient was re-operated on one year later due to obstruction of her small intestine. She also suffered from septicemia due to bacterial overgrowth in her intestine. The patient was given a *Lactobacillus*-containing carbohydrate-rich fruit drink when she was 3 years old. A couple of weeks later her parents saw her walking erratically and she had the smell of alcohol. A breath analyzer showed an alcohol level of 22 mmol/L (101 mg/dL). When the carbohydrate-rich fruit drink was discontinued, her symptoms resolved, but when the drink was reinstated, her symptoms returned and her blood alcohol level was 15 mmol/L (69 mg/dL). Liver enzymes and alcohol biomarkers were, however, normal. A culture of gastric fluid and feces showed the presence of *Candida kefyr*, and after she was treated with oral fluconazole for one week, all of her symptoms were resolved. A month later her symptoms reappeared and a high alcohol level was again detected in her blood. A new culture of gastric fluid showed the presence of *Saccharomyces cerevisiae*. Again the patient was treated with fluconazole and her symptoms were resolved. The cause of her alcohol intoxication was due to "Auto-Brewery Syndrome." A diet less rich in carbohydrates was selected and she had no such symptoms for the next 2 years [8].

Although blood alcohol is usually directly determined in a driver suspected of driving under the influence of alcohol, blood alcohol level can also be predicted by using the Widmark formula, which can be simplified as follows to calculate blood alcohol in percent (Equation 18.1):

$$C = (\text{Number of Drinks} \times 3.1/\text{Weight in Pounds} \times r) - 0.015\,t \quad (18.1)$$

Here, C is the blood alcohol in percent (mg/dL); r is 0.7 for men and 0.6 for women, and t is time (in hours).

Because most standard drinks contain approximately the same amount of alcohol, it is only important to know how many drinks one person consumes. The type of drink does not matter and that makes the calculation easy. For example, if a 160-lb. man drinks five beers in a 2-hour period, his blood alcohol at the end of the timeframe would be (Equation 18.2):

$$C = (5 \times 3.1/160 \times 0.7) - 0.015 \times 2$$
$$= 0.138 - 0.030 \quad (18.2)$$
$$= 0.108\% \text{ or blood alcohol of } 108 \text{ mg/dL}$$

18.6 ANALYSIS OF ALCOHOL IN BODY FLUIDS: LIMITATIONS AND PITFALLS

Alcohol is most commonly measured in whole blood or serum. Alcohol concentration is also measured in urine, but less frequently to demonstrate abstinence because alcohol can be detected a little longer in urine than in blood. Usually in blood no alcohol can be detected 24 h after heavy drinking. In hospital laboratories, ethyl alcohol is also analyzed using enzymatic methods and automated analyzers. There are several different automated analyzers available from various diagnostic companies that are capable of analyzing alcohol in serum or plasma. Enzyme-based automated methods are generally not applicable for analysis of whole blood, although modified methods are available for analysis of alcohol in urine specimens. Enzymatic automated analysis of alcohol is based on the following principles:

- Conversion of alcohol to acetaldehyde by alcohol dehydrogenase. In this process NAD is converted into NADH. NAD has no absorption of ultraviolet light at a wavelength of 340 nm, while NADH absorbs at 340 nm. Therefore, an absorption peak is seen when alcohol is converted into acetaldehyde because NAD is also converted into NADH.
- Peak intensity is proportional to the amount of alcohol present in the specimen. If no alcohol is present, no peak is absorbed.

Usually methanol, isopropyl alcohol, ethylene glycol, and acetone have negligible effects on alcohol determination using enzymatic methods, but propanol, if present, can cause 15–20% cross-reactivity with the alcohol assay.

Although isopropyl alcohol (rubbing alcohol) is common in households, propanol is used in much lesser frequency in household products. However, interference of lactate dehydrogenase (LDH) and lactate in enzymatic methods of alcohol determination is significant. Therefore, an enzymatic alcohol assay is unsuitable for determination of alcohol in postmortem blood because it contains high concentrations of lactate dehydrogenase and lactate. Postmortem blood alcohol must be determined by gas chromatography (GC), most commonly headspace GC. Lactate concentrations also tend to increase in trauma patients. Therefore, a false positive alcohol result may be observed in these patients if an enzymatic assay is used. Key points regarding interferences in enzymatic alcohol methods include:

- Enzymatic methods for alcohol determination are unsuitable for postmortem alcohol analysis due to high concentrations of LDH and lactate; only gas chromatographic methods must be used. Alternatively, negative urine alcohol, but positive blood alcohol, may indicate interference because LDH is absent in urine due to its high molecular weight and therefore cannot interfere with urine alcohol determination. However, for legal blood alcohol determination, the GC method is always used.
- Alcohol may be produced by the activity of microorganisms after death. Therefore, elevated blood alcohol in postmortem specimens may not confirm alcohol intake prior to death. The vitreous humor is a better source for determination of postmortem alcohol.
- Alternatively, the presence of ethyl glucuronide and ethyl sulfate (which are metabolites of alcohol) in postmortem blood or urine confirms alcohol abuse prior to death. However, if the postmortem blood alcohol level is positive but no ethyl glucuronide or ethyl sulfate can be detected in blood or urine, it is an indication of postmortem production of alcohol and it can be concluded that the deceased did not consume alcohol prior to death.

In urine, alcohol can be determined up to 48 h after drinking, depending on the amount of alcohol consumed. Usually no blood alcohol is detected 24 h after drinking, even with alcohol abuse. Although dividing the urine alcohol level by 1.3 can provide an approximate blood alcohol level, this approach has many limitations. In addition, alcohol production in vitro after urine collection is a major problem for interpretation of urine alcohol level. Uncontrolled diabetes mellitus can cause glycosuria, and if a yeast infection is present, in vitro production of alcohol can result due to contamination of urine containing glucose with *Candida albicans*. Women with urinary tract infections can have the same problem. There are several case reports of false positive alcohol in urine due to such problems. Storing urine at 4°C and using 1% sodium fluoride or potassium fluoride as a preservative can minimize the problem.

During the police investigation of a rape victim, blood alcohol was negative (<10 mg/dL) but urine alcohol was 82 mg/dL. No illicit drug was detected. Because of the long time interval between the rape and collection of the specimen (approximately 15 hours), the investigator thought that the girl was probably drunk at the time of assault, and with time her blood alcohol was cleared (alcohol can still be detected in urine because the window of detection of alcohol in urine is longer than blood). However, the girl denied drinking. Because the girl had Type 1 diabetes and no fluoride was used as preservative during urine collection, expert testimony was sought during the court hearing; reanalysis of the urine specimen showed high glucose. In addition, the ethanol value was increased to 550 mg/dL, indicating a false positive result. Further evidence of post-sampling of alcohol formation came from the observation that the ratio serotonin metabolites (5-hydroxytryptophol to 5-hydroindoleacetic acid) was low (14 nmol/mmol) in the girl. A value of 15 or less is normal, and it was concluded that alcohol was produced in her urine after collection due to conversion of glucose present in urine into alcohol [9].

18.7 BIOMARKERS OF ALCOHOL ABUSE

Biomarkers of alcohol abuse can be divided into two broad categories: state and trait (genetic predisposition) markers. Alcohol is metabolized by the liver enzymes alcohol dehydrogenase (ADH) and acetaldehyde dehydrogenase (ALDH); they can accelerate or slow down metabolism of alcohol. ALDH exists in two major forms, ALDH1 and ALDH2 (the more active of the two). People carrying different ADH and ALDH isoforms metabolize alcohol at different rates. ADH and ALDH isoforms arise from a polymorphism in the structures of genes that code these enzymes. Two alcohol dehydrogenase genes (ADH2 and ADH3) on chromosome 4 and one acetaldehyde dehydrogenase gene (ALDH2) on chromosome 12 are known to exhibit polymorphism, thus controlling activities of both enzymes. The frequency of these polymorphisms differs between ethnic groups. One of the best understood polymorphisms of alcohol-metabolizing enzymes is associated with the gene coding ALDH2 enzyme. One allele, known as ALDH2*2, which is found in approximately 40 percent of people of Far East Asian descent but rarely in Caucasians, produces a partially inactive enzyme because of a specific mutation in the gene that encodes this enzyme. In people carrying the ALDH2*2 allele, even moderate alcohol consumption results in acetaldehyde accumulation in the blood because acetaldehyde is only slowly removed from the blood due to a less active form of the enzyme. An elevated acetaldehyde level after drinking can lead to an unwanted reaction towards alcohol, such as flushing, nausea, and rapid heartbeat, thus deterring people from drinking.

The state markers of alcohol include:

- Liver enzymes, particularly gamma-glutamyltransferase (GGT).
- Mean corpuscular volume (MCV).
- Carbohydrate-deficient transferrin.
- Serum and urine hexosaminidase.
- Sialic acid.
- Acetaldehyde-protein adducts.
- Ethyl glucuronide and ethyl sulfate.
- Fatty acid ethyl ester.

Alcohol biomarkers are primarily used for screening patients for possible alcohol abuse. They are also used for identification of pregnant women who may be abusing alcohol because fetal alcohol syndrome is a totally preventable disorder. Alcohol biomarkers are also used in emergency room settings, psychiatric clinics, and internal medicine settings because self-reporting of alcohol use is not always accurate as some patients are reluctant to admit a problem with alcohol. The addition of biomarkers can help identify individuals who need treatment for alcohol abuse.

Traditional state biomarkers of alcohol use are indirect biomarkers, which are elevated in a person who consumes moderate to heavy amounts of alcohol. These biomarkers are elevated due to toxicity of alcohol on a particular organ; for example, liver enzyme gamma-glutamyltransferase (GGT) is elevated after heavy alcohol consumption. Mean corpuscular volume (MCV), as well as the first Food and Drug Administration (FDA) approved biomarker of alcohol abuse, carbohydrate-deficient transferrin, are also indirect markers. In addition, serum and urine hexosaminidase and sialic acid are also indirect biomarkers of alcohol abuse. In contrast, minor alcohol metabolites such as ethyl glucuronide, ethyl sulfate, or biomolecules derived from the interaction of alcohol with other molecules such as fatty acid ethyl ester and phosphatidyl ethanol, are direct biomarkers of alcohol consumption.

Because alcohol is produced by bacterial action after death, ethyl glucuronide and ethyl sulfate are postmortem markers of antemortem alcohol ingestion because neither one is formed after death. In one study involving 36 death investigations where postmortem ethanol production was suspected, ethyl glucuronide and ethyl sulfate were measured in both the urine and blood of the deceased. In 19 out of 36 deceased, the concentration of ethyl glucuronide in blood ranged from 0.1 to 23.2 µg/L, while urinary ethyl glucuronide concentrations ranged from 1.9 to 182 µg/L. For ethyl sulfate, the blood concentration ranged from 0.04 to 7.9 µg/L, while urine concentrations ranged from 0.3 to 99 µg/L. In 16 other individuals no ethyl glucuronide or ethyl sulfate was detected. The authors concluded that, in 36 cases, alcohol consumption before death was likely in 19 of the deceased who only showed positive ethyl glucuronide and ethyl sulfate concentrations in blood and urine [10].

Fatty acid ethyl esters are direct markers of alcohol abuse because they are formed due to a chemical reaction between fatty acids and alcohol (ethanol). Fatty acids are an integral part of triglyceride structure, but a small number of fatty acids, also known as free fatty acids, are found in circulation. The chemical reaction between alcohol and fatty acids is known as esterification, and is mediated by fatty acid ethyl ester synthase (FAEE synthase), an enzyme found in abundance in the liver and pancreas. Carboxylesterase, lipase, another enzyme that liberates free fatty acids from complex lipids, can also induce the reaction between alcohol and fatty acids to generate fatty acid ethyl esters. These compounds are found in circulation, but they are also incorporated into hair follicles through sebum and can be used as a biomarker of alcohol abuse. There are four major fatty acid ethyl esters: ethyl myristate, ethyl palmitate, ethyl stearate, and ethyl oleate. These compounds can be measured in blood or hair using gas chromatography/mass spectrometry. The results are usually expressed as the sum of all four fatty acid ethyl ester concentrations. The reference range and window of detection of alcohol abuse according to these various markers are summarized in Table 18.3.

Table 18.3 Reference Range and Detection Period of Alcohol Abuse by Slate Markers

Slate Marker	Type of Marker*	Cut-Off Value	Window of Detection
Gamma-glutamyltransferase (GGT)	Indirect	>63 U/L	2−3 weeks
Mean corpuscular volume (MCV)	Indirect	>100 fl	2−4 months
Carbohydrate-deficient transferrin	Indirect	>2−2.5%	2−3 weeks
Serum and urine beta-hexosaminidase	Indirect	Varies	1−2 weeks, serum 2−4 weeks, urine
Sialic acid	Indirect	>60 mg/dL	Variable
Acetaldehyde−hemoglobin adducts	Direct	Not established	1 week
Ethyl glucuronide	Direct	>1000 ng/mL in urine >25 pg/mg in hair	1 week, urine, Months in hair
Ethyl sulfate	Direct	Not established	1 week, urine
Fatty acid ethylester	Direct	>0.5 ng/mg of hair	Months in hair <1 day in serum

*Direct markers are either metabolites of alcohol or adducts formed with the alcohol molecule.

18.8 METHANOL ABUSE

Methanol (wood spirits) is found in many household chemicals (auto products, cleaning products, etc.), but methylated spirits is the most common household chemical that contains methanol. Methanol is easily absorbed, even through the skin, and may cause toxicity. Inhalation of methanol through carburetor cleaner is a major route of domestic exposure to methanol. Accidental ingestion of windshield washer fluid is also another common cause of methanol intoxication. Routine occupational exposure to methanol-containing products is relatively safe. Like ethanol, exposure to methanol during pregnancy is dangerous.

A small amount of methanol is found in alcoholic beverages as a part of the natural fermentation process. This small amount does not cause any harm because the ethanol present in the drink protects the human body from methanol toxicity. However, illicit drinks prepared from methylated spirits can cause severe and even fatal illness. Illegally prepared moonshine whiskey can contain much higher amounts of methanol. It is one of the major sources of the epidemic of methanol toxicity worldwide. Methanol is readily absorbed after ingestion or inhalation and subsequent entry into the blood stream. A small amount of methanol is excreted unchanged in urine and also through exhaled breath. The majority of methanol is metabolized by the same enzyme in the liver that metabolizes ethanol: alcohol dehydrogenase. In this process formaldehyde is generated and is further metabolized by another liver enzyme (acetaldehyde dehydrogenase) to formic acid (Equation 18.3):

$$\text{Methanol} \rightarrow \text{Formaldehyde} \rightarrow \text{Formic Acid} \qquad (18.3)$$

Methanol itself is relatively non-toxic and methanol toxicity is a classic example of "lethal synthesis," where metabolites of methanol in the body are the major cause of methanol toxicity. Formic acid, the end product of methanol metabolism, is the key factor in causing toxicity from methanol, including blindness and death. Important points regarding methanol intoxication include:

- The lethal dose of methanol in humans is not fully established. Although it is assumed that ingestion of anywhere from 30 to 100 mL of methanol may cause death, fatality from methanol can occur even after ingestion of 15 mL of 40% methanol, and blindness can result from consuming as little as 4 mL of methanol.
- If blood methanol concentration exceeds 20 mg/dL, treatment should be initiated. However, a clinician may treat a patient with much lower methanol concentration depending on the clinical picture of the patient.

The best way to establish the diagnosis of methanol toxicity is by direct measurement of methanol with gas chromatography. If that is not available, high anion gap and osmolar gap with suspected methanol ingestion can be used for diagnosis of methanol poisoning. Methanol poisoning can be treated with an infusion of ethanol (blood ethanol targeted as 100 mg/dL). The goal is to slow down production of formic acid, the toxic metabolite. In addition, 4-methylpyrazole (fomepizole), sodium bicarbonate, and even dialysis can be used for treating methanol poisoning.

18.9 ABUSE OF ETHYLENE GLYCOL AND OTHER ALCOHOLS

Ethylene glycol is a colorless and relatively non-volatile liquid that has a high boiling point and a sweet taste, which is why children and pets tend to ingest it (causing ethylene glycol toxicity). An adult may drink ethylene glycol as a substitute for ethanol or in a suicide attempt. Because of the low melting point and high boiling point, ethylene glycol is used as a major ingredient in automobile antifreeze. Ethylene glycol is used in de-icing fluid, and in industry ethylene glycol is widely used as a starting material for preparing various polyester products.

Because ethylene glycol is relatively non-volatile, inhalation exposure is not generally considered an occupational health hazard. Absorption of ethylene glycol through the skin can cause serious toxicity, especially if there are any skin lesions. The major route of exposure to ethylene glycol is ingestion of ethylene glycol-containing fluids. Ethylene glycol is rapidly and completely absorbed from the intestinal tract after oral ingestion. Ethylene glycol itself is relatively non-toxic (like methanol) but its metabolites are toxic. Ethylene glycol is primarily metabolized in the liver (approximately 80%) while another 20% is excreted in the urine unchanged. Metabolism of ethylene glycol by the liver is a four-step process. Ethylene glycol is first metabolized to glycoaldehyde by alcohol dehydrogenase and then glycoaldehyde is further metabolized by aldehyde dehydrogenase into glycolic acid. Finally, glycolic acid is transformed into oxalic acid through an intermediate glyoxylic acid. Oxalic acid then combines with calcium to cause deposition of calcium oxalate in the kidneys, which results in severe renal failure (Equation 18.4):

$$\text{Oxalic acid} + \text{Ca}^{2+} \rightarrow \text{Calcium oxalate crystals causing nephrotoxicity}$$

$$(18.4)$$

Major complications of ethylene glycol poisoning are metabolic acidosis and renal failure. These complications can even be fatal. The lethal dose of ethylene glycol is usually assumed to be 100 mL, but there are reports of fatality from ethylene glycol poisoning even from ingestion of only 30 mL [11]. Blood levels of ethylene glycol are usually measured by head space gas chromatography either singly or in combination with other volatile compounds such as methanol, acetone, and isopropyl alcohol. In addition, there are some enzymatic methods available for rapid determination of blood ethylene glycol levels using an automated analyzer in the clinical laboratory. Limitations of enzymatic methods for ethylene glycol determination include:

- Like the enzymatic method for alcohol, the method for ethylene glycol determination produces a false positive ethylene glycol level if lactate and lactate dehydrogenase are present in the serum specimen.
- Interestingly, in patients poisoned with ethylene glycol, falsely elevated lactate may be observed using blood gas analyzers, but chemistry analyzers usually do not show this false elevation.

CASE REPORT

A 29-year-old man with a history of psychosis and substance abuse presented to the emergency department in a confused state. His blood pressure was 180/100 mm of Hg and he had a score of 11/15 on the Glasgow Coma Scale. Testing of arterial blood in the emergency department using an ABL 725 analyzer (Radiometer) showed a highly elevated lactate level of 24 mmol/L, indicating severe life-threatening lactic acidosis. A urine toxicology screen showed the presence of cannabinoid metabolite. In addition, urine sediment showed calcium oxalate crystals, which indicated abuse of ethylene glycol. Further screening of his serum showed a highly elevated ethylene glycol level of 64 mg/dL. No methanol or ethanol was detected in the serum specimen. However, when a second serum specimen was analyzed in the main hospital laboratory using a DxC-800 automated analyzer (Beckman, Coulter, Brea, CA), the lactate level was normal (4.7 mmol/L). At that point it was determined that the patient had ethylene glycol poisoning rather than lactic acidosis. He was treated with ethanol infusion, bicarbonate infusion, and hemodialysis, and he completely recovered with no further sign of renal insufficiency. Patients with ethylene glycol poisoning may show false positive lactate by blood gas analyzers [12].

Ethylene glycol poisoning is treated similarly to methanol poisoning using bicarbonate, ethanol, fomepizole, or hemodialysis. Propylene glycol, which is similar to ethylene glycol, is used as an industrial solvent and can also be used in antifreeze formulations. Propylene glycol is significantly less toxic than ethylene glycol and is the preferred antifreeze used in motor homes and recreational vehicles. Propylene glycol is also used as a diluent for oral, topical, or intravenous pharmaceutical preparations so that active ingredients can be dissolved properly in the formulation. Isopropyl alcohol is also known as rubbing alcohol, and is a 70% aqueous solution of isopropyl alcohol. Isopropyl alcohol is slowly metabolized into acetone by alcohol dehydrogenase. Acetone is also found in many domestic products, for example, nail polish remover.

Neither isopropyl alcohol nor acetone can cause metabolic acidosis, and poisoning from these compounds can be less life-threatening than methanol or ethylene glycol poisoning. However, there are reports of death from severe isopropyl alcohol poisoning. Overdose with methanol and ethanol may cause metabolic acidosis, but overdose with isopropyl alcohol causes ketosis without acidosis because isopropyl alcohol is converted into acetone.

KEY POINTS

- Alcohol content of various alcoholic beverages varies widely, but different amounts are consumed for different drinks. Therefore, one standard drink contains approximately 0.6 ounces of pure alcohol or 14 grams of pure alcohol.
- Guidelines of moderate drinking: For men, no more than 2 drinks a day (up to 14 drinks per week); for women, not more than one drink a day (not exceeding 7 drinks per week); but for anyone 65 years or older, one drink per day regardless of gender.
- Alcohol metabolism follows zero-order kinetics (blood alcohol is cleared at a constant rate regardless of blood alcohol level; 0.018–0.020% for males and 0.015–0.018% for females). Metabolism depends on age, gender, body weight, amount of food consumed (less blood alcohol if consumed with food), and genetic makeup of the person.
- The legal limit of blood alcohol in all states in the U.S. is currently 0.08% whole blood alcohol (80 mg/dL). Serum alcohol concentration is higher than whole blood alcohol concentration. In order to convert serum alcohol level to whole blood alcohol level, serum alcohol level must be divided by 1.15. Therefore, a serum alcohol concentration of 100 mg/dL (0.1%) is equivalent to 87 mg/dL (0.087%) whole blood concentration.
- Blood alcohol can be calculated based on gender, number of standard drinks consumed by the person, and body weight using the Widmark formula: $C = $ (Number of drinks $\times$ 3.1/Weight in pounds $\times$ r) $-$ 0.015 t, where C is blood alcohol in percent (mg/dL) and r is 0.7 for men and 0.6 for women, and t is time (measured in hours).
- Enzymatic methods for alcohol determination are commonly used in hospital laboratories, but high lactate dehydrogenase and lactate, if present in the specimen, can falsely elevate the serum alcohol level even if alcohol is absent. Gas chromatography-based methods are free from such interferences. Enzymatic assays are unsuitable for alcohol determination in postmortem blood because high lactate dehydrogenase and lactate can be found in postmortem specimens due to cellular breakdown. In addition, alcohol can be produced by the activity of microorganisms after death. Therefore, elevated blood alcohol in postmortem specimens may not confirm alcohol intake prior to death. The vitreous humor is a

better source for determination of postmortem alcohol. Alternatively, the presence of ethyl glucuronide and ethyl sulfate (metabolites of alcohol in postmortem blood or urine) confirms alcohol abuse prior to death. However, if the postmortem blood alcohol level is positive but no ethyl glucuronide or ethyl sulfate can be detected in the blood or urine, it is an indication of postmortem production of alcohol, and it can be concluded that the deceased did not consume alcohol prior to death.

- Endogenous production of alcohol is minimal except for very rare instances of patients with auto-brewery syndrome, where, if a patient eats carbohydrate-rich food, endogenous alcohol production at a high level is possibly due to gut infection related to fungus.
- Although blood alcohol is more commonly measured, urine alcohol can also be measured in workplace drug testing. Again, gas chromatography is preferred, although enzymatic methods can also be used. Because lactate dehydrogenase is absent in urine (high molecular weight), urine alcohol determination is not subjected to the interference of high lactate dehydrogenase if present in serum (but lactic acid is present in urine). However, another problem with urine alcohol determination is in vitro production of alcohol if a female patient has poorly controlled diabetes and a yeast infection. In this case, urine may be contaminated with yeast, and glucose present in the urine can be converted into alcohol by the yeast.
- The state markers of alcohol include liver enzymes, particularly gamma-glutamyltransferase (GGT), mean corpuscular volume (MCV), carbohydrate-deficient transferrin, serum and urine hexosaminidase, sialic acid, acetaldehyde-protein adducts, ethyl glucuronide, ethyl sulfate, and fatty acid ethyl ester.
- Methanol is metabolized to toxic formaldehyde and formic acid, which can cause metabolic acidosis, blindness, and death. The lethal dose of methanol in humans is not fully established. Although it is assumed that ingestion of anywhere from 30 to 100 mL of methanol can cause death, fatality from methanol can occur even after ingestion of 15 mL of 40% methanol, and blindness can result from consuming as little as 4 mL of methanol.
- Ethylene glycol is metabolized finally into oxalic acid, which combines with calcium to form calcium oxalate. Calcium oxalate crystals deposit in the kidney and cause renal failure; a severe overdose of ethylene glycol can be fatal. Both methanol and ethylene glycol can be treated with ethanol infusion to slow down metabolism of methanol or ethylene glycol (ethanol is a preferred substrate for alcohol dehydrogenase, which also metabolizes methanol and ethylene glycol). Hemodialysis is used for treating life-threatening methanol or ethylene glycol overdoses. An overdose with methanol and ethanol can cause metabolic acidosis, but an overdose with isopropyl alcohol causes ketosis without acidosis.

REFERENCES

[1] Kerr WC, Greenfield TK, Tujague J, Brown SE. A drink is a drink? Variation in the amount of alcohol contained in beer, wine and spirits drinks in a US methodological sample. Alcohol Clin Exp Res 2005;29:2015−21.

[2] United States Department of Agriculture and United States Department of Health and Human Services. In: *Dietary guidelines for Americans*. Chapter 9 − Alcoholic Beverages. Washington, DC: US Government Printing Office; 2005. p. 43−6. Available at http://www.health.gov/DIETARYGUIDELINES/dga2005/document/html/chapter9.htm (Accessed 7/23/2013).

[3] Reid MC, Fiellin DA, O'Connor PG. Hazardous and harmful alcohol consumption in primary care. Arch Int Med 2008;159:1681−9.

[4] Tolstrup J, Jensen MK, Tjonneland A, Overvad K, et al. Prospective study of alcohol drinking patterns and coronary heart disease in women and men. Br Med J 332(7552):1244−8.

[5] Ruf JC. Alcohol, wine and platelet function. Biol Res 2004;37:209−15.

[6] Rainey P. Relation between serum and whole blood ethanol concentrations. Clin Chem 1993;39:2288−92.

[7] Logan BK, Jones AW. Endogenous ethanol auto brewery syndrome as a drunk-driving defense. Med Sci Law 2000;40:206−13.

[8] Jansson-Nettelbladt E, Meurling S, Petrini B, Sjolin J, et al. Endogenous ethanol fermentation in a child with short bowel syndrome. Acta Paediatr 2006;95:502−4.

[9] Jones AW, Eklund A, Helander A. Misleading results of ethanol analysis in urine specimens from rape victim suffering from diabetes. J Clin Forensic Med 2007;7:144−6.

[10] Hoiseth G, Karinen R, Christophersen A, Morland J. Practical use of ethyl glucuronide and ethyl sulfate in postmortem cases as markers of antemortem alcohol ingestion. Int J Legal Med 2010;124:143−8.

[11] Walder AD, Tyler CKG. Ethylene glycol antifreeze poisoning: Three case reports and a review of treatment. Anesthesia 1994;49:964−7.

[12] Sandberg Y, Rood PPM, Russcher H, Zwaans JJM, et al. Falsely elevated lactate in severe ethylene glycol intoxication. Netherland J Med 2010;68:320−3.

CHAPTER 19

Common Poisonings Including Heavy Metal Poisoning

19.1 POISONING FROM ANALGESICS

Poisonings from analgesics are due to overdoses or suicide attempts using over-the-counter (OTC) drugs such as acetaminophen and aspirin (acetyl salicylate). Determination of serum acetaminophen and salicylate concentrations is useful in clinical laboratories for the diagnosis of poisoning from these drugs. Important points regarding acetaminophen toxicity include:

- Severe acetaminophen toxicity usually occurs in an adult after consuming 7–10 g of acetaminophen (15–20 tablets, 500 mg each).
- Acetaminophen is normally metabolized to glucuronide and sulfate conjugates by liver enzymes, but a small amount is metabolized by the cytochrome P-450 mixed-function oxidase family of enzymes to a toxic metabolite (N-acetyl-p-benzoquinone imine), which is detoxified after conjugating with glutathione present in the liver.
- Acetaminophen toxicity is due to formation of excess toxic metabolite (N-acetyl-p-benzoquinone imine) during severe overdose because the glutathione supply of the liver is depleted and this metabolite is no longer conjugated with glutathione.
- The antidote for acetaminophen poisoning (N-acetylcysteine, Mucomyst) is a precursor of glutathione and can detoxify the toxic metabolite of acetaminophen. This antidote must be administered as soon as possible (certainly within 8 h of acetaminophen overdose) for maximum benefit.

Antidotes/treatments of various common poisonings are listed in Table 19.1. Although the therapeutic range of acetaminophen is 10–30 μg/mL and the toxic concentration is over 150 μg/mL (4 h after ingestion), alcoholics may experience severe liver toxicity after consuming a moderate dosage of acetaminophen because alcohol is known to deplete the liver's glutathione supply. Strikingly abnormal liver enzymes in alcoholics after consuming a moderate dosage of acetaminophen have been well documented in the literature [1].

CONTENTS

A. Dasgupta and A. Wahed: Clinical Chemistry, Immunology and Laboratory Quality Control
DOI: http://dx.doi.org/10.1016/B978-0-12-407821-5.00019-X

Table 19.1 Treatments/Antidotes for Common Poisonings

Poisoning	Treatment/Antidote
Acetaminophen	N-Acetylcysteine (Mucomyst)
Acetyl salicylate	Activated charcoal, sodium bicarbonate to correct acid–base disorder, and hemodialysis
Carbon monoxide	100% oxygen or hyperbaric oxygen therapy
Cyanide	Hydroxocobalamin or any combination of amyl nitrate, sodium nitrate, ferrous sulfate, or dicobalt edentate
Tricyclic antidepressants	Sodium bicarbonate
Benzodiazepines	Flumazenil
Opiates	Naloxone
Methanol	Ethyl alcohol or fomepizole (4-methylpyrazole)
Ethylene glycol	Ethyl alcohol or fomepizole (4-methylpyrazole)
Organophosphorus/ carbamate insecticides	Atropine, pralidoxime
Lead poisoning	Various chelating agents such as calcium sodium ethylenediamine tetraacetic acid (EDTA), D-penicillamine, or 2,3-dimercaptosuccinic acid (DMSA)
Mercury poisoning	British–Lewisite (BAL, dimercaprol) or 2,3-dimercapto-1-propanesulfonate (DMPS)
Arsenic poisoning	BAL, DMSA

Aspirin (acetyl salicylate) poisoning is also common. After ingestion, acetyl salicylate is hydrolyzed by liver and blood esterase into salicylic acid (salicylate, the pharmacologically active drug). The therapeutic range is $150-250\ \mu g/mL$, although an analgesic effect may be observed below the $100\ \mu g/mL$ level. Life-threatening toxicity is observed with salicylate levels at $500\ \mu g/mL$ or above. Important points regarding salicylate toxicity include:

- Aspirin can cause potentially life-threatening toxicity known as "Reyes syndrome" in children and adolescents with certain viral infections such as varicella and influenza. Therefore, salicylate use is contraindicated in these patients.
- Salicylate overdose results in direct stimulation of the central respiratory system, causing hyperventilation and respiratory alkalosis.
- Salicylate overdose also causes hyperthermia.
- Finally, salicylate overdose causes metabolic acidosis due to accumulation of organic acids as a result of inhibition of the Krebs cycle. Increased anion gaps and high serum osmolarity are indirect indications of salicylate overdose.

Activated charcoal can be given to an overdose patient to prevent further absorption. In addition, sodium bicarbonate can be used for correcting acid−base disorder. Finally, hemodialysis may be needed for a patient with salicylate levels in the near-fatal range.

19.2 METHYL SALICYLATE POISONING

Methyl salicylate is a major component of oil of wintergreen (prepared by distillation from wintergreen leaves). Methyl salicylate has an analgesic effect and is used in many over-the-counter analgesic creams or gels designed only for topical use. Methyl salicylate, if ingested, is very poisonous. Although aspirin is acetyl salicylate that is structurally close to methyl salicylate, after ingestion it is rapidly broken down into salicylic acid by blood and liver esterase; very little methyl salicylate is broken down to salicylic acid by the esterase in blood. Methyl salicylate is a relatively common cause of poisoning in children, and ingestion of one teaspoon of oil of wintergreen can be fatal [2]. The popular topical ointment Bengay® contains 15% methyl salicylate while Bengay Muscle Pin/Ultra Strength contains 30% methyl salicylate. Many Chinese medicines and medicated oils contain high amounts of methyl salicylate. Poisoning from methyl salicylate can occur due to abuse of topical creams containing methyl salicylate. Salicylate can be detected in blood after excessive topical application of creams containing methyl salicylate. Bell and Duggin reported the case of a 40-year-old man who became acutely ill after receiving treatment from an unregistered naturopath (herbal skin cream treatment for psoriasis). The herbal cream contained methyl salicylate, and in this case transcutaneous absorption of methyl salicylate was enhanced due to psoriasis [3].

19.3 CARBON MONOXIDE POISONING

Carbon monoxide poisoning can occur accidentally from a faulty ventilated home heating unit or intentionally by inhaling car exhaust in a suicide attempt. In general, carbon monoxide is produced due to incomplete combustion. It is also produced when burning wood and by charcoal grills, propane grills, and many gasoline-operated instruments. In addition, carbon monoxide is also present in cigarette smoke. When inhaled, carbon monoxide tightly binds to hemoglobin, producing carboxyhemoglobin. Because the binding affinity of carbon monoxide is 250 times more than the binding affinity of oxygen, in the presence of carbon monoxide, hemoglobin preferentially binds with carbon monoxide, thus causing severe hypoxia. Because carbon monoxide is odorless, it cannot be detected, and sometimes victims are not even aware that they are being exposed to a lethal carbon monoxide environment.

The blood level of carbon monoxide is usually determined spectrophotometrically by measuring carboxyhemoglobin levels with a CO-oximeter. Commercially available instruments can perform absorption measurements of blood specimens at various wavelengths to determine the concentration of oxyhemoglobin, deoxyhemoglobin, carboxyhemoglobin, and methemoglobin. More recently, non-invasive pulse CO-oximeters have become available for measurement of various hemoglobin components for screening in emergency departments to identify patients with carbon monoxide poisoning [4]. Important factors regarding carboxyhemoglobin levels in blood include:

- Non-smokers living in rural areas usually have a blood carboxyhemoglobin level of less than 0.5%, while urban non-smokers may have levels up to 2%.
- Smokers may have carboxyhemoglobin levels of 5−6%.
- Minor symptoms of carbon monoxide poisoning such as shortness of breath may be experienced with a carboxyhemoglobin concentration of 10% or higher, while at a 30% carboxyhemoglobin level, full-blown symptoms of carbon monoxide poisoning (severe headache, fatigue, nausea, vomiting, and difficulty in judgment) are observed.
- A carboxyhemoglobin level of 60% to 70% and above may cause respiratory failure and even death. A value over 80% may cause quick death.
- Fetal hemoglobin has slightly different spectrophotometric properties than adult hemoglobin. Therefore, a falsely high carboxyhemoglobin value has been reported in neonates using a CO-oximeter.

The treatment of a person with carbon monoxide poisoning is to administer 100% oxygen. The half-life of carboxyhemoglobin is reduced significantly when a victim breathes 100% oxygen rather than room air. If the victim is placed in a hyperbaric chamber (oxygen treatment at 2 to 3 times atmospheric pressure), the half-life of carboxyhemoglobin is approximately 15 min. However, this treatment is usually reserved for a victim experiencing severe carbon monoxide poisoning.

CASE REPORT

A previously healthy 3-month-old girl presented to the pediatric emergency department with smoke inhalation from a malfunctioning furnace. The exact duration of exposure was unknown. On examination, she was alert and afebrile with a Glasgow Coma Scale score of 15, normal vitals, and mildly elevated carboxyhemoglobin levels. Because of the concern of carbon monoxide poisoning, the baby received 100% oxygen. Six hours later, a repeat carboxyhemoglobin level was 11.2% even though the baby was doing well clinically. Despite continued normobaric oxygen therapy, her carboxyhemoglobin level was still elevated. This prolonged her emergency department stay. Further investigation revealed that her falsely elevated carboxyhemoglobin level was due to interference of fetal hemoglobin in spectrophotometric measurements of carboxyhemoglobin [5].

19.4 CYANIDE POISONING

The most lethal form of cyanide poisoning is inhalation of hydrocyanic acid (HCN, prussic acid). If an inorganic cyanide such as potassium cyanide is ingested (e.g. in a suicide attempt), it is converted into HCN, which causes toxicity and even fatality. Cyanide poisoning is relatively uncommon. The most common cause is smoke inhalation from burning common household substances such as plastics, silk, or rubber, which can produce cyanide smoke that contains HCN. Therefore, firefighters are also at high risk of cyanide poisoning. However, during fires, both carbon monoxide and cyanide poisoning can occur. Lundquist et al. studied blood cyanide and carboxyhemoglobin levels in 19 victims who were found dead in building fires. The results indicated that 50% of the victims had been exposed to toxic levels of HCN, and 90% to toxic levels of carbon monoxide [6].

Some stone fruits (fruits that contain a pit or solid core), such as apricots, cherries, peaches, pears, plums, and prunes, contain cyanogenic glycoside. Massive ingestion of these fruit pits can be dangerous, but eating the flesh of these fruits is not a concern. Cyanogenic glycoside is also present in cassava roots and fresh bamboo shoots. Therefore, these foods should be cooked before consumption.

After exposure to HCN, cyanide ion tightly binds with hemoglobin to produce cyanhemoglobin, which can cause severe hypoxia. Although the body can transform cyanide into relatively non-toxic thiocyanate, this process is slow and not effective in avoiding life-threatening cyanide poisoning. Cyanide poisoning produces non-specific symptoms. Therefore, determining blood cyanide level is useful for diagnosis. Usually spectroscopic methods are used for cyanide determination. The normal blood cyanide level is less than 0.2 μg/mL. A cyanide level above 2 μg/mL can produce severe toxicity, and a level above 5 μg/mL may be lethal if not treated immediately.

There are several antidotes for cyanide poisoning, including inhalation of amyl nitrate and administration of sodium thiosulfate, sodium nitrate, ferrous sulfate, dicobalt edentate, or hydroxocobalamin. Oxygen therapy is also useful. If a cyanide-poisoning victim receives prompt medical care, a life can be saved. Borron et al. concluded in their study that 67% of patients with confirmed cyanide poisoning after smoke inhalation survived after successful administration of a hydroxocobalamin antidote [7].

19.5 OVERDOSE WITH TRICYCLIC ANTIDEPRESSANTS

Although overdose with tricyclic antidepressants was a serious problem in the past, the number of cases related to tricyclic overdose is on the decline

Table 19.2 Common Tricyclic Antidepressants and their Active Metabolites	
Tricyclic Antidepressant	**Active Metabolite**
Imipramine	Desipramine
Amitriptyline	Nortriptyline
Doxepin	Nordoxepin
Desipramine	Not applicable
Nortriptyline	Not applicable
Protriptyline	Not applicable
Trimipramine	Not applicable

because these antidepressants are prescribed less frequently today due to the availability of newer psychoactive drugs with improved safety margins (especially selective serotonin reuptake inhibitors, SSRIs). Some of the tricyclic antidepressants are metabolized to their active metabolites. Interestingly, active metabolites are also available as drugs for therapy (Table 19.2). Two major manifestations of tricyclic antidepressants include:

- Central nervous system anticholinergic effects such as dry mouth, dry skin, flushing, and urinary retention.
- Cardiovascular effects.

Mortality from tricyclic antidepressant overdose is usually due to cardiovascular toxicity, including cardiac arrhythmia. In the electrocardiogram, QRS prolongation of more than 100 milliseconds (msec) is typically a sign of severe toxicity, and is usually associated with a total tricyclic antidepressant level of 1,000 ng/mL or more. Heart rates above 100 are also associated with severe toxicity [8]. Tricyclic antidepressant overdose can be treated by alkalinization using sodium bicarbonate. Other therapies may also be initiated depending on the clinical condition of the patient (as determined by the physician).

19.6 BENZODIAZEPINE AND OPIATE OVERDOSE

Benzodiazepines are one of the most commonly observed prescription medications (along with opioids). Flumazenil, a specific benzodiazepine antagonist, is useful in reversing the sedation and respiratory depression that are characteristics of benzodiazepine overdose. However, some controversy exists in using flumazenil in treating mixed benzodiazepine/tricyclic antidepressant overdose due to possible precipitation of seizure activity [9].

Naloxone (Narcan) is a synthetic derivative of oxymorphone that can antagonize pharmacological effects of opiates. Therefore, naloxone is an excellent antidote for opiate overdose not only with morphine and heroin, but also with other opioids such as hydrocodone, oxycodone, and oxymorphone. This antidote is administered by intravenous injection (it is poorly absorbed after oral administration), and onset of action is very rapid (within 2−5 min). However, it is important to note that opioid addiction is treated with methadone.

19.7 ALCOHOL POISONING

Poisoning with alcohol (ethanol), methanol, and ethylene glycol are also common. In addition to standard care (such as gastric irrigation, etc.), both methanol and ethylene glycol poisoning can be treated with intravenous administration of ethyl alcohol to achieve a blood ethanol level of 0.1% (100 mg/dL). This minimizes metabolism of methanol and ethylene glycol to their toxic metabolites (formic acid for methanol, and oxalic acid for ethylene glycol). This is possible because methanol and ethylene glycol are indeed metabolized by alcohol dehydrogenase, but ethanol is the preferred substrate. Fomepizole (4-methylpyrazole, Antizol, administered intravenously) is also a competitive inhibitor of alcohol dehydrogenase that can be used instead of ethyl alcohol to treat methanol and ethylene glycol poisoning. However, both fomepizole and ethyl alcohol cannot be used simultaneously because fomepizole also significantly increases the elimination of ethyl alcohol. Depending on the blood levels of methanol and ethylene glycol, dialysis may also be initiated. Treatment of alcohol poisoning is mostly standard, but ethanol is also dialyzable (see Chapter 18).

It is important to note that alcohol abuse is treated in detoxification centers with disulfiram (Antabuse), which inhibits alcohol dehydrogenase. A person who is supposed to be alcohol-free experiences uncomfortable physical reactions (nausea, flushing, vomiting, and headaches) due to accumulation of acetaldehyde after drinking even one drink. Other medications used in treating alcohol abuse include naltrexone, which blocks the good feeling after drinking, and acamprosate, which helps combat alcohol cravings. However, in contrast to disulfiram, naltrexone and acamprosate do not produce an uncomfortable feeling after drinking alcohol in a person who is supposed to be sober.

19.8 POISONING FROM ORGANOPHOSPHORUS AND CARBAMATE INSECTICIDES

Pesticides are widely used in agriculture to control insects, fungi, weeds, and microorganisms in order to increase crop yield. Chemicals in pesticides may cause toxicity in humans if the chemical comes in contact with the skin or eyes, or is inhaled or swallowed (most commonly in a suicide attempt). Pesticides can be absorbed through the skin after exposure, but can be absorbed more readily through the eyes or lungs after breathing contaminated air since many pesticides form droplets or fumes. Pesticides can be grouped according to chemical classes. The most common include organophosphates (also called organophosphorus), organochlorines, and carbamates. Thiocarbamate and dithiocarbamate are subclasses of carbamate pesticides.

Organophosphates, and to some extent carbamate pesticides, inhibit cholinesterase. Two major human cholinesterases are acetylcholinesterase, found primarily in nerve tissues and erythrocytes, and pseudocholinesterase, which is found in both serum and the liver. Poisoning with carbamate is usually less severe because carbamate binds to acetylcholinesterase reversibly, but organophosphates bind to acetylcholinesterase irreversibly. Nevertheless, both organophosphorus and carbamate toxicity can cause serious life-threatening situations, including fatalities.

Currently, red cell (erythrocyte) acetylcholinesterase activity or serum plasma cholinesterase (also known as pseudocholinesterase or butyrylcholinesterase) activity can be measured for diagnosis as well as for monitoring the progress of therapy in patients poisoned with organophosphate or carbamate insecticides. The antidote for organophosphorus and carbamate poisoning is atropine to block muscarinic action of excess acetylcholine. Pralidoxime is also given to reactivate cholinesterase.

19.9 LEAD POISONING

Lead is a heavy metal and a divalent cation that has been used in human civilization for a long time. In 1976, lead-based paints in toys were banned in the U.S., and lead-based household paints were banned in 1978. In 1986 lead-based gasoline was phased out. Today, major sources of lead exposure are the following:

- Deteriorated lead-based paint in older housing.
- Old water pipes with lead-based soldering.
- Soil contaminated with lead.
- Stained glass and glazed ceramics used by artists.
- Moonshine liquors.
- Many Asian herbal remedies, which can be contaminated with heavy metals such as lead, arsenic, and mercury.
- Lead shots from ingestion of hunted birds (an unusual source).
- Lead bullet left in the body after a firearm injury (a rare source).

Lead enters the body either through inhalation or ingestion. It is distributed in the body in three main compartments: blood, soft tissue, and bone. Approximately 99% of the lead in blood is bound to red blood cells and 1% is free in plasma. The skeleton is the main lead depot in the human body, and may represent approximately 90–95% of the body's burden of lead in adults. Important issues regarding lead toxicity include:

- Lead inhibits delta-aminolevulinic acid dehydrogenase, one of the enzymes that catalyzes the synthesis of heme from porphyrin (thus increasing the erythrocyte concentration of protoporphyrin). Lead also inhibits the enzyme ferrochelatase, which leads to inhibition of iron incorporation in protoporphyrin. As a result, zinc protoporphyrin (ZPP) is produced and remains in circulation for 120 days.
- Lead exposure can cause microcytic anemia, and acute lead toxicity can cause hemolytic anemia. Lead is a nephrotoxic agent.
- The neurotoxicity of lead produces more adverse effects in children than in adults.

Although the presence of erythrocytic zinc protoporphyrin can be used for the diagnosis of lead poisoning, ZPP levels are also elevated in anemia. The ZPP-to-heme ratio can in this case also be used for diagnosis of lead poisoning. In addition, determination of blood lead level is also useful for assessing lead exposure. Heparinized whole blood is the appropriate specimen for this test since most of the lead in blood is bound to erythrocytes. Lead testing on a dried filter-paper blood spot is also used routinely by some laboratories in screening for lead poisoning, especially in small children and neonates. The lead level in blood can be measured by atomic absorption. However, other techniques such as anodic stripping voltammetry and inductively coupled plasma mass spectrometry (ICP-MS) can also be used. Determination of lead in 24-h urine is also useful for monitoring therapies for lead toxicity. For collection of 24-h urine, the patient should void directly into a lead-free container (borosilicate glass or polyethylene).

The Centers for Disease Control and Prevention (CDC) recommends that blood lead levels in children should be less than $10\,\mu g/dL$. In many cases of lead poisoning, removing the person from the source of exposure is sufficient to reduce the blood lead level. The World Health Organization (WHO) defines lead levels over $30\,\mu g/dL$ as indicative of significant exposure to lead. Chelation therapy is usually performed for blood lead concentrations over $60\,\mu g/dL$. Calcium sodium EDTA is an effective chelating agent, but D-penicillamine is also effective in treating lead poisoning. More recently, 2,3-dimercaptosuccinic acid (DMSA) has been introduced as an orally given chelating agent to replace EDTA.

CASE REPORT

A 58-year-old woman from India, currently residing in the U.S., presented to the emergency department with a 10-day history of progressively worsening post-prandial lower abdominal pain, nausea, and vomiting. Her past medical history included well-controlled non-insulin-dependent diabetes mellitus and hypertension. A physical exam was notable only for abdominal tenderness in the lower quadrants. Laboratory studies revealed normochromic, normocytic anemia with a hemoglobin level of 7.7 g/dL, hematocrit of 22.6%, MCV of 87 fL, and a normal iron level. A CT scan of the abdomen and pelvis showed no specific abnormalities. The patient was discharged with a prescription of antiemetics and instructions to follow up with her primary care physician. Five days later, the patient returned to the emergency department with worsening abdominal pain, nausea, and bilious vomiting. Laboratory test results showed hemoglobin of 8.8 g/dL, hematocrit of 23.5%, MCV of 87 fL, and a corrected reticulocyte count of 7%. The patient was admitted, and a review of her peripheral blood smear demonstrated normochromic, normocytic anemia with extensive, coarse basophilic stippling of the erythrocytes, an indicator of lead poisoning. This triggered a screening for heavy metals, which revealed an elevated blood lead level of 102 µg/dL (normal is <10 µg/dL). Zinc protoporphyrin (ZPP) was subsequently found to be elevated at 912 µg/dL (normal is <35 µg/dL), indicating severe lead poisoning. Upon further questioning, the patient disclosed that she had been taking an Indian Ayurvedic medicine called Jambrulin over a period of 5 to 6 weeks for her diabetes. The patient was instructed not to take Jambrulin, and was treated with 2,3-dimercaptosuccinic acid (DMSA, an oral lead chelator) for two weeks. At the end of chelation therapy, her blood lead level was decreased to 46 µg/dL, and her abdominal pain was resolved. The Jambrulin pill tested was found to contain approximately 21.5 mg of lead, which explained her severe lead poisoning [10].

19.10 MERCURY POISONING

Mercury is a heavy metal with known toxicity. Most human exposure to mercury is from outgassing of mercury from dental amalgam or from eating fish and seafood contaminated with mercury. Occupational exposure to mercury vapor also causes mercury toxicity. The target organ for inhaled mercury vapor is the brain. Mercury exposure can also occur from inorganic mercury (mercurous and mercuric salts) as well as from organic mercury (mostly methyl mercury). Mercury in all forms is toxic because it alters the tertiary and quaternary structure of proteins by binding with sulfhydryl groups. Consequently, mercury can potentially impair the function of any organ, but neurological functions are the most often affected.

Exposure to methyl mercury from eating fish, shellfish (both fresh and saltwater), or sea mammals is dangerous for pregnant women and children. Over 3,000 lakes in the U.S. have been closed to fishing due to mercury contamination [11]. Methyl mercury may be present in higher amounts in predatory fish. Pregnant women, women planning to be pregnant, and children, should not eat fish (shark, swordfish, king mackerel, golden bass, and snapper) where high amounts of mercury could accumulate. Other people can eat up to 7 ounces of these fish per week. Salmon, cod, flounder, catfish, and other seafoods such as crabs and scallops may also contain mercury in lower amounts. Determination of blood, urine, and even hair levels of mercury is useful for determination of exposure to mercury. WHO experts have determined that a 24-h urine mercury level over 50 µg is indicative of excessive exposure to mercury.

Mercury poisoning is treated with chelating therapy. British–Lewisite (BAL, dimercaprol), a chelating agent developed to treat arsenic poisoning, is also effective in chelating mercury. However, this chelating agent is toxic, and, more recently, safer chelating agents such as 2,3-dimercapto-1-propanesulfonate (DMPS) have been developed for clinical use.

CASE REPORT

A 2-year-old boy showed increased aggressive behavior for the past 6 months. His father was diagnosed with mercury poisoning by a physician two months prior after complaining of allergies, rashes, abdominal pain, and diarrhea. The family ate fish (salmon, barramundi, or snapper) at least five times a week, and they had also used herbal medicines in the past. The blood mercury level of the boy was 31.6 µg/L, indicating mercury poisoning. Elevated levels of mercury were also found in his urine and hair. In addition, elevated levels of mercury were detected in hair specimens of his parents. The boy was therefore treated with chelation therapy [12].

19.11 ARSENIC POISONING

Soluble inorganic arsenic (as arsenic salts) is acutely toxic, and ingestion of a large dose may cause gastrointestinal, neurological, and cardiovascular toxicity, and can eventually even be fatal. Long-term chronic exposure to arsenic can occur from drinking water from wells where groundwater is contaminated with arsenic. This can be a particular problem in many developing countries, and long-term exposure can cause cancer in the skin, lungs, bladder, and kidneys. Hypertension and cardiovascular disorders are also common after chronic exposure to arsenic. In addition, early chronic exposure to arsenic can cause skin changes such as hyperkeratosis and pigmentation changes. WHO guidelines indicate that the maximum limit of arsenic in drinking water should be 10 µg/L. Of note, arsenic levels are greater than 200 µg/L in some drinking water wells in Bangladesh [13].

Other sources of arsenic poisoning are some Asian herbal remedies and Indian Ayurvedic medicines. In some Indian Ayurvedic medicines, arsenic and other heavy metals are used as components. Some kelp supplements may contain high quantities of arsenic. Serum arsenic concentrations are elevated for a short time only after exposure. Therefore, hair analysis of arsenic is useful in investigating chronic exposure to arsenic. Hair arsenic levels greater than 1 µg/g in dry hair are indicative of excessive exposure to arsenic.

Severe arsenic toxicity can be treated with chelation therapy. Originally, BAL was widely used as a chelating agent, but it has for the most part been replaced by safer chelating agents (such as DMSA).

19.12 POISONING FROM OTHER METALS/SOURCES

Metal toxicity can also occur from exposure to other heavy metals such as aluminum, antimony, beryllium, cadmium, chromium, cobalt, copper, iron, manganese, nickel, selenium, silicon, and thallium. However, toxicities from these heavy metals are observed less frequently than from lead, mercury, and arsenic poisoning. Aluminum toxicity has been documented after excessive use of aluminum-based over-the-counter antacids (secondary to aluminum-containing phosphate binding agents), but such toxicity is rare. Patients undergoing dialysis are also exposed to aluminum, but toxicity is very rare. Therefore, routine plasma monitoring of dialysis patients is not necessary, and it should only be conducted if excessive aluminum exposure is suspected.

The classic accumulation of copper in the human body is due to the genetic disease known as "Wilson's disease." Iron supplements are used for treating anemia, but excess ingestion of iron supplements can cause iron overload and toxicity. Asbestos fiber toxicity was a public health concern when asbestos-containing products were used as insulating materials in houses. Asbestos dust inhalation is particularly harmful. However, asbestos is no longer used in most instances. Silicone-based implants (containing elastomers of silicon) are another source of silicon exposure, especially if there is a rupture of an implant. Axillary silicone lymphadenopathy (presence of silicone in the lymph node) is a potential complication of breast augmentation if there is a rupture of the silicone-based implant, but such phenomena are rare. Omakobia et al. reported the case of a woman who presented with a painless swelling in the left supraclavicular region. She had a previous cosmetic breast augmentation using silicone-containing implants. Radiological imaging and biopsy of the swelling demonstrated the presence of a silicone foreign body secondary to breast implant rupture [14].

KEY POINTS

- Usually severe acetaminophen toxicity occurs in an adult after consuming 7–10 g of acetaminophen (15–20 tablets, 500 mg each). Acetaminophen is normally metabolized to glucuronide and sulfate conjugates by liver enzymes, but a small amount is metabolized by the cytochrome P-450 mixed-function oxidase family of enzymes to a toxic metabolite (N-acetyl-p-benzoquinone imine), which is detoxified after conjugation with glutathione present in the liver. In severe acetaminophen overdose, the glutathione supply in the liver is exhausted; this toxic metabolite causes liver damage. The antidote for acetaminophen poisoning is N-acetylcysteine (Mucomyst), which is a precursor of glutathione and can also detoxify the toxic metabolite of acetaminophen. This antidote must be administered as soon as possible (certainly within 8 h of acetaminophen overdose) for maximum benefit.

- Aspirin can cause a potentially life-threatening toxicity known as "Reyes syndrome" in children and adolescents with certain viral infections (such as varicella and influenza). Therefore, salicylate use is contraindicated for these groups of patients.

- Salicylate overdose results in direct stimulation of the central respiratory system, thus causing hyperventilation, respiratory alkalosis, and hyperthermia, followed by metabolic acidosis.

- Methyl salicylate should be used only as a topical analgesic. If ingested, it is very toxic.

- Carbon monoxide poisoning causes an increase in carboxyhemoglobin levels. However, smokers may have carboxyhemoglobin levels of 5–6%. A carboxyhemoglobin level of 60% to 70% and above can cause respiratory failure and even death. Fetal hemoglobin has slightly different spectrophotometric properties than adult hemoglobin. Therefore, falsely high carboxyhemoglobin values have been reported in neonates (using a CO-oximeter). Treatment for carbon monoxide poisoning is administration of 100% oxygen or oxygen therapy using a hyperbaric chamber.

- Cyanide poisoning can be life-threatening. Hydrogen cyanide (HCN) is the most lethal form. The most common source is smoke inhalation from a fire in the house from burning common household substances such as plastics, silk, and rubber (which can produce cyanide smoke containing HCN). After exposure to HCN, the cyanide ion tightly binds with hemoglobin to produce cyanhemoglobin, which results in severe hypoxia. Usually, spectroscopic methods are used for cyanide determination; a normal blood cyanide level is less than 0.2 μg/mL. A cyanide level above 2 μg/mL can produce severe toxicity, and a level above 5 μg/mL can be lethal if not treated immediately. There are several antidotes for treating cyanide poisoning, including inhalation of amyl nitrate and administration of sodium thiosulfate, sodium nitrate, ferrous sulfate, dicobalt edentate, and hydroxocobalamin.

- Overdose with tricyclic antidepressants causes central nervous system anticholinergic toxic effects such as dry mouth, dry skin, flushing, urinary retention, and cardiovascular toxicity. Mortality from tricyclic antidepressant overdose is usually due to cardiovascular toxicity, including cardiac arrhythmia. Usually in the electrocardiogram, QRS prolongation of more than 100 milliseconds (msec) is a sign of severe toxicity and is typically associated with a total tricyclic antidepressant level of 1,000 ng/mL or more. Heart rates above 100 or more are also associated with severe toxicity. A tricyclic antidepressant overdose can be treated by alkalinization using sodium bicarbonate.

- Benzodiazepine overdose can be treated with flumazenil, while naloxone is a good antidote for treating opioid overdose.
- Methanol and ethylene glycol overdoses are treated with either ethyl alcohol or fomepizole (4-methylpyrazole).
- Both organophosphorus and carbamate toxicity can cause serious life-threatening situations (including fatality), but, in general, carbamate insecticides are less toxic. Red cell (erythrocyte) acetylcholinesterase activity or serum plasma cholinesterase (also known as pseudocholinesterase or butyrylcholinesterase) activity can be measured for diagnosis as well as for monitoring progress of therapy in patients poisoned with organophosphate or carbamate insecticides.
- The Centers for Disease Control and Prevention (CDC) recommends that blood lead levels of children should be less than 10 μg/dL. Blood lead measurement is a good way of diagnosing lead toxicity. Lead inhibits delta-aminolevulinic acid dehydrogenase, one of the enzymes that catalyzes the synthesis of heme from porphyrin (thus increasing the erythrocyte concentration of protoporphyrin). Lead also inhibits the enzyme ferrochelatase, which leads to inhibition of the incorporation of iron in protoporphyrin. As a result, zinc protoporphyrin (ZPP) is produced, and remains in circulation for 120 days. Lead exposure can cause microcytic anemia, and acute lead toxicity can cause hemolytic anemia. Lead is a nephrotoxic agent. In addition, the neurotoxicity of lead produces more adverse effects in children than adults.
- Normal blood mercury levels should be lower than 10 μg/L (1 μg/dL); blood levels over 50 μg/L (5 μg/dL) are indicative of severe methyl mercury toxicity. Higher levels may be observed after poisoning with inorganic mercury. Eating fish from contaminated water can increase the risk of methyl mercury exposure. A 24-h urine mercury level over 50 μg is indicative of excessive exposure to mercury (as determined by WHO).
- Arsenic poisoning can occur from drinking well water contaminated with arsenic. Herbal supplements (especially some Indian Ayurvedic medicines) may be contaminated with heavy metals, including lead, arsenic, and mercury.

REFERENCES

[1] Seeff LB, Cuccherini BA, Zimmerman HJ, Adler E, et al. Acetaminophen hepatotoxicity in alcoholics: a therapeutic misadventure. Ann Intern Med 1986;104:399–404.

[2] Davis JE. Are one or two dangerous? Methyl salicylate exposure in toddlers. J Emerg Med 2007;32:63–9.

[3] Bell AJ, Duggin G. Acute methyl salicylate toxicity complicating herbal skin treatment for psoriasis. Emerg Med (Fremantle) 2002;14:188–90.

[4] Suner S, Partridge R, Sucov A, Valente J, et al. Non-invasive pulse CO-oximetry screening in the emergency department identifies occult carbon monoxide toxicity. J Emerg Med 2008;34:441–50.

[5] Mehrorta S, Edmonds M, Lim RK. False elevation of carboxyhemoglobin: a case report. Pediatr Emerg Care 2011;27:138–40.

[6] Lundquist P, Rammer L, Sorbo B. The role of hydrogen cyanide and carbon monoxide in fire casualities: a prospective study. Forensic Sci Int 1989;43:9—14.

[7] Borron SW, Baud FJ, Barriot P, Imbert M, et al. Prospective study of hydroxocobalamin for acute cyanide poisoning in smoke inhaler. Ann Emerg Med 2007;49:794—90.

[8] Lavoie FW, Gansert GG, Weiss RE. Value of initial ECG findings and plasma drug levels in cyclic antidepressant overdose. Ann Emerg Med 1990;19:696—700.

[9] Krisanda TJ. Flumazenil: an antidote for benzodiazepine toxicity. Am Fam Physician 1993;47:891—5.

[10] Gunturu KS, Nagarjan P, McPhedran P, Goodman TR, et al. Ayurvedic herbal medicine and lead poisoning. J Hematol Oncol 2011;20(4):51.

[11] Bernhoff RA. Mercury toxicity and treatment: a review of literature. J Environ Public Health 2012;2012:460508.

[12] Corbett SJ, Poon CC. Toxic levels of mercury in Chinese infants eating fish congee. Med J Aust 2008;188:59—60.

[13] Bolt HM. Arsenic: an ancient toxicant of continuous public health impact, from Iceman to Otzi until now. Arch Toxicol 2012;86:825—30.

[14] Omakobia E, Porter G, Armstrong S, Denton K. Silicone lymphadenopathy: an unexpected case of neck lump. J Laryngol Otol 2012;126:970—3.

CHAPTER 20

Pharmacogenomics

20.1 INTRODUCTION TO PHARMACOGENOMICS

The goal of pharmacogenomics is to understand polymorphisms of drug metabolizing enzymes, transporters, and/or receptors that ultimately determine the outcome of drug therapy. The first pharmacogenomics discovery was made over 50 years ago when it was demonstrated that patients with a genetic polymorphism that led to a deficiency of glucose-6-phosphate dehydrogenase developed hemolysis after treatment with primaquine [1]. With the completion of the Human Genome Project and the availability of pharmacogenomics tests, there are currently over 100 drugs for which testing can benefit patients. In reality, pharmacogenomics may be most beneficial in patients receiving warfarin therapy, chemotherapy with certain anticancer drugs, and pain management with certain opioids. Examples of drugs where pharmacogenomics testing is useful are listed in Table 20.1.

There is a great deal of variability at the DNA level between individuals that governs many characteristics of the person, including his or her ability to respond to a particular drug therapy. Single nucleotide polymorphisms (SNPs) account for over 90% of the genetic variations in the human genome. The rest of the genetic variations include insertions and deletions, tandem repeats, and microsatellites. The effect of butylcholinesterase genetic polymorphisms on the metabolism of the neuromuscular blocking agents succinylcholine and mivacurium used during general anesthesia has been well documented.

Based on the response to a drug, individuals can be classified as poor or extensive metabolizers. Molecular genetic testing can characterize an enzyme's gene to demonstrate which alleles (genetic polymorphisms) are present, and how such alleles may affect enzymatic activity. Some of these alleles may be associated with loss or reduction of gene function (alleles are denoted by an asterisk (*) and a number). In general, *1 usually means a

CONTENTS

353

A. Dasgupta and A. Wahed: Clinical Chemistry, Immunology and Laboratory Quality Control
DOI: http://dx.doi.org/10.1016/B978-0-12-407821-5.00020-6

Table 20.1 Examples of Drugs Where Pharmacogenomics Testing Has Clinical Significance

Drug Class	Individual Drug	Polymorphism
Anticoagulant	Warfarin	CYP2C19/VKORC1
Antineoplastic drugs	Irinotecan	UTG1A1
	6-Mercaptopurine,	TPMT
	Thiopurine, azathioprine	TPMT
	Tamoxifen	CYP2D6
Antidepressants	Amitriptyline/nortriptyline	CYP2C19/CYP2D6
	Nortriptyline	CYP2D6
	Doxepin	CYP2D6
	Paroxetine	CYP2D6
	Sertraline	CYP2C19
Narcotic analgesic	Codeine	CYP2D6
	Tramadol	CYP2D6
Immunosuppressant	Tacrolimus	CYP3A5

normally functioning gene, and hence a normally functioning enzyme. Different metabolizers of a drug that depend on genetic makeup include:

- Extensive Metabolizers (EM): Individuals who have two normal genes metabolize a drug normally.
- Poor Metabolizers (PM): Individuals with two non-functional genes metabolize a drug very slowly compared to a normal individual (EM).
- Ultra-Rapid Metabolizers (UM): These individuals may have multiple copies of active genes and may metabolize a particular drug so fast that the drug doesn't have any pharmacological effect.
- Intermediate Metabolizers (IM): These individuals may have one active and one non-active allele for the same gene.

20.2 POLYMORPHISM OF ENZYMES RESPONSIBLE FOR DRUG METABOLISM

Most drugs undergo phase I metabolism, which involves oxidation, reduction, or hydrolysis. Such reactions transform the drug into a more polar water-soluble metabolite. In addition, some drugs can undergo phase II metabolism, which entails conjugation of a polar group to the drug molecule to make it more polar. Enzymes responsible for such transformations may show a wide variation in enzymatic activities due to genetic polymorphisms. The goal of pharmacogenomics is to understand such genetic variations in order to predict the response of a particular drug in a particular patient. The cytochrome P-450 mixed-function oxidase (CYP), the most important family of enzymes responsible for drug metabolism, comprises a large group of heme-containing enzymes. These enzymes are found in abundance in the liver and other organs. The major CYP isoforms responsible for the metabolism of drugs include CYP1A2, CYP2B6, CYP2C9, CYP2C19, CYP2D6, CYP2E1, and CYP3A4/CYP3A5. Key points regarding CYP enzymes:

- CYP3A4 is the predominant isoform of the CYP family (almost 30%), and is responsible for the metabolism of many drugs.
- Genetic polymorphisms of CYP2D6, CYP2C9, and CYP2C19 have been well studied and account for some wide interindividual responses to various drugs. If the enzymatic activity is lost or significantly reduced due to a genetic polymorphism, then the individual may not be able to metabolize a particular drug (that is typically metabolized through that enzyme) effectively, and can suffer from drug toxicity.

Other polymorphically expressed drug-metabolizing enzymes are N-acetyltransferase (NAT1 and NAT2) and thiopurine-S-methyltransferase (TPMT). The slow acetylator phenotype of the NAT1/2 polymorphism results in isoniazid-induced peripheral neuropathy and sulfonamide-induced hypersensitivity reactions, while TPMT catalyzes inactivation of various anticancer and anti-inflammatory drugs. In addition, a polymorphism of uridine-5 diphosphate glucuronyl transferase (UDP-glucuronyl transferase) may also play a vital role in metabolism of certain drugs (e.g. irinotecan, an anticancer drug). This enzyme is responsible for conjugation of glucuronic acid with the drug molecule in phase II metabolism, thus inactivating the drug. This enzyme is mostly found in the liver, but may also be present in other organs. There are two main families of UDP-glucuronyl transferase: UGT1 and UGT2. Polymorphisms of UGT1A1 and UGT2B7 play important roles in the phase II metabolism of certain drugs.

20.3 POLYMORPHISM OF TRANSPORTER PROTEINS AND RECEPTORS

Most drug responses are determined by the interplay of several gene products that influence pharmacokinetics and pharmacodynamics (i.e. drug metabolizing enzymes, drug transporters, and drug targets). With the complete sequencing of the human genome, it has been estimated that approximately 500–1,200 genes code for drug transporters, and today the best-characterized drug transporter is the multidrug-resistant transporter P-glycoprotein/MDR1 (the gene product of the multiple drug-resistant protein, MDR1). Compared to drug-metabolizing enzymes, much less is known about the genetic polymorphisms of drug targets and receptors, but molecular research has revealed that many of the genes that encode drug targets demonstrate genetic polymorphisms.

20.4 PHARMACOGENOMICS AND WARFARIN THERAPY

Warfarin, the most widely used oral anticoagulant, is a synthetic compound available as a racemic mixture of 50% R-warfarin and 50% S-warfarin. Usually warfarin (Coumadin) therapy is monitored using the International Normalization Ratio (INR). However, considerable variability in the warfarin dose−response relationship between individuals is explained mainly by genetic polymorphisms of CYP2C9, the major enzyme that metabolizes warfarin and its target receptor, vitamin K epoxide reductase complex (VKORC1). Patients carrying CYP2C9*2 and CYP2C9*3 alleles require lower warfarin maintenance doses than patients with normal CYP2C9 activity. In addition, patients with VKORC1A haplotype require lower warfarin maintenance doses [2]. Federal Drug Administration (FDA)-approved devices are available for warfarin pharmacogenomics testing in individual patients. They test for CYP2C9 alleles as well as polymorphisms in VKORC1.

20.5 PHARMACOGENOMICS OF SELECTED ANTICANCER DRUGS

Cancer chemotherapy is characterized by wide variations of efficacy and toxicity among different patients due to wide interindividual variability in pharmacokinetics. In addition, most anticancer agents also have narrow therapeutic windows. Today, pharmacogenomics testing is useful in pharmacotherapy with certain anticancer drugs, most noticeably thiopurine, irinotecan, and tamoxifen.

Thiopurine drugs such as 6-mercaptopurine (6-MP), thioguanine, and azathioprine are metabolized by thiopurine S-methyltransferase (TPMT). These drugs are used for treating various conditions, including acute lymphoblastic leukemia, inflammatory bowel disease, rheumatoid arthritis, and organ rejection in transplant recipients. There are 18 mutations of TPMT genes, but three alleles, TPMT*2, *3A, and *3C, account for polymorphisms in a majority of patients with low to intermediate TPMT activities. Although polymorphisms were observed in all ethnic groups studied, approximately 10% of all Caucasians exhibit low TPMT activities, with approximately 1 out of 300 individuals having virtually no activity. Patients who inherit very low levels of TPMT are at risk of thiopurine-induced toxicity (such as myelosuppression) even when treated with a standard dosage of such drugs [3].

Irinotecan is metabolized to an active metabolite, 7-ethyl-10-hydroxycamptothecin (SN-38), which is then detoxified by UDP-glucuronosyltransferase (UGT1A1). As expected, decreased activity of UGT1A1 caused by polymorphisms in genes controlling enzymatic activity (in particular, the UGT1A1*28 allele) can cause severe toxicity in an individual after treating with a standard dose due to accumulation of the active metabolite SN-38. In response to these findings, the FDA has supported clinical pharmacogenetic testing by revising the package inserts for these anticancer drugs [4].

Tamoxifen is used for treating estrogen receptor-positive breast cancer. Tamoxifen is a prodrug that is converted into the active metabolite endoxifen (4-hydroxy-N-desmethyl-tamoxifen), mostly by CYP2D6. Poor metabolizers have lower levels of endoxifen in blood and respond poorly to tamoxifen therapy [5]. Approximately 5–14% of Caucasians, 0–5% of Africans, and 0–1% of Asians lack CYP2D6 activity. These individuals are poor metabolizers because they carry two defective alleles [6]. Lower serum endoxifen levels are observed in poor metabolizers compared to extensive metabolizers after a standard dose.

20.6 PHARMACOGENOMICS OF SELECTED OPIOID DRUGS

In general, the majority of therapeutic drugs used in pain management – including codeine, dihydrocodeine, fentanyl, hydrocodone, methadone, morphine, oxycodone, tramadol, and tricyclic antidepressants – are metabolized by polymorphic CYP450 enzymes such as CYP2D6, CYP3A4, and/or uridine diphosphate glucuronyl transferase 2B7 (UGT2B7). The wide range of genetic polymorphisms of CYP2D6 leads to four distinct groups of metabolizers, including ultra-rapid metabolizers containing multiple copies of the *CYP2D6* gene, extensive metabolizers with a single wild-type copy of the *CYP2D6* gene, intermediate metabolizers showing decreased enzymatic activity, and poor metabolizers with almost no detectable activity. Differences in drug metabolism due to polymorphism of these genes can lead to therapeutic failure or toxicity depending on the individual drug [7]. Codeine is metabolized by the liver into the more active drug morphine. Approximately 5–10% of Caucasians and 1–4% of most other ethnic groups have decreased CYP2D6 activity; these patients may not get adequate pain control following codeine therapy. In contrast, extensive metabolizers are at risk of toxicity due to fast accumulation of morphine in the blood. Similarly, tramadol is also transformed into O-desmethyltramadol (the active metabolite) by CYP2D6, and a fast metabolizer may be at risk of increased toxicity.

CASE REPORT

A full-term healthy male infant delivered vaginally showed intermittent periods of difficulty in breastfeeding and lethargy starting at Day 7 of his life. On Day 11 the baby underwent a well-baby pediatric visit, but on Day 12 the baby showed gray skin and his milk intake was down. He was found dead on Day 13. Postmortem analysis showed no anatomical abnormality that would explain the cause of death. However, analysis of his blood showed a morphine concentration of 70 ng/mL. The mother received codeine tablets (500 mg acetaminophen and 30 mg codeine) after giving the birth for episiotomy pain. She took two tablets on the first day, reduced the dose to one tablet per day from the 2nd day, but confirmed that she continued codeine tablets for two weeks. Because of poor feeding of the baby, she stored her breast milk on Day 10; her blood showed a morphine level of 87 ng/mL (the typical range in breast milk during codeine therapy is 1.9–20.5 ng/mL). Genotype analysis of the woman demonstrated that she was an ultra-rapid metabolizer due to gene duplication, thus causing increased formation of morphine after taking codeine tablets. The baby was exposed to a high amount of morphine from the breast milk, but was unable to eliminate it from his body due to an impaired capacity for conjugating morphine for inactivation and elimination. This case demonstrates that polymorphism of CYP2D6 in mother can be life-threatening for the breastfed baby [8].

20.7 PHARMACOGENOMICS OF SELECTED PSYCHOACTIVE DRUGS

Pharmacogenomics play an important role in determining serum and plasma levels of selected psychoactive drugs. The tricyclic antidepressant amitriptyline is metabolized by CYP2C19 to the active metabolite nortriptyline. CYP2D6 is needed for deactivation of nortriptyline. Adverse drug reactions tend to be associated with nortriptyline concentrations, and poor metabolizers of CYP2D6 are more likely to suffer from adverse effects due to the build-up of nortriptyline concentrations. Smith and Curry reported the case of a comatose woman who intentionally overdosed with amitriptyline. She demonstrated a rising total tricyclic antidepressant concentration over the 6 days after admission to the hospital. The level started declining on Day 7. Genotyping showed the patient to be homozygous for the CYP2D6*4 allele, the most common cause of CYP2D6 enzymatic deficiency among Caucasians [9]. Paroxetine, a selective serotonin reuptake inhibitor (SSRI), is metabolized by CYP2D6, and is also an inhibitor of CYP2D6. Poor metabolizers are at higher risk of adverse effects from paroxetine [10].

20.8 PHARMACOGENOMICS OF MISCELLANEOUS OTHER DRUGS

Organ transplant recipients receive immunosuppressants in order to prevent organ rejection. These drugs are metabolized by the cytochrome P-450 family of enzymes, including CYP3A4 and CYP3A5. Although polymorphisms of CYP3A4 do not alter the enzyme activity significantly, polymorphisms of

CYP3A5 may be clinically more significant because enzyme activities can vary significantly between different alleles. Pharmacogenomics testing can be used to identify polymorphisms of CYP3A5 to predict optimal initial dosage of tacrolimus [11].

Recent developments in the understanding of severe immunological adverse reactions following therapy with certain drugs indicate that major histocompatibility class I type B gene complex (HLA-B) is a biomarker that strongly predicts the onset of serious skin rashes such as Stevens–Johnson syndrome. HLA-B57 and HLA-B58 are major histocompatibility class I allotypes that are predictive of clinically important immune phenotypes. It has been demonstrated that HLA-B*1502, which is more abundant among Asians, is associated with severe skin rashes (including Stevens–Johnson syndrome) following treatment with carbamazepine. The FDA has released a warning suggesting HLA-B testing in Asians be performed before carbamazepine therapy is given. Although a strong correlation between carbamazepine-induced Stevens–Johnson syndrome has been found among Han Chinese patients, HLA-B*1511, a member of HLA-B75 found among Japanese, is also a risk factor for such adverse reactions from carbamazepine therapy [12]. HLA-B*5701 is strongly associated with hypersensitivity towards the anti-HIV drug abacavir. In addition, HLA-B*5801 is associated with hypersensitivity to allopurinol [13].

20.9 METHODS FOR PHARMACOGENOMICS TESTING

A variety of methods have been used to identify DNA polymorphisms that affect genes involved in enzymes responsible for drug metabolism or drug target receptor genes. These techniques either focus on analysis of a selected number of genes (e.g. classic Sanger sequencing, pyrosequencing, real-time polymerase chain reaction, and melting curve analysis) or on highly multiplexed analysis of a large number of genes (various microarray, microchip, and microbead techniques). There are two different ways a pharmacogenomics test can be introduced in the toxicology laboratory. The first and most convenient way is to use an FDA-approved test kit. The first test for which the FDA granted market approval using a DNA microarray was the AmpliChip CYP450 (Roche), which genotypes cytochrome P450 (CYP2D6 and CYP2C19). The test was approved for use with a scanner (Affymetrix Genetic Chip Microarray Instrument), and can predict phenotypes by testing for 27 CYP2D6 alleles. The assay utilizes a patient's blood or a specimen from a buccal swab. In addition to the DNA microarray, currently available pharmacogenomics tests include one that detects variations in the *UGT1A1* gene (which produces the enzyme UDP-glucuronosyltransferase), and another one that detects genetic variants of the CYP2C9 and vitamin K epoxide reductase (VKORC1) enzymes. Currently there are only nine FDA-approved pharmacogenomics tests.

Compared to therapeutic drug monitoring, pharmacogenomics testing is very expensive. An individual laboratory can, however, develop a pharmaco-genomics test (home-brew test). Although these tests do not require formal FDA approval, the quality of such tests is closely monitored by various accrediting agencies that certify clinical laboratories following rules from the Clinical Laboratory Improvement Act of 1988 (CLIA). Under CLIA rules, home-brew pharmacogenomics testing is considered high-complexity testing and must follow the strict guideline qualifications of the laboratory director and testing personnel.

KEY POINTS

- Single nucleotide polymorphisms (SNPs) account for over 90% of the genetic variations in the human genome, including genes that code enzymes responsible for drug metabolism.
- Individuals who have two normal genes are extensive metabolizers (EM) who metabolize a drug normally, but individuals with two non-functional genes are poor metabolizers (PM). Intermediate metabolizers (IM) may have one active allele and one non-active allele for the same gene while ultra-rapid metabolizers (UM) may have multiple copies of active genes; these individuals metabolize a particular drug so fast that the drug may not have any pharmacological effect.
- CYP3A4 is the predominant isoform of the CYP family (almost 30%), and is responsible for metabolism of many drugs. However, genetic polymorphisms of CYP2D6, CYP2C9, and CYP2C19 account for some wide interindividual responses to various drugs; pharmacogenetics testing usually focuses on polymorphisms of these isoenzymes.
- N-acetyltransferase (NAT1 and NAT2) and thiopurine-S-methyltransferase (TPMT) are also important drug-metabolizing enzymes that show significant genetic polymorphisms. In addition, polymorphisms of uridine-5-diphosphate glucuronyl transferase (UDP-glucuronyl transferase, the enzyme responsible for conjugation in the phase II part of drug metabolism) may also play a vital role in metabolism of certain drugs, for example, irinotecan, an anticancer drug.
- Warfarin pharmacogenomics have been well established where a genetic polymorphism of CYP2C9, the enzyme that metabolizes warfarin, and a polymorphism of warfarin's target receptor (vitamin K epoxide reductase complex, VKORC1) play important roles in the pharmacological actions of warfarin.
- Thiopurine anticancer drugs such as 6-mercaptopurine (6-MP), thioguanine, and azathioprine are metabolized by thiopurine S-methyltransferase (TPMT). Polymorphism of TPMT determines the pharmacological response of these anticancer drugs.

- Irinotecan is metabolized to an active metabolite, 7-ethyl-10-hydroxycamptothecin (SN-38), which is then detoxified by UDP-glucuronosyltransferase (UGT1A1). Therefore, decreased activity of UGT1A1 caused by polymorphisms in genes controlling enzymatic activity (in particular, the UGT1A1*28 allele) may cause severe toxicity in an individual after treatment with a standard dose due to accumulation of the active metabolite SN-38.
- Polymorphism of CYP2D6 is important for pharmacological activities of certain drugs that are metabolized by this enzyme. Tamoxifen is a prodrug that is converted into the active metabolite endoxifen (4-hydroxy-N-desmethyl-tamoxifen), mostly by CYP2D6. Poor metabolizers result in lower levels of endoxifen in the blood, and respond poorly to tamoxifen therapy. The major pharmacological activity of codeine is due to its conversion to morphine by CYP2D6. Poor metabolism may not yield adequate pain control due to a low level of morphine.
- It has been demonstrated that HLA-B*1502, which is more abundant among Asians, is associated with severe skin rashes, including Stevens—Johnson syndrome, following treatment with carbamazepine.
- The first FDA-approved test using a DNA microarray is the AmpliChip CYP450 (Roche), which genotypes cytochrome P-450 (CYP2D6 and CYP2C19). Now many pharmacogenomics test kits are commercially available.

REFERENCES

[1] Beutler E. Drug induced hemolytic anemia. Pharmacol Rev 1969;21:73—103.

[2] Li J, Wang S, Barone J, Malone B. Warfarin pharmacogenomics. Pharmacy Therapeutics 2009;34:422—7.

[3] Zhou S. Clinical pharmacogenomics of thiopurine S-methyltransferase. Curr Clin Pharmacol 2006;1:119—28.

[4] Fujita K, Sasaki Y. Pharmacogenomics in drug-metabolizing enzymes catalyzing anticancer drugs for personalized cancer chemotherapy. Curr Drug Metab 2007;8:554—62.

[5] Ingle JN. Pharmacogenomics of tamoxifen and aromatase inhibitors. Cancer 2008;112 (Suppl. 3):695—9.

[6] Frueh FW, Amur S, Mummaneni P, Epstein RS, et al. Pharmacogenomics biomarkers information in drug labels approved by the United States Food and Drug Administration: prevalence and related drug use. Pharmacotherapy 2008;28:992—8.

[7] Jannetto PJ, Bratanow NC. Utilization of pharmacogenomics and therapeutic drug monitoring for opioid pain management. Pharmacogenomics 2009;10:1157—67.

[8] Koren G, Cairns J, Chitayat D, Gaedigk A, et al. Pharmacogenomics of morphine poisoning in a breastfed neonate of a codeine prescribing mother. Lancet 2006;368:704.

[9] Smith JC, Curry SC. Prolonged toxicity after amitriptyline overdose in a patient deficient in CYP2D6 activity. J Med Toxicol 2011;7:220—3.

[10] Sheffield LJ, Ohillimore HE. Clinical use of pharmacogenomics tests in 2009. Clin Biochem Rev 2009;30:55—65.

[11] Warne N, MacPhee IA. Current progress in pharmacogenomics and individualized immunosuppressive drug dosing in organ transplantation. Curr Opin Mol Ther 2010;12:270—83.

[12] Kaniwa N, Saito Y, Aihara M, Matsunaga K, et al. HLA-B*1511 is a risk factor for carbamazepine induced Stevens–Johnson syndrome and toxic epidermal necrolysis in Japanese patients. Epilepsia 2010;51:2461–5.

[13] Kostenko L, Kjer-Nielsen L, Nicholson I, Hudson F, et al. Rapid screening for the detection of HLA-B57 and HLA-B58 in prevention of drug hypersensitivity. Tissue Antigens 2011;78:11–20.

CHAPTER 21

Hemoglobinopathy

21.1 HEMOGLOBIN STRUCTURE AND SYNTHESIS

Hemoglobin, the oxygen-carrying pigment of erythrocytes, consists of a heme portion (iron-containing chelate) and four globin chains. Six distinct species of normal hemoglobin are found in humans, three in normal adults, and three in fetuses. The globulins associated with the hemoglobin molecule (both embryonic stage and after birth) include alpha chain (α-chain), beta chain (β-chain), gamma chain (γ-chain), delta chain (δ-chain), epsilon chain (ε-chain), and zeta chain (ζ-chain). In the embryonic stage hemoglobin Gower and hemoglobin Portland are found, but these are replaced by hemoglobinF (HbF: two alpha chains and two gamma chains) in the fetus. Interestingly, HbF has a higher oxygen affinity than adult hemoglobin and is capable of transporting oxygen in peripheral tissues in the hypoxic fetal environment. In the third trimester the genes responsible for beta and gamma globulin synthesis are activated, and, as a result, adult hemoglobin, such as hemoglobin A (HbA: two alpha chains and two beta chains) and hemoglobin A_2 (HbA$_2$: two alpha chains and two delta chains) may also be found in neonates; however, HbF is still the major component. Newborn babies and infants up to 6 months old do not depend on HbA synthesis; the switch from Hb F to Hb A occurs around three months of age. Therefore, disorders due to beta chain defects, such as sickle cell disease, tend to manifest clinically after 6 months of age, although diseases due to alpha chain defects are manifested in utero or following birth. Embryonic, fetal, and adult hemoglobins are summarized in Table 21.1. The different types of naturally occurring embryonic, fetal, and adult hemoglobin vary in their tetramer–dimer subunit interface strength (stability) in the liganded (carboxyhemoglobin or oxyhemoglobin) state [1].

The normal hemoglobin (HbA) in adults contains two alpha chains and two beta chains. Each alpha chain contains 141 amino acids and each beta chain contains 146 amino acids. Hemoglobin A_2 (HbA$_2$) contains two alpha chains and two delta chains. The gene for the alpha chain is located in chromosome 16 (two genes in each chromosome, a total of four genes), while genes for the beta (one

CONTENTS

A. Dasgupta and A. Wahed: Clinical Chemistry, Immunology and Laboratory Quality Control
DOI: http://dx.doi.org/10.1016/B978-0-12-407821-5.00021-8

Table 21.1 Embryonic, Fetal, and Adult Hemoglobins

Period of Life	Hemoglobin Species	Globulin Chains	% In Adult
Embryonic	Gower-1	Two zeta, two epsilon	
	Gower-2	Two alpha, two epsilon	
	Portland-1	Two zeta, two gamma	
	Portland-2	Two zeta, two beta	
Fetal	Hemoglobin F	Two alpha, two gamma	
Adult	Hemoglobin A	Two alpha, two beta	92–95%
	Hemoglobin A_2	Two alpha, two delta	<3.5%
	Hemoglobin F	Two alpha, two gamma	<1%

gene on each chromosome, a total of two genes), gamma, and delta chains are located on chromosome 11. Adults have mostly HbA and small amounts of HbA_2 (less than 3.5%) and HbF (less than 1%). A small amount of fetal hemoglobin persists in adults due to a small clone of cells called F cells. When hemoglobin is circulating with erythrocytes, glycosylation of the globin chains may take place. These are referred to as X1c (X being any hemoglobin, e.g. HbA1c). When a hemoglobin molecule ages, glutathione binds to the cysteine at the 93rd position of the beta chain. This is HbA3 or HbA1d. Just like HbA1c and HbA1d, HbC1c, HbC1d, HbS1c, HbS1d may also exist in circulation in individuals with HbC and HbS respectively.

Heme is synthesized in a complex way involving enzymes in both the mitochondrion and cytosol. In the first step, glycine and succinyl-CoA combine in the mitochondria to form delta-aminolevulinic acid, which is transported into the cytoplasm and converted into porphobilinogen by the action of the enzyme aminolevulinic acid dehydrogenase. Then porphobilinogen is converted into coproporphyrinogen III through several steps involving multiple enzymes. Coproporphyrinogen III is transported into the mitochondria and converted into protoporphyrinogen III by coproporphyrinogen III oxidase. Then protoporphyrinogen III is converted into protoporphyrin IX by protoporphyrinogen III oxidase, and protoporphyrin IX is converted into heme by ferrochelatase. Finally, heme is transported into the cytosol and combines with globulin to form the hemoglobin molecule.

21.2 INTRODUCTION TO HEMOGLOBINOPATHIES

Hemoglobinopathies can be divided into three major categories:

- Quantitative disorders of hemoglobin synthesis: Production of structurally normal globin chains, but in decreased amounts (thalassemia syndrome).

- Qualitative disorders of hemoglobin structure: Production of structurally abnormal globulin chains such as hemoglobin S, C, O, or E. Sickle cell syndrome is the most common example of such disease.
- Failure to switch globin chain synthesis after birth: For example, hereditary persistence of fetal hemoglobin (HbF), a relatively benign condition. It can co-exist with thalassemia or sickle cell disease, and will result in decreased severity of such diseases (protective effect).

Hemoglobinopathies are transmitted in autosomal recessive fashion. Therefore, carriers who have one affected chromosome and one normal chromosome are usually healthy or only slightly anemic. When both parents are carriers, then children have a 25% chance of being normal, a 25% chance of being severely affected by the disease, or a 50% chance of being mostly normal. Hemoglobinopathies are caused by inherent mutations of genes coding for globin synthesis. Mutations can disrupt gene expression and cause reduced production of alpha or beta chain globin (thalassemias) or point mutations of the gene in the coding region (exons). This in turn can cause production of defective globin that results in formation of abnormal hemoglobin (hemoglobin variants) [2].

It has been estimated that approximately 5% of the world population are carriers of hemoglobin disorders. Moreover, hemoglobinopathies affect approximately 370,000 newborns worldwide each year. The hemoglobin variants of most clinical significance are hemoglobin S, C, and E. In West Africa, approximately 25% of the individuals are heterozygous for the hemoglobin S (HbS) gene, which is related to sickle cell diseases. In addition, high frequencies of HbS gene alleles are also found in people of the Caribbean, South and Central Africa, the Mediterranean, the Arabian Peninsula, and East India. Hemoglobin C (HbC) is found mostly in people living in or originating from West Africa. Hemoglobin E (HbE) is widely distributed between East India and Southeast Asia, with the highest prevalence in Thailand, Laos, and Cambodia; it is sporadically observed in parts of China and Indonesia as well. Thalassemia syndrome is not due to a structural defect in the globin chain, but to the lack of sufficient synthesis of globin chains; it is an inherited disease. Thalassemia syndrome can be divided into alpha (α) and beta (β) types. In general, β-thalassemia is observed in the Mediterranean, the Arabian Peninsula, Turkey, Iran, West and Central Africa, India, and other Southeast Asian countries, while α-thalassemia is commonly seen in parts of Africa, the Mediterranean, the Middle East, and throughout Southeast Asia [3]. Out of over 1,000 hemoglobinopathies reported, most of the disorders are asymptomatic. However, in other cases, significant clinical disorders can be seen, including:

- Thalassemias (both alpha and beta).
- Sickling disorders (HbSS, HbSC, HbSD, HbSO).

- Cyanosis (such as Hb Kansas).
- Hemolytic anemias (such as HbH).
- Erythrocytosis (such as Hb Malmo).

21.3 ALPHA-THALASSEMIA

There are two genetic loci for the alpha gene that result in four genes (alleles) for alpha-hemoglobin (α/α, α/α) on chromosome 16. Two alleles are inherited from each parent. Alpha-thalassemia occurs when there is a defect or deletion in one or more of the four genes responsible for alpha-globin production. Alpha-thalassemia can be divided into four categories:

- Silent Carriers: These are characterized by only one defective or deleted gene, but three functional genes. These individuals have no health problems. An unusual case of silent carrier is seen in individuals carrying one defective Constant Spring mutation but three functional genes (no health problems).
- Alpha-Thalassemia Trait: Characterized by two deleted or defective genes and two functional genes. These individuals may have mild anemia.
- Alpha-Thalassemia Major (Hemoglobin H Disease): Characterized by three deleted or defective genes and only one functional gene. These patients have persistent anemia and significant health problems. When hemoglobin H disease is combined with hemoglobin Constant Spring, the severity is greater than hemoglobin H disease alone. However, if the child inherits one hemoglobin Constant Spring gene from the mother and one from the father, then the child is homozygous for hemoglobin Constant Spring and the severity of the disease is similar to hemoglobin H disease.
- Hydrops Fetalis: Characterized by no functional alpha gene; these individuals have hemoglobin Bart. This condition is seriously life-threatening unless an intrauterine transfusion is initiated.

When an alpha gene is functional, it is denoted as "α." If it is not functional or is deleted, it is designated as "$-$." There is not much difference in impaired alpha-globin synthesis between a deleted gene and a non-functioning defective gene. With a deletion or a defect of one gene ($-/\alpha$, α/α), little clinical effect is observed because three alpha genes are enough to allow normal hemoglobin production. These patients are sometimes referred to as "silent carriers" because there are no clinical symptoms but mean corpuscular volume (MCV) and mean corpuscular hemoglobin (MCH) may be slightly decreased. These individuals are diagnosed by deduction only when they have children with a thalassemia trait or hemoglobin H disorder. An unusual case of the silent carrier state is an individual carrying one hemoglobin Constant Spring mutation but three functional genes. Hemoglobin

Constant Spring (hemoglobin variant isolated from a family of ethnic Chinese background from the Constant Spring district of Jamaica) is a hemoglobin variant where the mutation of the alpha-globin gene produces an abnormally long alpha chain (172 amino acids instead of the normal 146 amino acids). Hemoglobin Constant Spring is due to a non-deletion mutation of the alpha gene that results in the production of unstable alpha-globin. Moreover, this alpha-globin is produced in very low quantity (approximately 1% of the normal expression level) and is found in people living or originating in Southeast Asia.

When two genes are defective or deleted, the alpha-thalassemia trait is present. There are two forms of the alpha-thalassemia trait. Alpha-thalassemia 1 $(-/-, \alpha/\alpha)$ results from the *cis*-deletion of both alpha genes on the same chromosome. This mutation is found in Southeast Asian populations. Alpha-thalassemia 2 $(-/\alpha, -/\alpha)$ results from the *trans*-deletion of the alpha gene on two different chromosomes. This mutation is found in African and African-American populations (prevalence of the disease is 28% in African-Americans). In the alpha-thalassemia trait two functioning alpha genes are present and as a result erythropoiesis is almost normal in these individuals, with only mild microcytic hypochromic anemia (low MCV and MCH). This form of the disease can mimic iron deficiency anemia. Therefore, distinguishing alpha-thalassemia from iron deficiency anemia is essential.

If three genes are affected $(-/-, -/\alpha)$, the disease is called hemoglobin H disease, which is a severe form of alpha-thalassemia; patients with this severe anemia require a blood transfusion. Because only one alpha gene is responsible for production of alpha-globin in HbH disease, a high β-globin-to-α-globin ratio (a 2- to 5-fold increase in β-globin production) can result in formation of a tetramer containing only the β; this form of hemoglobin is called HbH (four β chains). This form of hemoglobin cannot deliver oxygen in peripheral tissues because hemoglobin H has a very high affinity for oxygen. A microcytic hypochromic anemia with target cells and Heinz bodies (which represents precipitated HbH) is present in the peripheral blood smear of these patients. Moreover, red cells that contain hemoglobin H are sensitive to oxidative stress, and may be more susceptible to hemolysis, especially when oxidants such as sulfonamides are administered. More mature erythrocytes also contain increased amounts of precipitated hemoglobin H (Heinz bodies). These are removed from the circulation prematurely, which can also cause hemolysis. Therefore, clinically these patients experience varying severities of chronic hemolytic anemia. Due to the subsequent increase in erythropoiesis, erythroid hyperplasia can result and cause bone structure abnormalities with marrow hyperplasia, bone thinning, maxillary hyperplasia, and pathologic fractures. When hemoglobin Constant Spring is associated with HbH disease, a more severe form of anemia is observed that requires frequent transfusions [4]. However, when a child inherits one hemoglobin Constant Spring gene from the father and one from the mother, then hemoglobin Constant Spring disease is present, which is less severe than hemoglobin H—hemoglobin Constant Spring disease; severity is comparable to HbH disease. Patients with hemoglobin H and related diseases require transfusion and chelation therapy to remove excess iron.

When four genes are defective or deleted $(-/-, -/-)$, the result is Hemoglobin Bart's disease, where alpha-globin is absent because no gene is present to promote alpha-globin synthesis. As a result, four gamma (γ) chains form a tetramer. As in HbH, the hemoglobin in Hb Bart's is unstable, which impairs the ability of the red cells to release oxygen to the surrounding tissues. The fetus cannot usually survive gestation, which causes stillbirth with hydrops fetalis. However, more recently, with support through intrauterine transfusion and a neonatal intensive care unit, survival may be possible. Still, survivors may have severe transfusion-dependent anemia like patients suffering from beta-thalassemia major. Bone marrow or cord blood transplants may be helpful.

CASE REPORT

The patient was diagnosed with hemoglobin Bart's disease at 22 weeks gestational age and gene analysis revealed a 2-alpha-globin-4-gene deletion. The patient's Laotian parents were both discovered to have an alpha-2-thalassemia trait after two previous pregnancies ended in severe hydrops fetalis. The patient received intrauterine transfusion at 25, 28, and 32 weeks. The baby was born after 35-6/7 weeks of gestation (birth weight 2,550 g, hemoglobin 11.2 g/dL) and a double volume exchange transfusion was performed. The baby boy subsequently received transfusion every 2–3 weeks to maintain his hemoglobin level at > 9.5 g/dL until he was eight months' old. Then he received hematopoietic cell transplantation. Because a suitable marrow donor could not be identified, unrelated cord blood with a 4 out of 6 human leukocyte antigen (HLA) match was used. A resumed transfusion and chelation therapy continued until he was 18 months' old when a 6 of 6 HLA matched unrelated donor hematopoietic cell transplantation was performed. At one year post-transplant, the baby was transfusion-independent and retained 90% of the donor engraftment in his bone marrow. At the time of the report the boy was 7 years old and in an age-appropriate school grade, but was receiving special education. However, he was short-statured and had hypothyroidism [5].

21.4 BETA-THALASSEMIA

Beta-thalassemia is due to deficit or absent production of beta-globin, resulting in excess production of alpha-globin. Synthesis of beta-globin can vary from near complete to absent, thus causing a range of severity of beta-thalassemia due to mutation of genes (one gene each on chromosome 11); more than 200 point mutations have been reported. However, deletion of both genes is rare. Beta-thalassemia can be broadly divided into three categories:

- Beta-Thalassemia Trait: Characterized by one defective gene and one normal gene. Individuals may experience mild anemia but may not necessarily be transfusion-dependent.
- Beta-Thalassemia Intermedia: Characterized by two defective genes, but some beta-globin production is still observed in these individuals. However, some individuals may have significant health problems that require intermittent transfusion.
- Beta-Thalassemia Major (Cooley's Anemia): Characterized by two defective genes, and almost no function of either gene, leading to no synthesis of beta-globin. These individuals have a severe form of disease that requires lifelong transfusions and they may have a shortened lifespan.

If a defective gene is incapable of producing any beta-globin, it is characterized as "β^0," causing a more severe form of beta-thalassemia. However, if the mutated gene can retain some function, it is characterized as "β^+." In the case of one gene defect, beta-thalassemia minor (trait: patients are β^0/β or β^+/β) is observed and individuals are either normal or mildly anemic. These patients have increased HbA_2. In addition, HbF may also be elevated. MCV and MCH are low, but these patients are not transfusion-dependent. If both genes are affected, resulting in severely impaired production of beta-globin (β^0/β^0 or β^+/β^0), the disease is severe and is called beta-thalassemia major (also known as Cooley's anemia). However, due to the presence of fetal hemoglobin, symptoms of beta-thalassemia major are not observed prior to 6 months of age. Patients with beta-thalassemia major have elevated HbA_2 and HbF (although in some individuals HbF may be normal). If production of beta-globin is moderately hampered, then the disease is called beta-thalassemia intermedia (β^0/β or β^+/β^+). These individuals have less severe disease complications than with beta-thalassemia major. In patients with beta-thalassemia major, an excess alpha-globin chain precipitates, leading to hemolytic anemia. These patients require lifelong transfusion and chelation therapy. Interestingly, having β^0 or β^+ does not predict the severity of disease because patients with both types have been diagnosed with beta-thalassemia major or intermedia. Major features of alpha- and beta-thalassemia are summarized in Table 21.2.

21.5 DELTA-THALASSEMIA

Delta-thalassemia is due to mutation of the genes responsible for synthesis of the delta chain. A mutation that prevents formation of the delta chain is

Table 21.2 Major Features of Alpha- and Beta-Thalassemias

Disease	Number of Deleted Genes	Comments
Alpha-thalassemia silent carrier	One of four gene deletions	Asymptomatic. May have low MCV, MCH.
Alpha-thalassemia trait	Two of four gene deletions	Asymptomatic/mild symptoms. Mild microcytic hypochromic anemia.
Hemoglobin H disease	Three of four gene deletions	Microcytic hypochromic anemia and HbH found in adults and Hb Bart's found in neonates. HbH may co-exist with Hb Constant Spring, a more severe disease than HbH.
Hydrops fetalis	Four of four gene deletions	Hemoglobin Bart's disease. Most severe form may cause stillbirth/hydrops fetalis.
Beta-thalassemia trait	One gene defect	Asymptomatic
Beta-thalassemia intermedia	Both genes defective	Variable degree of severity as some beta-globin is still produced.
Beat-thalassemia major	Both genes defective	Severe impairment or no beta-globin synthesis. Severe disease with anemia, splenomegaly, requiring lifelong transfusions.

called delta^0, and if some delta chain is formed, the mutation is called delta$^+$. If an individual inherits two delta^0 mutations, no delta chain is produced and no HbA$_2$ can be detected in the blood (normal level <3.5%). However, if an individual inherits two delta$^+$ mutations, a decrease in HbA$_2$ is observed. All patients with delta-thalassemia have normal hematological consequences although the presence of the delta mutation can obscure diagnosis of the beta-thalassemia trait because in beta-thalassemia, HbA$_2$ is increased but the presence of delta may reduce HbA$_2$ concentration, thus masking diagnosis of the beta-thalassemia trait.

Delta–beta thalassemia is a rare hemoglobinopathy characterized by decreased production (or total absence) of delta- and beta-globin. As a compensatory mechanism, gamma chain synthesis is increased, resulting in a significant amount of fetal hemoglobin (HbF) in the blood, which is homogenously distributed in red blood cells. This condition is found in many ethnic groups, but is especially seen in individuals with Greek or Italian ancestry. Heterozygous individuals are asymptomatic with normal HbA$_2$ but rarely reported homozygous individuals experience mild symptoms.

21.6 SICKLE CELL DISEASE

The term "sickle cell disease" includes all manifestations of abnormal hemoglobin S (HbS): sickle cell trait (HbAS), homozygous sickle cell disease

(HbSS), and a range of mixed heterozygous hemoglobinopathies such as hemoglobin SC disease, hemoglobin SD disease, hemoglobin SO Arab disease, and HbS combined with beta-thalassemia. Sickle cell disease affects millions of people throughout the world and is particularly common in people in, or people migrating from, sub-Saharan Africa, the Caribbean, Central America, Saudi Arabia, India, and Mediterranean countries such as Turkey, Greece, and Italy. Sickle cell disease is the most commonly observed hemoglobinopathy in the U.S. and affects 1 in every 500 African-American births and 1 in every 36,000 Hispanic-American births. Sickle cell disease is a dangerous hemoglobinopathy. Symptoms of sickle disease start before age 1 with chronic hemolytic anemia, developmental disorders, crises that include extreme pain (sickle cell crisis), high susceptibility to various infections, spleen crisis, acute thoracic syndrome, and increased risk of stroke. Optimally, treated individuals can have a lifespan of 50 to 60 years [6].

In sickle cell disease, the normal round shape of red blood cells (RBCs) is changed into a crescent shape, hence the name "sickle cell." In the heterozygous form (HbAS), sickle cell disease protects against infection from *Plasmodium falciparum* malaria, but this is not the case in the more severe form of homozygous sickle cell disease (HbSS). The genetic defect that produces sickle hemoglobin is a single nucleotide substitution at codon 6 of the beta-globin gene on chromosome 11 that results in a point mutation in the beta-globin chain of hemoglobin (substitution of valine for glutamic acid at the sixth position). Hemoglobin S is formed when two normal alpha-globins combine with two mutant beta-globins. Because of this hydrophobic amino acid substitution, HbS polymerizes upon deoxygenation and multiple polymers bundle into a rod-like structure, resulting in a deformed RBC. Various possible diagnoses of patients with HbS hemoglobinopathy include sickle cell trait (HbAS), sickle cell disease (HbSS), and sickle cell disease status post RBC transfusion/exchange. Patients with sickle cell trait may also have concomitant alpha-thalassemia; a diagnosis of HbS/beta-thalassemia (0/+, +/+) is also occasionally made. Double heterozygous states of HbSC, HbSD, and HbSO Arab are important sickling states that should not be missed. Hemoglobin C is formed due to substitution of a glutamic acid residue with a lysine residue at the sixth position of beta-globin. Individuals who are heterozygous with HbC disease are asymptomatic with no apparent disease, but homozygous individuals have almost all hemoglobin (>95%) that is HbC and experience chronic hemolytic anemia and pain crisis. However, individuals who are heterozygous with both hemoglobin C and hemoglobin S (HbSC disease) have weaker symptoms than sickle cell disease because HbC does not polymerize as readily as HbS.

Patients with HbSS disease may have increased HbF. The distribution of HbF amongst the haplotypes of HbSS are HbF (5−7% in Bantu, Benin, or

Cameroon), HbF (7−10% in Senegal), and HbF (10−25% in the Arab-Indian area). Hydroxyurea also causes an increase in HbF. This is usually accompanied by macrocytosis. HbF can also be increased in HbS/HPFH (HPFH: hereditary persistence of fetal hemoglobin). HbA_2 values are typically increased in sickle cell disease and more so upon high-performance liquid chromatographic (HPLC) analysis. This is because the post-translational modification form of HbS (HbS1d) produces a peak in the A_2 window. This elevated value of HbA_2 may produce diagnostic confusion with HbSS disease and HbS/beta-thalassemia. It is important to remember that microcytosis is not a feature of HbSS disease, and patients with HbS/beta-thalassemia typically exhibit microcytosis.

HbSS patients and HbS/β^0-thalassemia patients do not have any HbA unless the patient has been transfused or has undergone red cell exchange. Glycated HbS has the same retention time in HPLC (approximately 2.5 minutes) as HbA. This produces a small peak in the A window and raises the possibility of HbS/β^+-thalassemia. HbS/α-thalassemia is considered a possibility when the percentage of HbS is lower than expected. Classical cases are 60% of HbA and approximately 35−40% of HbS. Cases of HbS/α-thalassemia should have lower values of HbS, typically below 30% with microcytosis. A similar picture will also be present in patients with sickle cell trait and iron deficiency. Various features of sickle cell disease are summarized in Table 21.3.

Table 21.3 Major Features of Sickle Cell Diseases

Disease	Hemoglobin Variants	Clinical Features
Sickle cell trait (heterozygous)	HbAS	HbS: 35−40%; $HbA_2 < 3.5\%$; normal hemoglobin; no apparent illness.
Sickle cell disease	HbSS	HbS > 90%; $HbA_2 < 3.5\%$; no HbA2; hemoglobin: 6−8 g/dL; severe disease with chronic hemolytic anemia.
Sickle cell-β^0-thalassemia	HbSβ^0	HbS > 80%; $HbA_2 > 3.5\%$; no HbA; hemoglobin: 7−9 g/dL; severe sickle cell disease.
Sickle cell-β^+-thalassemia	HbSβ^+	HbS > 60%; $HbA_2 > 3.5\%$, HbA: 5−30 %; hemoglobin: 9−12 g/dL; variable mild to moderate sickle cell disease.
Hemoglobin SC disease	HbSC	HbS: 50%; HbC: 50%; hemoglobin: 10−12 g/dL; moderate sickling disease, but chronic hemolytic anemia may be present.
Hemoglobin S/HPFH		HbS: 60%; $HbA_2 < 3.5\%$; HbF: 30−40%; hemoglobin: 11−14 g/dL; no HbA; mild sickling disease.

21.7 HEREDITARY PERSISTENCE OF FETAL HEMOGLOBIN

In individuals with hereditary persistence of fetal hemoglobin, significant amounts of fetal hemoglobin (HbF) can be detected well into adulthood. In normal adults HbF represents less than 1% of total hemoglobin, but in hereditary persistence of fetal hemoglobin (HPFH) the percentage of HbF can be significantly elevated; HbA_2, however, is normal. HPFH is divided into two major groups: deletional and non-deletional. Deletional HPFH is caused by a variable length deletion in the beta-globin gene cluster. This leads to decreased (or absent) beta-globin synthesis and a compensatory increase in gamma-globin synthesis with a pancellular or homogenous distribution of HbF in red blood cells. Non-deletional HPFH is a broad category of related disorders with increased HbF typically distributed heterocellularly. Heterocellular distribution is also seen in beta-thalassemia and delta–beta-thalassemia.

Both homozygous and heterozygous HPFH are asymptomatic with no clinical or significant hematological change, although individuals with homozygous HPFH may show up to 100% HbF, while the heterozygous version typically shows 20–28% HbF. If HPFH is associated with sickle cell, it can reduce the severity of the disease. Compound heterozygotes for sickle hemoglobin (HbS) and HPFH have high levels of HbF, but these individuals experience few (if any) sickle cell disease-related complications [7]. If HPFH is associated with thalassemia, individuals also experience less severe disease complications.

21.8 OTHER HEMOGLOBIN VARIANTS

Hemoglobin D (hemoglobin D Punjab, also known as hemoglobin D Los Angeles) is formed due to the substitution of glutamine for glutamic acid. HbD Punjab is one of the most commonly observed abnormalities worldwide, and is not only found in the Punjab region of India, but also in Italy, Belgium, Austria, and Turkey. Hemoglobin D disease can occur in four different forms, including heterozygous HbD trait, HbD-thalassemia, HbSD disease, and, very rarely, homozygous HbD disease. Heterozygous HbD disease is a benign condition with no apparent illness, but when HbD is associated with HbS or beta-thalassemia, clinical conditions such as sickling disease and moderate hemolytic anemia may be observed. Heterozygous HbD is rare, and usually presents with mild hemolytic anemia and mild-to-moderate splenomegaly [8].

Hemoglobin E is due to a point mutation of beta-globin that results in the substitution of lysine for glutamic acid in position 26. As a result, production of beta-globin is diminished. HbE also has a structural defect and is a thalassemia-like phenotype. HbE is unstable and can form Heinz bodies under oxidative stress. The HbE trait is associated with moderately severe microcytosis, but usually no anemia is present. However, individuals who are homozygous for HbE often present with modest anemia that is similar to the thalassemia trait. When beta-thalassemia is combined with HbE (e.g. in HbE/β^0-thalassemia), patients may have significant anemia that requires transfusion, similar to patients with beta-thalassemia intermedia.

Hemoglobin O-Arab (Hb-O-Arab, also known as Hb Egypt) is a rare abnormal hemoglobin variant where, at position 121 of beta-globin, the normal glutamic acid is replaced by lysine. Hb-O-Arab is found in people from the Balkans, the Middle East, and Africa. Patients who are heterozygous for Hb-O-Arab may experience mild anemia and microcytosis similar to patients with beta-thalassemia minor. Patients with the homozygous version (which is extremely rare) may have anemia, but, despite an abnormal hemoglobin pattern, these patients may be mostly asymptomatic. However, patients with HbS/Hb-O-Arab may experience severe clinical symptoms similar to individuals with HbS/S. Similarly, patients with Hb-O-Arab/beta-thalassemia may experience severe anemia, with hemoglobin levels between 6 and 8 g/dL and splenomegaly [9].

Hemoglobin Lepore is an unusual hemoglobin molecule composed of two alpha chains and two delta—beta chains as a result of the fusion of delta and beta genes. The delta—beta chains have the first 87 amino acids of the delta chain and 32 amino acids of the beta chain. There are three common variants of hemoglobin Lepore: Hb Lepore Washington (also known as Hb Lepore Boston), Hb Lepore Baltimore, and Hb Lepore Hollandia. Hemoglobin Lepore is seen in individuals of Mediterranean descent. Individuals with HbA/Hb Lepore are asymptomatic, with Hb Lepore representing 5—15% of hemoglobin, and slightly elevated HbF (2—3%). However, homozygous Lepore individuals suffer from severe anemia similar to patients with beta-thalassemia intermedia, with Hb Lepore representing 8—30% of hemoglobin and the remainder of the hemoglobin as HbF. Patients with Hb Lepore/beta-thalassemia experience severe disease complications similar to patients with beta-thalassemia major.

Hemoglobin G-Philadelphia (HbG) is the most common alpha chain defect, is observed in 1 in 5,000 African-Americans, and is associated with alpha-thalassemia 2 deletions. Therefore, these individuals have only three functioning alpha genes; HbG represents one-third of total hemoglobin. HbS is the most common beta chain defect, and is often observed in the African-American population. HbG is the most common alpha chain defect, and, again, occurs most often in African-American people. Therefore, it is possible that an African-American individual may have HbS/HbG where the hemoglobin molecule contains one normal alpha chain, one alpha G chain, one normal beta chain, and one beta S chain. This can result in detection of various hemoglobins in the blood, including HbA (alpha 2, beta 2), HbS (alpha 2, beta S2), HbG (alpha G_2, beta 2), and HbS/G (alpha G_2, beta 2). In addition, HbG_2 (alpha 2, delta 2), which is the counterpart of HbA_2, is also present.

Increase in the percentage of fetal hemoglobin is associated with multiple pathologic states. These include beta-thalassemia, delta−beta-thalassemia, and hereditary persistence of fetal hemoglobin (HPFH). While beta-thalassemia is associated with high HbA_2, the latter two states are associated with normal HbA_2 values. Hematologic malignancies are associated with increased hemoglobin F and include acute erythroid leukemia (AML, M6) and juvenile myelomonocytic leukemia (JMML). Aplastic anemia is also associated with an increase in percentage of HbF. In elucidating the actual cause of high HbF, it is important to consider the actual percentage of HbF and HbA_2 values, as well as the correlation with complete blood count (CBC) and peripheral smear. It is also important to note that drugs (hydroxyurea, sodium valproate, erythropoietin) and stress erythropoiesis can also result in high HbF. Hydroxyurea is used in sickle cell disease patients to increase the amount of HbF, the presence of which may help to reduce the clinical effects of the disease. Measuring the level of HbF can be useful in determining the appropriate dose of hydroxyurea. In 15−20% of cases of pregnancy, HbF can be raised to values as high as 5%.

Other rarely reported hemoglobinopathies involve hemoglobin I, hemoglobin J, and hemoglobin Hope. Rarely reported unstable hemoglobins are hemoglobin Koln, hemoglobin Hasharon, and hemoglobin Zurich (for these, an isopropanol test is positive). Certain rarely reported hemoglobin variants are hemoglobin Malmo, hemoglobin Andrew, hemoglobin Minneapolis, hemoglobin British Columbia, and hemoglobin Kempsey. Patients with these rare hemoglobin variants experience erythrocytosis. Hemoglobin I is due to a single alpha−globin substitution (lysine substitution at position 16 for glutamic acid). HbI is clinically insignificant except in rare occasions when it is associated with alpha-thalassemia, where approximately 70% of the hemoglobin is HbI. Hemoglobin J is characterized as a fast-moving band in hemoglobin electrophoresis (band found close to the anode, the farthest point from application of the sample), and more than 50 variants have been reported, including HbJ Capetown and HbJ Chicago. However, heterozygous hemoglobinopathy involving HbJ is clinically insignificant. In hemoglobin Hope, aspartic acid is substituted for glycine at position 136 of the beta chain. Other important hemoglobinopathies are summarized in Table 21.4.

Table 21.4 Various Other Hemoglobinopathies

Diagnosis	Hemoglobin/Hematology	Comments
HbC trait (HbAC)	HbA: 60%; HbC: 40%. Normal/microcytic	HbC implies ancestry from West Africa, clinically insignificant.
HbCC disease	No HbA, HbC almost 100%. Mild microcytic	Mild chronic hemolytic anemia.
HbC trait/α-thalassemia	HbA major, HbC < 30%	
HbC/β-thalassemia	Microcytic, hypochromatic	Moderate to severe anemia with splenomegaly.
HbE trait (HbAE)	HbA major, HbE: 30−35%. Normal/microcytic	No clinical significance, found in Cambodia, Laos, and Thailand (Hb E triangle, where HbE trait is 50−60% of population) and Southeast Asia.
HbE disease	No HbA; mostly HbE. Microcytic hypochromic red cells +/- anemia	Usually asymptomatic.
HbE trait with α-thalassemia	Majority is A: HbE < 25%	
HbO trait (HbAO)	Majority is A; HbO: 30−40%. Normal CBC	Clinically insignificant; HbS/O is a sickling disorder.
HbD trait (HbAD)	HbA > HbD. Normal CBC	Clinically insignificant; HbS/D is a sickling disorder.
HbG trait (HbAG)	HbA > HbG. Normal CBC	Clinically insignificant.

21.9 LABORATORY INVESTIGATION OF HEMOGLOBIN DISORDERS

Multiple methodologies exist to detect hemoglobinopathies and thalassemias. Three methods that are routinely employed are gel electrophoresis, high-performance liquid chromatography (HPLC), and capillary electrophoresis. If any one method detects an abnormality, a second method must be used to confirm it. In addition, relevant clinical history, review of the complete blood count (CBC), and a peripheral smear provide important correlations in the pursuit of an accurate diagnosis.

21.9.1 Gel Electrophoresis

In hemoglobin electrophoresis red cell lysates are subjected to electric fields under alkaline (alkaline gel) and acidic (acid gel) pH. This can be carried out on filter paper, a cellulose acetate membrane, a starch gel, a citrate agar gel, or an agarose gel. Separation of different hemoglobins is largely (but not solely) dependent on the charge of the hemoglobin molecule. A change in the amino acid composition of the globin chains results in alteration of the

Table 21.5 Migration of Various Hemoglobin Bands in Alkaline Gel and Acid Gel Electrophoresis

Region	Hemoglobin Present
Alkaline Gel Electrophoresis	
Top band (farthest from origin: H Lane)	HbH, HbI
J Lane	HbJ
	Hb Bart's and HbN are between HbJ and HbH lane
Fast hemoglobin	Hb Hope
A Lane	HbA
F Lane	HbF
S Lane	HbS, HbD, HbG, Hb Lepore
C Lane	HbC, HbE, HbO, HbA_2, HbS/G hybrid
Carbonic anhydrase band (faint)	HbG_2, HbA_2', HbCS
Acid Gel Electrophoresis	
Top band: C Lane	HbC
S Lane	HbS, HbS/G hybrid; HbO and HbH are between S and A lane
A lane	HbA, HbE, HbA_2, HbD, HbG, HbLepore, HbJ, HbI, HbN, HbH
F Lane	HbF, Hb Hope, Hb Bart's

charge of the hemoglobin molecule, resulting in a change of the speed of migration. In gel electrophoresis, different hemoglobins migrate at different speeds. The top lane is called the H lane and is mainly composed of HbH and HbI, while the point of origin is before the carbonic acid band (Table 21.5). On the alkaline gel in hemoglobin electrophoresis, the H is fast-migrating, and the band on the gel should be the same distance from J as A is from J in the opposite direction. On the acid gel, the H migrates between the S and hemoglobins. The patterns of various bands in acid gel electrophoresis are summarized in Table 21.5.

21.9.2 High-Performance Liquid Chromatography

High-Performance Liquid Chromatography (HPLC) systems utilize a weak cation exchange column system. A sample of an RBC lysate in buffer is injected into the system followed by application of a mobile phase so that various hemoglobins can partition (interact) between the stationary phase and mobile phase. The time required for different hemoglobin molecules to elute is referred to as the retention time. The eluted hemoglobin molecules are detected by light

Table 21.6 Approximate Retention Times of Peaks in HPLC Analysis of Hemoglobins

Approximate Retention Time	Hemoglobin
0.7 min (Peak 1)	Acetylated HbF, HbH, Hb Bart's, bilirubin
1.1 min	HbF
1.3 min (Peak 2)	Hb A1c, Hb Hope
1.7 min (Peak 3)	Aged HbA (HbA1d), HbJ, HbN, HbI
2.5 min	HbA, HbS1c
3.7 min	HbA_2, HbE, Hb Lepore, HbS1d
3.9−4.2 min	HbD, HbG
4.5 min	HbS, HbA_2', HbC1c, Hb-O-Arab has a broad range from 4.5 to 5 min
4.6−4.7 min	HbG_2
4.9 min	HbC (preceding the main peak is a small peak, HbC1d), HbS/G hybrid, HbCS (3 peaks: 2−3%)

absorbance. HPLC permits the provisional identification of many more variant types of hemoglobins that cannot be distinguished by conventional gel electrophoresis. When HPLC is used, a recognized problem is carryover of a specimen from one to the next. For example, if the first specimen belongs to a patient with sickle cell disease (HbSS), then a small peak may be seen at the "S" window in the next specimen. This can lead to diagnostic confusion as well as the necessity of re-running the sample. Approximate retention times of common hemoglobins in a typical HPLC analysis are summarized in Table 21.6.

21.9.3 Capillary Electrophoresis

In capillary electrophoresis a thin capillary tube made of fused silica is used. When an electric field is applied, the buffer solution in the capillary tube generates an electro-endosmotic flow that moves toward the cathode. Separation of individual hemoglobins takes place due to differences in overall charges. Different hemoglobins are represented in different zones. Capillary zone electrophoresis has an advantage over HPLC in that hemoglobin adducts (glycated hemoglobins and the aging adduct HbX1d) do not separate from the main hemoglobin peak in capillary electrophoresis, which makes interpretation easier. Common hemoglobin zones in capillary electrophoresis are given in Table 21.7.

Other less commonly used methodologies include isoelectric focusing, DNA analysis, and mass spectrometry. It is important to note that hemoglobinopathies can interfere with the measurement of glycosylated hemoglobin

Table 21.7 Zones in Which Common Hemoglobins Appear on Capillary Electrophoresis

Zone	Hemoglobin
Zone 1	HbA_2'
Zone 2	HbC, HbCS
Zone 3	HbA_2, Hb-O-Arab
Zone 4	HbE, Hb Koln
Zone 5	HbS
Zone 6	HbD-Punjab/Los Angeles/Iran, HbG-Philadelphia
Zone 7	HbF
Zone 8	Acetylated HbF
Zone 9	HbA
Zone 10	Hb Hope
Zone 11	Denatured HbA
Zone 12	Hb Bart's
Zone 13	
Zone 14	
Zone 15	HbH

(HbA1C), providing unreliable results. When an HbA1C result is inconsistent with a patient's clinical picture, the possibility of hemoglobinopathy must be considered. Depending on the methodology used for measurement of HbA1C (e.g. HPLC, immunoassay, etc.), results may be falsely elevated or lower. Patients with the HbC trait (particularly) show variable results. In such cases, a test that is not affected by hemoglobinopathy, such as fructosamine measurement (representative average blood glucose: 2−3 weeks), may be used [10].

21.10 DIAGNOSTIC TIPS FOR THALASSEMIA, SICKLE CELL DISEASE, AND OTHER HEMOGLOBINOPATHIES

For diagnosis of alpha-thalassemia, routine blood count (CBC) analysis is the first step. Mean corpuscular volume (MCV), mean corpuscular hemoglobin (MCH), and red cell distribution width (RDW) can provide important clues not only in the diagnosis of thalassemias, but also other hemoglobin disorders. Thalassemias are characterized by hypochromatic and microcytic anemia, and it is important to differentiate thalassemia from iron deficiency anemia because iron supplements have no benefits in patients with thalassemia. Often silent carriers of alpha-thalassemia are diagnosed incidentally when their CBC shows a mild microcytic anemia. However, serum iron and serum ferritin levels are normal in a silent carrier of alpha-thalassemia, but are reduced in a patient with iron deficiency anemia. In addition, microcytic anemia with normal RDW also indicates the thalassemia trait. In hemoglobin H disease, MCV is further reduced, but in iron deficiency anemia, MCV is rarely less than 80 fL. In addition, MCH is also reduced. For children, an MCV of <80 fL can be common, and a Mentzer index (MCV/RBC) is useful in differentiating thalassemia from iron deficiency anemia. In iron deficiency anemia this ratio is usually greater than 13, but in thalassemia, this value is less than 13. However, for accurate diagnosis of alpha-thalassemia, genetic testing is essential. Hemoglobin electrophoresis is not usually helpful for diagnosis of alpha-thalassemia except in infants where the presence of Hb Bart's or HbH indicates alpha-thalassemia. Hemoglobin electrophoresis is usually normal in an individual with the alpha-thalassemia trait. However, in an individual with HbH disease, the presence of hemoglobin H in electrophoresis along with Hb Bart's provides useful diagnostic clues. In hydrops fetalis, newborns often die or are born with gross abnormalities. Circulating erythrocytes are markedly hypochromic and anisopoikilocytosis is common. In addition, many nucleated erythroblasts are present in peripheral blood smears. Most of the hemoglobin observed in electrophoresis is Hb Bart's. Parental genetic testing is essential for counseling parents who may potentially give birth to a baby with hydrops fetalis.

A patient with beta-thalassemia major disease can be identified during infancy; after 6 months of age these patients present with irritability, growth retardation, abnormal swelling, and jaundice. Individuals with microcytic anemia, but milder symptoms that start later in life, are likely suffering from beta-thalassemia intermedia. Hemoglobin electrophoresis of individuals with the beta-thalassemia trait usually have reduced or absent HbA, elevated levels of HbA_2, and elevated levels of HbF. Therefore, for the diagnosis of the beta-thalassemia trait, the proportion of HbA_2 relative to the other hemoglobins is an important indicator. In certain cases, HbA_2 variants may also be present. In such cases the total HbA_2 (HbA_2 and HbA_2 variant) need to be considered for the diagnosis of beta-thalassemia. HbA_2' is the most common of the known HbA_2 variants, and is reported in 1−2% of African-Americans; it is detected in heterozygous and homozygous states, and in combination with other Hb variants and thalassemia. The major clinical significance of HbA_2' is that, for the diagnosis or exclusion of beta-thalassemia minor, the sum of HbA_2 and HbA_2' must be considered. HbA_2', when present, accounts for a small percentage (1−2%) in heterozygotes and is difficult to detect by gel electrophoresis. It is, however, easily detected by capillary electrophoresis and HPLC. In HPLC HbA_2' elutes in the "S" window. In the HbAS trait and HBSS disease, HbA_2' could be masked by the presence of HbS. In the HbAC trait and HbCC disease, glycosylated HbC will also elute in the "S" window.

Table 21.8 Hematological Features of Alpha- and Beta-Thalassemias

Disease	CBC	Hemoglobin Electrophoresis
*Alpha-Thalassemia**		
Silent carrier	Hb: Normal; MCH <27 pg	Normal
Trait	Hb: Normal; MCH <26 pg, MCV <75 fL	Normal
HbH disease	Hb: 8–10 g/dL; MCH <22 pg; MCV low	HbH: 10–20%
Hydrops fetalis	Hb <6 g/dL; MCH <20 pg	Hb Bart's: 80–90%
		HbH <1%
*Beta-Thalassemia**		
Minor	Hb: Normal or low; MCV: 55–75 fL# MCH: 19–25 pg	HbA$_2$ >3.5%
Intermedia	Hb: 6–10 g/dL; MCV: 55–70 fL	HbA$_2$: Variable
Mild or compound heterozygous	MCH: 15–23 pg	HbF: up to 100%
Major	Hb <7 g/dL; MCV: 50–60 fL; MCH: 14–20 pg	HbA$_2$: Variable HbF: High

Mentzer Index for children is <13 for both alpha- and beta-thalassemias. #MCV (Abnormal): adults <80 fL; children (ages 7–12) <76; children (6 months to 6 years) <70.

In these conditions, HbA$_2$′ will remain undetected. Conversely, sickle cell patients on a chronic transfusion protocol or recent efficient RBC exchange can result in a very small percentage of HbS that the pathologist may interpret as HbA$_2$′. It has been documented that the HbA$_2$ concentration may be raised in HIV during treatment. Severe iron deficiency anemia can reduce HbA$_2$ levels and this can obscure diagnosis of the beta-thalassemia trait. Hematological features of alpha- and beta-thalassemia are given in Table 21.8 [6,11].

HbF quantification is useful in the diagnosis of beta-thalassemia and other hemoglobinopathies. Quantification of HbF may be an issue when HPLC is used. Fast variants (e.g. HbH or Hb Bart's) may not be quantified as they can elute off the column before the instrument begins to integrate in many systems designed for adult samples. This affects the quantity of HbF. If an alpha-globin variant separates from HbA, then there should be an HbF variant that will separate from normal HbF, but it may not separate from other hemoglobin adducts present. In this case the total HbF will not be adequately quantified. HbF variants can also be due to mutation of the gamma-globin chain, and again this can result in a separate peak and incorrect quantification. Some beta chain variants and adducts will not separate from HbF and this can lead to incorrect quantification. If HbF appears to be greater than 10% on HPLC, its nature should be confirmed

Table 21.9 Diagnostic Approach to Sickle Cell Disease

Hemoglobin Pattern	Diagnosis/Comments
Patient has HbA and HbS	HbAS trait or HbSS disease (post-transfusion) or HbS/β^+-thalassemia or a normal person transfused from a donor with HbAS trait. Transfusion history is essential for diagnosis. For patient with HbAS trait, HbA is majority and HbS is 30–40%; if donor was HbS trait then S% is usually between 0.8% to 14% of the total hemoglobin. In HbS/β^+-thalassemia HbA$_2$ is expected to be high and there should be microcytosis and hypochromia of the red cells. HbA% is typically 5–25% depending on severity of genetic defect.
Patient has HbS, but no HbA	HbSS disease; HbS/β^0-thalassemia, HbA$_2$ is elevated, with low MCV and MCH.
Patient has HbS and high HbF	HbS/HPFH and HbSS disease while patient is on hydroxyurea. High MCV favors hydroxyurea therapy; medication history will be required.

by an alternative method to exclude misidentification of HbN or HbJ as HbF. Characterization of patients with high HbF includes evaluation of the following:

- Consideration of whether HbF is physiologically appropriate for the age.
- Beta-thalassemia trait, intermedia (20–40%), or major (60–98%). Here HbA$_2$ will also be raised. Patients should have microcytic hypochromic anemia with normal RDW and a disproportionately high RBC count. A peripheral smear should exhibit target cells.
- Delta–beta-thalassemia: Here HbA$_2$ is normal, but HbF is increased due to an increase in gamma chains. However, the increase in gamma chains does not entirely compensate for the decreased beta chains. Moreover, the alpha chain is present in excess. The trait shows microcytosis without anemia. Homozygous patients have high severity of disease compared to thalassemia intermedia.

Hemoglobin electrophoresis is useful in diagnosis of sickle cell disease by identification of HbS. The diagnostic approach for sickle cell disease is summarized in Table 21.9. However, a solubility test can also aid in diagnosis of sickle cell disease. When a blood sample containing HbS is added to a test solution containing saponin (to lyse cells) and sodium hydrosulfite (to deoxygenate the solution), a cloudy turbid suspension is formed if HbS is present. If no HbS is present, the solution remains clear. A false negative result may be observed if HbS is <10%, as is often the case in infants younger than 3 months [12].

For diagnosis of the HbS/G hybrid on alkaline gel electrophoresis, one band is expected in the A lane, one band in the S lane (due to HbS and HbG), one band in the C lane (due to S/G hybrid), and one band in the carbonic anhy-drase area (due to HbG_2). Therefore, a total of four bands should be observed. If the band in the carbonic anhydrase is not prominent, at least three bands should be seen. On the acid gel electrophoresis, one band is expected in the A lane (due to HbA, HbG, and HbG_2) and one band in the S lane (due to HbS and HbS/G hybrid). In electrophoresis a band should be seen in zone 5 (HbS) and zone 6 (HbG). It is important to emphasize that for hemoglobinopathies, gel electrophoresis results must be confirmed by a second method, either HPLC or capillary electrophoresis.

In the presence of HbS, if a higher value of HbF is observed, then HbS/HPFH can be suspected. In this case CBC should be normal and HbF should be between 25and35%. However, with HbS/beta-thalassemia, HbF could also be high. In HPFH and HbS/HPFH, distribution of HbF in red cells is normocellular, but in delta—beta-thalassemia and HbSS with high HbF, it is heterocellular. Kleihauer—Betke tests or flow cytometry with anti-F antibody will illustrate the difference. Interpretations of various other hemoglobinopa-thies are given in Table 21.10. The logical approach for diagnosis of hemo-globinopathies where an initial band is present in the C lane of an alkaline gel is given in Figure 21.1. Approaches where the initial band is present in the E lane are given in Figure 21.2.

Universal newborn screening for hemoglobinopathies is now required in all 50 states and the District of Columbia. In addition, the American College of Obstetricians and Gynecologists provides guidelines for screening of couples that may be at risk of having children with hemoglobinopathy. Diagnostic approaches for various hemoglobinopathies are summarized in Table 21.10. Persons of Northern European, Japanese, Native American, or Korean descent are at low risk for hemoglobinopathies, but people with ancestors from Southeast Asia, Africa, or the Mediterranean are at high risk. A complete blood count should be done to accurately measure hemoglobin. If all para-meters are normal and the couple belongs to a low-risk group, no further testing may be necessary. For higher risk couples, hemoglobin analysis by electrophoresis or another method is recommended. A solubility test for sickle cell may be helpful. Genetic screening can help physicians identify cou-ples at risk of having children with hemoglobinopathy. Molecular protocols for hemoglobinopathies started in the 1970s using Southern blotting and restriction fragment length polymorphism analysis for prenatal sickle cell dis-ease. With the development of polymerase chain reaction (PCR) molecular testing for hemoglobinopathies, much less DNA is now required for analysis [13]. Currently, however, molecular testing for diagnosis of hemoglobinopa-thies that is certain to establish a firm diagnosis, especially for the alpha-thalassemia trait (direct gene analysis), is available in large academic medical centers and reference laboratories only.

Table 21.10 Diagnostic Approaches to Various Hemoglobinopathies

Diagnosis	Features
HbC	Band in C lane in the alkaline gel: possibilities are C, E, or O.
	Band in C lane in acid gel.
	HPLC shows peak around 5 min with small peak just before main peak (HbC1d). Small peak may also be observed at 4.5 min (HbC1c).
	OR
	Capillary electrophoresis shows peak in Zone 2.
HbE	Band in C lane in alkaline gel: possibilities are C, E, or O.
	Band in A lane in acid gel.
	HPLC show a peak at 3.5 minutes and is >10%.
	OR
	Capillary electrophoresis shows peak in Zone 4.
HbO	Band in C lane in alkaline gel: possibilities are C, E, or O.
	Band between A and S lane in acid gel.
	HPLC shows peak between 4.5 and 5 minutes.
	OR
	Capillary electrophoresis shows peak in Zone 3 (O-Arab).
HbS	Band in S lane in alkaline gel; possibilities: S, D, G, Lepore.
	Band in S lane in acid gel.
	HPLC shows peak at 4.5 minutes.
	OR
	Capillary electrophoresis shows peak in Zone 5.
HbD	Band in S lane in alkaline gel: possibilities are S, D, G, Lepore.
	Band in A lane in acid gel.
	HPLC shows peak at 3.9 to 4.2 minutes; no additional peak.
	OR
	Capillary electrophoresis shows a peak in Zone 6.
HbG	Band in S lane in alkaline gel: possibilities are S, D, G, Lepore.
	Band in A lane in acid gel.
	HPLC shows peak at 3.9−4.2 min and small additional peak (G_2).
	OR
	Capillary electrophoresis shows peak in Zone 6.
Hb Lepore*	Band (faint) in S lane in alkaline gel; possibilities: S, D, G, Lepore.
	Band in A lane in acid gel.
	HPLC shows peak at 3.7 min (A_2 peak); quantity is lower than D, G, or E. Small increase in % HbF.
	OR
	Capillary electrophoresis shows peak in Zone 6.

*Hb Lepore band in the alkaline gel is faint.

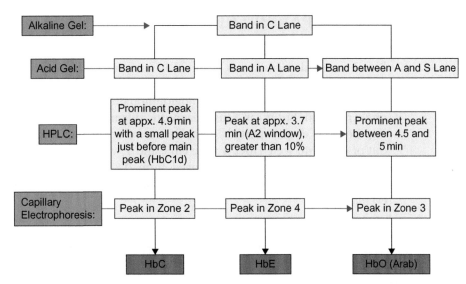

FIGURE 21.1

Interpretation of hemoglobinopathy when a band is present in the C lane in the alkaline gel. This figure is reproduced in color in the color plate section. *(Courtesy of Andres Quesda, M.D., Department of Pathology and Laboratory Medicine, University of Texas, Houston Medical School.)*

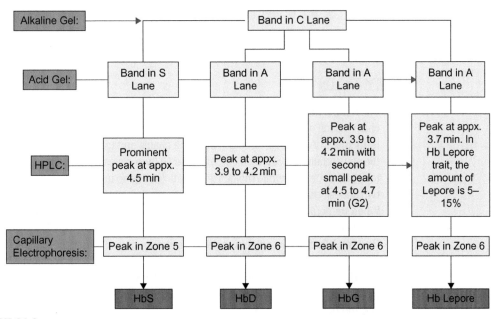

FIGURE 21.2

Interpretation of hemoglobinopathy when a band is present in the E lane in the alkaline gel. This figure is reproduced in color in the color plate section. *(Courtesy of Andres Quesda, M.D., Department of Pathology and Laboratory Medicine, University of Texas, Houston Medical School.)*

21.11 APPARENT HEMOGLOBINOPATHY AFTER BLOOD TRANSFUSION

Blood transfusion history is essential in interpreting abnormal hemoglobin patterns because small peaks of abnormal hemoglobin can appear from blood transfusions. Apparent hemoglobinopathy after blood transfusion is rarely reported, but it can cause diagnostic dilemmas that require repeat testing. Kozarski et al. reported 52 incidences of apparent hemoglobinopathies out of which 46 were HbC, 4 were HbS, and 2 were Hb-O-Arab. The percentage of abnormal hemoglobin ranged from 0.8% to 14% (median: 5.6%). The authors recommended identifying and notifying the donor in such events [14].

CASE REPORT

A 2-year-old male with thalassemia major showed a peak of 18.5% in HPLC analysis in the sickle window at 4.36 min retention time, thus creating a diagnostic dilemma. There was also a slow-moving band in the HbS region of alkaline electrophoresis. The boy showed on HPLC analysis an HbS of 18.5%, an HbA of 66%, and an HbF of 1%. His MCV was 71.7 fL and hemoglobin was 11 g/dL. On examination of old records, it was noted that no HbS was present. The patient had received a blood transfusion, probably from a donor with the sickle cell trait [15].

KEY POINTS

- The normal hemoglobin (HbA) in adults contains two alpha chains and two beta chains. Each alpha chain contains 141 amino acids and each beta chain 146. Hemoglobin A_2 (HbA$_2$) contains two alpha chains and two delta chains. The gene for the alpha chain is located on chromosome 16 (two genes on each chromosome, a total of four genes), while the genes for beta (one gene in each chromosome, a total of two genes), gamma, and delta chains are located on chromosome 11.
- When hemoglobin is circulating with erythrocytes, glycosylation of the globin chains may take place. These are referred to as X1c (X being any hemoglobin, e.g. HbA1c). When a hemoglobin molecule is aging, glutathione is bound to cysteine at the 93rd position of the beta chain. This is HbA3 or HbA1d. Just like HbA1c and HbA1d, there can exist HbC1c, HbC1d, HbS1c, and HbS1d.
- Heme is synthesized in a complex way involving enzymes in both mitochondria and the cytosol.
- Hemoglobinopathies can be divided into three major categories: (1) quantitative disorders of hemoglobin synthesis where production of structurally normal but decreased amounts of globin chains (thalassemia syndrome) occurs; (2) qualitative

disorders in hemoglobin structure where there is production of structurally abnormal globulin chains such as hemoglobin S, C, O, or E (sickle cell syndrome is the most common example of such disease); and (3) failure to switch globin chain synthesis after birth. Here, hereditary persistence of fetal hemoglobin (HbF), a relatively benign condition, can co-exist with thalassemia or sickle cell disease, but with decreased severity of such diseases (a protective effect).

- Hemoglobinopathies are transmitted in autosomal recessive fashion.
- Disorders due to beta chain defects such as sickle cell disease tend to manifest clinically after 6 months of age, although diseases due to alpha chain defects are manifested in utero or following birth.
- The hemoglobin variants of most clinical significance are hemoglobin S, C, and E. In West Africa, approximately 25% of individuals are heterozygous for the hemoglobin S (HbS) gene, which is related to sickle cell diseases. In addition, high frequencies of HbS gene alleles are also found in people from the Caribbean, South and Central Africa, the Mediterranean, the Arabian Peninsula, and East India. Hemoglobin C (HbC) is found mostly in people living or originating from West Africa. Hemoglobin E (HbE) is widely distributed between East India and Southeast Asia, with the highest prevalence in Thailand, Laos, and Cambodia, but may be sporadically observed in parts of China and Indonesia. Thalassemia syndrome is not due to a structural defect in the globin chain, but is due to lack of sufficient synthesis of the globin chain; it is a genetically inherited disease. Thalassemia syndrome can be divided into alpha-thalassemia and beta-thalassemia. In general, β-thalassemia is observed in the Mediterranean, the Arabian Peninsula, Turkey, Iran, West and Central Africa, India, and Southeast Asian countries, while α-thalassemia is commonly seen in parts of Africa, the Mediterranean, the Middle East, and throughout Southeast Asia.
- Alpha-thalassemia occurs when there is a defect or deletion in one or more of four genes responsible for alpha-globin production. Alpha-thalassemia can be divided into four categories:
  - The Silent Carriers: Characterized by only one defective or deleted gene but three functional genes. These individuals have no health problem. An unusual case of silent carrier is individuals carrying one defective Constant Spring mutation but three functional genes. These individuals also have no health problem.
  - Alpha-Thalassemia Trait: Characterized by two deleted or defective genes and two functional genes. These individuals may have mild anemia.
  - Alpha-Thalassemia Major (Hemoglobin H Disease): Characterized by three deleted or defective genes and only one functional gene. These patients have persistent anemia and significant health problems. When hemoglobin H disease is combined with hemoglobin Constant Spring, the severity of disease is more than just hemoglobin H disease. However, if a child inherits one hemoglobin Constant Spring gene from its mother and one from its father,

then the child has homozygous hemoglobin Constant Spring and the severity of disease is similar to hemoglobin H disease.

- Hydrops Fetalis: Characterized by no functional alpha gene. These individuals have hemoglobin Bart. This condition is severely life-threatening unless intrauterine transfusion is initiated.

- Hemoglobin Constant Spring (hemoglobin variant isolated from a family of ethnic Chinese background from the Constant Spring district of Jamaica) is a hemoglobin variant where mutation of the alpha-globin gene produces an abnormally long alpha chain (172 amino acids instead of the normal 146). Hemoglobin Constant Spring is due to a non-deletion mutation of the alpha gene that results in production of unstable alpha-globin. Moreover, this alpha-globin is produced in very low quantity (approximately 1% of the normal expression level) and is found in people living or originating in Southeast Asia.

- Beta-thalassemia can be broadly divided into three categories:
  - Beta-Thalassemia Trait: Characterized by one defective gene and one normal gene. Individuals may experience mild anemia but are not transfusion-dependent.
  - Beta-Thalassemia Intermedia: Characterized by two defective genes; some beta-globin production is still observed in these individuals. However, some individuals may have significant health problems requiring intermittent transfusions.
  - Beta-Thalassemia Major (Cooley's Anemia): Characterized by two defective genes; almost no function of either gene, leading to no synthesis of beta-globin. These individuals have a severe form of disease requiring lifelong transfusion and may have shortened lifespans.

- Patients with beta-thalassemia major have elevated HbA_2 and HbF (although in some individuals HbF may be normal).

- In the heterozygous form (HbAS), the sickle cell trait protects from infection of *Plasmodium falciparum* malaria, but not in the more severe form of homozygous sickle cell disease (HbSS). The genetic defect that produces sickle hemoglobin is a single nucleotide substitution at codon 6 of the beta-globin gene on chromosome 11; it results in a point mutation in the beta-globin chain of hemoglobin (substitution of valine for glutamic acid at the sixth position).

- Double heterozygous states of HbSC, HbSD, HbS-O-Arab are important sickling states that should not be missed.

- Hemoglobin C is formed due to substitution of a glutamic acid residue with a lysine residue at the sixth position of beta-globin. Hemoglobin E is due to a point mutation of beta-globin that results in substitution of lysine for glutamic acid in position 26.

- Hemoglobin Lepore is an unusual hemoglobin molecule composed of two alpha chains and two delta—beta chains as a result of a fusion gene of delta and beta genes. The delta—beta chains have the first 87 amino acids of the delta chain and 32 amino acids of the beta chain.

- Individuals with HbA/Hb Lepore are asymptomatic, with Hb Lepore representing 5—15% of hemoglobin and, slightly elevated HbF (2—3%). However, homozygous Lepore individuals suffer from severe anemia similar to patients with beta-thalassemia intermedia, with Hb Lepore representing 8—30% of hemoglobin and the remainder hemoglobin F.

- Hemoglobin G-Philadelphia (HbG) is the most common alpha chain defect, observed in 1 in 5,000 African-Americans, and is associated with alpha-thalassemia 2 deletions.

- It is possible that an African-American individual may have HbS/HbG where the hemoglobin molecule contains one normal alpha chain, one alpha G chain, one normal beta chain, and one beta S chain. This can result in the detection of various hemoglobins in the blood, including HbA (alpha 2, beta 2), HbS (alpha 2, beta S2), HbG (alpha G_2, beta 2), and HbS/G (alpha G_2, beta S2). In addition, HbG_2 (alpha 2, delta 2), which is the counterpart of HbA_2, is also present.

- Increase in fetal hemoglobin percentage is associated with multiple pathologic states. These include beta-thalassemia, delta—beta-thalassemia, and hereditary persistence of fetal hemoglobin (HPFH). Beta-thalassemia is associated with high HbA_2, and the latter two states are associated with normal HbA_2 values. Hematologic malignancies are associated with increased hemoglobin F, and include acute erythroid leukemia (AML, M6) and juvenile myelomonocytic leukemia (JMML). Aplastic anemia is also associated with an increase in percentage of HbF. In elucidating the actual cause of high HbF, it is important to consider the actual percentage of HbF and HbA_2 values as well as the correlation with complete blood count (CBC) and peripheral smear findings. It is also important to note that drugs (hydroxyurea, sodium valproate, erythropoietin) and stress erythropoiesis can also result in high HbF. Hydroxyurea is used in sickle cell disease patients to increase the amount of HbF, the presence of which can help reduce the clinical effects of the disease. Measuring the level of HbF can be useful in determining the appropriate dose of hydroxyurea. In 15—20% of cases of pregnancy, HbF may be raised to values as high as 5%.

REFERENCES

[1] Manning LR, Russell JR, Padovan JC, Chait BT, et al. Human embryonic, fetal and adult hemoglobins have different subunit interface strength. Correlation with lifespan in the red cell. Protein Sci 2007;16:1641—58.

[2] Giordano PC. Strategies for basic laboratory diagnostics of the hemoglobinopathies in multi-ethnic societies: interpretation of results and pitfalls. Int J Lab Hematol 2012;35:465—79.

[3] Rappaport VJ, Velazquez M, Williams K. Hemoglobinopathies in pregnancy. Obset Gynecol Clin N Am 2004;31:287—317.

[4] Sriiam S, Leecharoenkiat A, Lithanatudom P, Wannatung T, et al. Proteomic analysis of hemoglobin H Constant Spring (HB H-CS) erythroblasts. Blood Cells Mol Dis 2012;48:77–85.

[5] Yi JS, Moertel CL, Baker KS. Homozygous α-thalassemia treated with intrauterine transfusion and unrelated donor hematopoietic cell transplantation. J Pediatr 2009;154:766–8.

[6] Kohne E. Hemoglobinopathies: clinical manifestations, diagnosis and treatment. Dtsch Arztebl Int 2011;108:532–40.

[7] Ngo DA, Aygun B, Akinsheye I, Hankins JS. Fetal hemoglobin levels and hematological characteristics of compound heterozygotes for hemoglobin S and deletional hereditary persistence of fetal hemoglobin. Br J Haematol 2012;156:259–64.

[8] Pandey S, Mishra RM, Pandey S, Shah V, et al. Molecular characterization of hemoglobin D Punjab traits and clinical-hematological profile of patients. Sao Paulo Med J 2012;130:248–51.

[9] Dror S. Clinical and hematological features of homozygous hemoglobin O-Arab [beta 121 Glu – LYS]. Pediatr Blood Cancer 2013;60:506–7.

[10] Smaldone A. Glycemic control and hemoglobinopathy: when A1C may not be reliable. Diabetes Spectrum 2008;21:46–9.

[11] Muncie H, Campbell JS. Alpha and beta thalassemia. Am Fam Physician 2009;339:344–71.

[12] Lubin B, Witkowska E, Kleman K. Laboratory diagnosis of hemoglobinopathies. Clin Biochem 1991;24:363–74.

[13] Benson JA, Therell BL. History and current status of newborn screening for hemoglobinopathies. Semin Perinatol 2010;34:134–44.

[14] Kozarski TB, Howanitz PJ, Howanitz JH, Lilic N, et al. Blood transfusion leading to apparent hemoglobin C, S and O-Arab hemoglobinopathies. Arch Pathol Lab Med 2006;130:1830–3.

[15] Gupta SK, Sharma M, Tyagi S, Pati HP. Transfusion induced hemoglobinopathy in patients with beta-thalassemia major. Indian J Pathol Microbiol 2011;54:609–11.

CHAPTER 22

Protein Electrophoresis and Immunofixation

22.1 MONOCLONAL GAMMOPATHY

Serum protein electrophoresis, urine electrophoresis, serum immunofixation, and urine immunofixation are all performed primarily to investigate suspicion of monoclonal gammopathy in a patient. Monoclonal gammopathy is present in a patient when a monoclonal protein (M protein or paraprotein) is identified in a patient's serum, urine, or both. Monoclonal gammopathy may be of undetermined significance or due to myeloma or a B cell lymphoproliferative disorder that produces this paraprotein. Multiple myeloma, a malignant disorder of bone marrow, is the most common form of myeloma. This disease is called multiple myeloma because it affects multiple organs in the body. In multiple myeloma, plasma cells (which proliferate at a low rate) become malignant with a massive clonal expression that results in a high rate of production of monoclonal immunoglobulin in the circulation. Monoclonal gammopathy of undetermined significance was first described in 1978 and is a pre-cancerous condition that affects approximately 3% of people over 50 years of age [1]. This condition may progress to multiple myeloma among 1% of these individuals every year. A variant of monoclonal gammopathy of undetermined significance is asymptomatic or smoldering plasma cell myeloma in which the diagnostic criteria for multiple myeloma are present but no related organ damage is observed. Circulating micro RNAs can be a potential biomarker for distinguishing normal people from patients with multiple myeloma or related conditions [2]. The risk of malignant transformation of monoclonal gammopathy of undetermined significance into multiple myeloma is higher in females than males and is also higher in individuals showing the presence of IgA paraprotein as compared to individuals with IgG paraprotein [3].

The paraprotein can be found as an intact immunoglobulin, only light chains (light chain myeloma, light chain deposition disease, or amyloid light chain amyloidosis), or, rarely, only as heavy chains (heavy chain disease). Paraproteins can be detected in the serum and can also be excreted into the

391

A. Dasgupta and A. Wahed: Clinical Chemistry, Immunology and Laboratory Quality Control
DOI: http://dx.doi.org/10.1016/B978-0-12-407821-5.00022-X

urine. Sometimes if the paraprotein is only light chain (light chain disease) it is detected in the urine alone but not in the serum. The serum may paradoxically exhibit only hypogammaglobulinemia. It is important to note that the presence of paraprotein in serum, urine, or both indicates monoclonal gammopathy and not necessarily the presence of multiple myeloma in a patient. Multiple myeloma is one of the causes of monoclonal gammopathy. Transient monoclonal gammopathy may be observed in an immunocompromised patient suffering from infection due to an opportunistic pathogen such as cytomegalovirus [4]. Monoclonal gammopathy is usually observed in patients over 50 years of age and is rare in children. Gerritsen et al. studied 4,000 pediatric patients over a 10-year period and observed monoclonal gammopathy only in 155 children, but such gammopathies were found most frequently in patients suffering from primary and secondary immunodeficiency, hematological malignancies, autoimmune disease, and severe aplastic anemia. Follow-up analysis revealed that most of these monoclonal gammopathies were transient [5].

Agarose gel electrophoresis and capillary electrophoresis are two principal methods employed in screening for paraproteins. Both methods are applicable for both serum and urine specimens. Once a paraprotein is detected, confirmation and the isotyping of paraprotein are essential. This is usually achieved by immunofixation. For urine immunofixation, the best practice is to utilize a 24-hour urine specimen that has been concentrated; such technique allows for detection of even a faint band.

22.2 SERUM PROTEIN ELECTROPHORESIS

Serum protein electrophoresis (SPEP) is an inexpensive, easy-to-perform screening procedure for initial identification of monoclonal bands. Monoclonal bands are usually seen in the gamma zone, but may be seen in proximity to the beta band or rarely in the alpha-2 region. Blood can be collected in a tube with a clot activator, and after separation from blood components, serum is then placed on special paper treated with agarose gel followed by exposure to an electric current in the presence of a buffer solution (electrophoretic cell). Various serum proteins are then separated based on charge. After a predetermined time of exposure to an electric field, the special paper is removed, dried, placed on a fixative to prevent further diffusion of specimen components, and then stained to visualize various protein bands. Coomassie brilliant blue is a common staining agent to visualize bands in serum protein electrophoresis. Then, using a densitometer, each fraction is quantitated. The serum protein components are separated into five major fractions:

- Albumin
- Alpha-1 globulins (alpha-1 zone)

- Alpha-2 globulins (alpha-2 zone)
- Beta globulins (beta zone often splits into beta-1 and beta-2 bands)
- Gamma globulins (gamma zone).

Albumin and globulins are two major fractions of the electrophoresis pattern. Albumin, the largest band, lies closest to the positive electrode (anode) and has a molecular weight of approximately 67 kDa (67,000 Daltons). Reduced intensity of this band is observed in inflammation, liver dysfunction, uremia, nephrotic syndrome, and other conditions that lead to hypoalbuminemia, such as critical illness and pregnancy. A smear observed in front of the albumin band may be due to hyperbilirubinemia or the presence of certain drugs. A band in front of the albumin band may be due to prealbumin (a carrier for thyroxine and vitamin A) that is commonly seen in cerebrospinal fluid specimens or serum specimens in patients with malnutrition. Two (rather than one) albumin bands may represent bisalbuminemia. This is a familial abnormality with no clinical significance. Analbuminemia is a genetically inherited metabolic disorder that was first described in 1954. This disorder is rare, and affects less than 1 in one million births. This condition is benign because low albumin levels are compensated for by high levels of non-albumin proteins and circulatory adaptation. Hyperlipidemia is usually observed in these patients. Pseudo-analbuminemia due to the presence of a slow-moving albumin variant appearing in the alpha-1 region of serum protein electrophoresis has also been reported [6].

Moving towards the negative portion of the gel (cathode), the alpha zone is the next band after albumin. The alpha zone can be sub-divided into two zones: the alpha-1 band and alpha-2 band. The alpha-1 band mostly consists of alpha-1-antitrypsin (AT, 90%), alpha-1-chymotrypsin, and thyroid-binding globulin. Alpha-1-antitrypsin is an acute-phase reactant and its concentration is increased in inflammation and other conditions. The alpha-1-antitrypsin band is decreased in patients with alpha-1-antitrypsin deficiency or decreased production of globulin in patients with severe liver disease. At the leading edge of this band, a haze due to high density lipoprotein (HDL) may be observed, although different stains are used (Sudan Red 7B or Oil Red O) for lipoprotein analysis using electrophoresis. The alpha-2 band consists of alpha-2-macroglobulin, haptoglobin, and ceruloplasmin. Because both haptoglobin and ceruloplasmin are acute-phase reactants, this band is increased in inflammatory states. Alpha-2-macroglobulin is increased in nephrotic syndrome and cirrhosis of liver.

The beta zone may consist of two bands, beta-1 and beta-2. Beta-1 is mostly composed of transferrin and low density lipoprotein. An increased beta-1 band is observed in iron deficiency anemia due to an increased level of free transferrin. This band may also be elevated in pregnant women. Very low density lipoprotein usually appears in the pre-beta zone. The beta-2 band is mostly composed of complement proteins. If two bands are observed in the beta-2 region, it implies either electrophoresis of the plasma specimen (fibrinogen band) instead of the serum specimen or IgA paraprotein.

Much of the clinical interest of serum protein electrophoresis is focused on the gamma zone because immunoglobulins mostly migrate to this region. Usually the C-reactive protein band is found between the beta and gamma regions. Serum protein electrophoresis is most commonly ordered when multiple myeloma is suspected and observation of a monoclonal band (M band, paraprotein) indicates that monoclonal gammopathy may be present in the patient. If the M band or paraprotein is observed in serum protein electrophoresis, the following steps are performed:

- The monoclonal band is measured quantitatively using a densitometric scan of the gel.
- Serum and/or urine immunofixation is conducted to confirm the presence of the paraprotein as well as to determine the isotype of the paraprotein.
- A serum light chain assay is conducted or recommended to the ordering clinician.

Monoclonal gammopathy can be due to various underlying diseases, including multiple myeloma. In about 5% of cases two paraproteins may be detected. This is referred to as biclonal gammopathy. A patient may also have non-secretory myeloma, as in the case of a plasma cell neoplasm in which the clonal cells are neither producing nor secreting M proteins. The most commonly observed paraprotein is IgG followed by IgA, light chain, and, rarely, IgD. When a monoclonal band is identified using serum protein electrophoresis, serum immunofixation and 24-hour urine immunofixation is typically recommended. There are certain situations where a band may be apparent, but in reality it is not a monoclonal band. Examples include:

- Fibrinogen is seen as a discrete band when electrophoresis is performed on the plasma instead of the serum specimen. This fibrinogen band is seen between the beta and gamma regions. If the electrophoresis is repeated after the addition of thrombin, this band should disappear. In addition, an immunofixation study should be negative.
- Intravascular hemolysis results in the release of free hemoglobin in circulation (which binds to haptoglobin). The hemoglobin–haptoglobin complex may appear as a large band in the alpha-2 area. Serum immunofixation studies should be negative in such cases.
- In patients with iron deficiency anemia, concentrations of transferrin may be high, which can result in a band in the beta region. Again, immunofixation should be negative.

- Patients with nephrotic syndrome usually show low albumin and total protein, but this condition can also produce increased alpha-2 and beta fractions. Bands in either of these regions may mimic a monoclonal band.
- When performing gel electrophoresis, a band may be visible at the point of application. Typically this band is present in all samples performed at the same time using the same agarose gel support material.

Common problems associated with interpretation of serum protein electrophoresis are summarized in Box 22.1. A low concentration of a paraprotein may not be detected by serum electrophoresis. There are also certain situations where a false negative interpretation could be made on serum electrophoresis. These situations include:

- A clear band is not seen in cases of alpha heavy chain disease (HCD). This is presumably due to tendency of these chains to polymerize or to their high carbohydrate content. HCDs are rare B cell lymphoproliferative neoplasms characterized by the production of a monoclonal component consisting of monoclonal immunoglobulin heavy chain without associated light chain.
- In mu HCD a localized band is found in only 40% of cases. Panhypogammaglobulinemia is a prominent feature in such patients.
- In occasional cases of gamma HCD, a localized band may not be seen.
- When a paraprotein forms dimers, pentamers, polymers, or aggregates with each other, or when forming complexes with other plasma components, a broad smear may be visible instead of a distinct band.

Box 22.1 COMMON PROBLEMS ASSOCIATED WITH SERUM PROTEIN ELECTROPHORESIS

- Serum protein electrophoresis performed using plasma instead of serum produces an additional distinct band between the beta and gamma zones due to fibrinogen, but such a band is absent in subsequent immunofixation studies.
- A band may be seen at the point of application. Typically this band is present in all samples performed at the same time.
- If concentration of transferrin is high (e.g. iron deficiency), a strong band in the beta region is observed.

- In nephrotic syndrome, prominent bands may be seen in alpha-2 and beta regions that are not due to monoclonal proteins.
- Hemoglobin—haptoglobin complexes (seen in intravascular hemolysis) may produce a band in the alpha-2 region.
- Paraproteins may form dimers, pentamers, polymers, or aggregates with each other, resulting in a broad smear rather than a distinct band.
- In light chain myeloma, light chains are rapidly excreted in the urine and no corresponding band may be present in serum protein electrophoresis.

- Some patients may produce only light chains, which are rapidly excreted in the urine, and no distinct band may be present in the serum protein electrophoresis. Urine protein electrophoresis is more appropriate for diagnosis of light chain disease. When light chains cause nephropathy and result in renal insufficiency, excretion of the light chains is hampered and a band may be seen in serum electrophoresis.
- In some patients with IgD myeloma, the paraprotein band may be very faint.

Hypogammaglobulinemia can be congenital or acquired. Amongst the acquired causes are multiple myeloma and primary amyloidosis. Panhypogammaglobulinemia can occur in about 10% of cases of multiple myeloma. Most of these patients have a Bence−Jones protein in their urine, but lack intact immunoglobulins in their serum. Bence−Jones proteins are monoclonal free kappa or lambda light chains in the urine. Detection of Bence−Jones protein may be suggestive of multiple myeloma or Waldenström's macroglobulinemia. Panhypogammaglobulinemia can also be seen in 20% of cases of primary amyloidosis. It is important to recommend urine immunofixation studies when panhypogammaglobulinemia is present in serum protein electrophoresis.

Although monoclonal gammopathy is the major reason for serum protein electrophoresis, polyclonal gammopathy may also be observed in some patients. Monoclonal gammopathies are associated with a clonal process that is malignant or potentially malignant. However, polyclonal gammopathy, in which there is a non-specific increase in gamma globulins, may not be associated with malignancies. Many conditions can lead to polyclonal gammopathies. Serum protein electrophoresis can also exhibit changes that imply specific underlying clinical conditions other than monoclonal gammopathy. Common features of serum protein electrophoresis in various disease states other than monoclonal gammopathy include:

- Inflammation: Increased intensity of alpha-1 and alpha-2 with a sharp leading edge of alpha-1 may be observed, but with chronic inflammation the albumin band may be decreased with increased gamma zone due to polyclonal gammopathy.
- Nephrotic syndrome: The albumin band is decreased due to hypoalbuminemia. In addition, the alpha-2 band may be more distinct.
- Cirrhosis or chronic liver disease: A low albumin band due to significant hypoalbuminemia with a prominent beta-2 band and beta−gamma bridging are characteristic features. In addition, polyclonal hypergammaglobulinemia is observed.

Various clinical conditions other than monoclonal gammopathy that lead to abnormal patterns in serum protein electrophoresis are listed in Table 22.1.

Table 22.1 Abnormal Serum Protein Electrophoresis Pattern Due to Various Diseases Other than Monoclonal Gammopathy

Disease	Abnormal Pattern				
	Albumin	Alpha-1	Alpha-2	Beta	Gamma Zone
Acute ilnflammation	Reduced	Increased	Increased	No change	No change
Chronic inflammation	Reduced	Increased	Increased	No change	Increased
Nephrotic syndrome*	Reduced	No change	Increased	Increased#	No change
Liver disease/cirrhosis*	Reduced	No change	No change	Beta-gamma bridging	Increased
Polyclonal gammopathy	No change	No change	No change	No change	Increased

*Total protein is reduced.
#Increased beta zone due to secondary hyperlipoproteinemia.

22.3 URINE ELECTROPHORESIS

Urine protein electrophoresis is analogous to serum protein electrophoresis and is used to detect monoclonal proteins in the urine. Ideally it should be performed on a 24-hour urine sample (concentrated 50−100 times). Molecules less than 15 kDa are filtered through a glomerular filtration process and are excreted freely into urine. In contrast, only selected molecules with molecular weights between 16 and 69 kDa can be filtered by the kidney and may appear in the urine. Albumin is approximately 67 kDa. Therefore, trace albumin in urine is physiological.

Molecular weight of the protein, concentration of the protein in the blood, charge, and hydrostatic pressure all regulate passage of a protein through the glomerular filtration process.

Proteins that pass through glomerular filtration include albumin, alpha-1-acid glycoprotein (orosomucoid), alpha-1-microglobulin, beta-2-microglobulin, retinol-binding protein, and trace amounts of gamma globulins. However, 90% of these are reabsorbed and only a small amount may be excreted in the urine. Normally, total urinary protein is <150 mg/24 h and consists of mostly albumin and Tamm−Horsfall protein (secreted from the ascending limb of the loop of Henle). The extent of proteinuria can be assessed by quantifying the amount of proteinuria as well as by expressing it as the protein-to-creatinine ratio. Normal protein-to-creatinine ratio is <0.5 in children 6 months to 2 years of age, <0.25 in children above 2 years, and <0.2 in adults.

Proteinuria with minor injury (typically only albumin is lost in urine) can be related to vigorous physical exercise, congestive heart failure, pregnancy, alcohol abuse, or hyperthermia. Overflow proteinuria can be seen in patients with myeloma, or massive hemolysis of crush injury (myoglobin in urine).

In addition, beta-2-microglobulin, eosinophil-derived neurotoxin, and lysozymes can produce bands in urine electrophoresis. Therefore, immunofixation studies are required to document true M proteins and rule out the presence of other proteins in urine electrophoresis.

Proteinuria can be classified as glomerular, tubular, or combined proteinuria. Glomerular proteinuria can be sub-classified as selective glomerular proteinuria (urine has albumin and transferrin bands) or non-selective glomerular proteinuria (urine has all different types of proteins). In glomerular proteinuria the dominant protein present is always albumin. In tubular proteinuria, albumin is a minor component. The presence of alpha-1-microglobulin and beta-2-microglobulin are indicators of tubular damage. Please see Chapter 11 for more detail.

22.4 IMMUNOFIXATION STUDIES

In immunofixation, electrophoresis of one specimen from a patient suspected of monoclonal gammopathy is performed using five separate lanes. Then, each sample is overlaid with different monoclonal antibodies: anti-gamma (to detect gamma heavy chain), anti-mu (to detect mu heavy chain), anti-alpha (to detect alpha heavy chain), anti-kappa (to detect kappa light chain), and anti-lambda (to detect lambda light chain). An antigen–antibody reaction should take place. After washing to remove unbound antibodies, the gel paper is stained, which allows identification of a specific isotope of the monoclonal protein. A normal serum protein electrophoresis does not exclude diagnosis of myeloma because approximately 11% of myeloma patients may have normal serum protein electrophoresis. Therefore, serum and urine immunofixation studies should be performed regardless of serum electrophoresis results if clinical suspicion is high. It is also important to note that an M band or paraprotein in serum protein electrophoresis may not be a true band unless it is identified by using serum or urine immunofixation since these tests are more sensitive than serum protein electrophoresis. In addition, immunofixation techniques can also determine the particular isotype of the monoclonal protein. However, immunofixation techniques cannot estimate the quantity of the M protein. In contrast, serum protein electrophoresis is capable of estimating the concentration of an M protein.

Sometimes in multiple myeloma only free light chains are produced. The concentration of the light chains in serum may be so low that these light chains remain undetected using serum protein electrophoresis and even serum immunofixation. In such cases immunofixation on a 24-hour urine sample is useful. Another available test is detection of serum free light chains by immunoassay, which allows for calculation of the ratio of kappa-to-lambda free light chains. This test is more sensitive than urine immunofixation.

One source of possible error in urine immunofixation studies is the "step ladder" pattern. Here multiple bands are seen in the kappa (more often) or lambda lanes, and are indicative of polyclonal spillage rather than monoclonal spillage into the urine. During urine immunofixation, five or six faint, regular, diffuse bands with hazy background staining between bands may be seen. This is more often seen in the kappa lane than the lambda lane. This is referred to as the step ladder pattern and is a feature of polyclonal hypergammaglobulinemia with spillage into the urine.

CASE REPORT

A 38-year-old male presented with complaints of backache over 2 months (associated with low-grade fever). He also had poor appetite and experienced weight loss. Laboratory investigation revealed anemia with a hemoglobin level of 5.8 g/dL, elevated serum creatinine (3.9 mg/dL), and urea (108 mg/dL). He also showed hypercalcemia with a serum calcium level of 12.8 mg/dL and had a low serum albumin level of 2.7 g/dL. His serum protein electrophoresis showed a discrete suspicious band in the gamma region that was conventionally placed as an M band. However, a serum immunofixation study revealed a sharply localized discrete kappa band corresponding to the suspicious band in the serum protein electrophoresis. Urine Bence–Jones protein was positive and urine protein electrophoresis showed a sharp, densely stained band in the gamma region. Urine immunofixation confirmed this band as the kappa band. These findings confirmed the diagnosis of monoclonal gammopathy involving production of only light chain. Therefore, the patient was suffering from light chain disease multiple myeloma with extramedullary involvement [7].

22.5 CAPILLARY ZONE ELECTROPHORESIS

Capillary zone electrophoresis is an alternative method of performing serum protein electrophoresis. Protein stains are not required, and a point of application is not observed. It is considered to be faster and more sensitive compared to agarose gel electrophoresis where a classical case of monoclonal gammopathy produces a peak, typically in the gamma zone. However, subtle changes in the gamma zones may also represent underlying monoclonal gammopathy. Interpretation can be subjective, and a relatively high percentage of cases may be referred for ancillary studies such as immunofixation, depending on the preference of the pathologist who is interpreting the results. However, disregarding a subtle change in capillary zone electrophoresis may potentially result in missing a case. The capillary electrophoresis pattern of a normal serum is given in Figure 22.1.

22.6 FREE LIGHT CHAIN ASSAY

Patients with monoclonal gammopathy can have negative serum protein electrophoresis and serum immunofixation studies. This may be due to very low levels

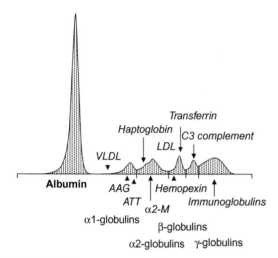

FIGURE 22.1

Capillary electrophoresis of serum proteins from a healthy individual [8]. *VLDL: very low density lipoprotein, AAG: α-1 acid glycoprotein, ATT: α-1 antitrypsin, α2-M: α-2-macroglobulin.©American Association for Clinical Chemistry. Reprinted with permission.*

of paraproteins and light chain gammopathy in which the light chains are rapidly cleared from the serum by the kidneys. Because of this, urine electrophoresis and urine immunofixation are part of the work-up for cases where monoclonal gammopathy is a clinical consideration. Urine electrophoresis and urine immunofixation studies are also performed to document the amount (if any) of potentially nephrotoxic light chains being excreted in the urine in a case of monoclonal gammopathy.

Quantitative serum assays for kappa and lambda free light chain disease have increased the sensitivity of serum testing strategies for identifying monoclonal gammopathies, especially the light chain diseases. Cases that may appear as non-secretory myeloma can actually be cases of light chain myeloma. Free light chain assays allow disease monitoring as well as providing prognostic information for monoclonal gammopathy of undetermined significance (MGUS), smoldering myeloma.

The rapid clearance of light chains by the kidney is reduced in renal failure. Levels may be 20–30 times higher than normal in end-stage renal disease. In addition, the kappa:lambda ratio may be as high as 3:1 in renal failure (normal is 0.26/1.65). Therefore, patients with renal failure may be misdiagnosed as having kappa light chain monoclonal gammopathy. If a patient has lambda light chain monoclonal gammopathy, with the relative increase in kappa light chain in renal failure, the ratio may become normal. Thus, a case of lambda light chain monoclonal gammopathy may be missed.

22.7 PARAPROTEIN INTERFERENCES IN CLINICAL LABORATORY TESTS

Interference of paraprotein can produce both false positive and false negative test results depending on the analyte and the analyzer. However, the magnitude of interference may not correlate with the amount of paraprotein present in serum. The most common interferences include falsely low high density lipoprotein (HDL) cholesterol, falsely high bilirubin, and altered values of inorganic phosphate. Other tests in which altered results can occur include low density lipoprotein (LDL) cholesterol, cholesterol, C reactive protein, creatinine, glucose, urea nitrogen, inorganic calcium, and blood count. Roy reported a case of a 65-year-old man with IgG kappa (3.5 g/dL) multiple myeloma who showed undetectable HDL cholesterol. However, his HDL cholesterol level 2 years prior was 42 mg/dL, indicating that his undetectable HDL cholesterol was due to interference of paraprotein [9].

CASE REPORT

A 51-year-old man diagnosed with IgG lambda multiple myeloma showed an elevated bilirubin level of 19.6 mg/dL without jaundice, but his direct bilirubin level in serum was within normal range. This indicated that his hyperbilirubinemia could be artifactual. The patient had a monoclonal gammopathy of 10 g/dL and his total serum protein was 14.3 g/dL. Therapy with lenalidomide and dexamethasone resulted in a reduction of monoclonal gammopathy, and during the last follow-up by the authors prior to publishing this report, his monoclonal protein was 4.8 g/dL and his total bilirubin was 0.35 g/dL, a value that was within normal limits, indicating that prior bilirubin of 19.6 mg/dL was due to interference of paraprotein in the bilirubin assay [10].

22.8 CEREBROSPINAL FLUID ELECTROPHORESIS

Qualitative assessment of cerebrospinal fluid by electrophoresis (CSF electrophoresis) for oligoclonal bands is an important diagnostic for multiple sclerosis. Multiple sclerosis is an inflammatory demyelinating disease of the central nervous system (CNS). CSF fluid is used in the diagnosis of multiple sclerosis to identify intrathecal IgG synthesis as reflected qualitatively by the presence of oligoclonal bands in CSF electrophoresis and quantitatively by IgG index or IgG synthesis ratio. The immunoglobulin increase in CSF fluid in multiple sclerosis is predominantly IgG, although the synthesis of IgM and IgA may also be increased.

Oligoclonal bands are defined as at least two bands seen in the CSF lane with no corresponding band present in the serum lane. Thus, it is crucial to perform CSF and serum electrophoresis simultaneously. Oligoclonal bands may be found in 90% or more of patients clinically diagnosed with multiple sclerosis. However, oligoclonal bands may be seen in patients with CNS infections such as Lyme disease, and in patients with autoimmune diseases, brain tumors, and lymphoproliferative disorders. Thus, it is important to note that detection of oligoclonal bands is not always due to multiple sclerosis. The characteristic feature of the CSF electrophoresis pattern is the presence of a prealbumin band and a band in the beta-2 region due to desialated transferrin (also known as beta-1-transferrin). These bands are not present in the serum lane.

An abnormality of CSF IgG production can be expressed as a percentage of the total protein, as a percentage of albumin, or by the use of the IgG index. The IgG index is defined in Equation 22.1 (CSF is cerebrospinal fluid):

$$\text{IgG index} = \frac{[\text{CSF IgG/CSF albumin}]}{\text{Serum IgG/Serum albumin}} \qquad (22.1)$$

A normal value for an IgG index in adults is 0.23−0.64. However, it is important to correlate clinical findings with IgG index and the findings of the electrophoresis. In multiple sclerosis, CSF is grossly normal and the pressure is also normal. The total leukocyte count is normal in the majority of patients. If the white blood cell count is elevated, it rarely exceeds 50 cells/μL. Lymphocytes are the predominant cells found, and the total protein concentration of CSF is also within normal range. If there is a systemic immune reaction or a monoclonal gammopathy is also present, then bands should be present in both the serum and CSF lanes and should correspond with each other. These are not oligoclonal bands. Although detection of oligoclonal bands is one of the major reasons to order CSF electrophoresis, other characteristic features of CSF electrophoresis include:

- Prealbumin band (transthyretin band) is seen anodal (positive electrode) to albumin (a band above the albumin band). This band is not present in serum electrophoresis.
- Albumin band in CSF electrophoresis is slightly anodal to the albumin band present in the corresponding serum electrophoresis.
- The alpha-2 band is significantly denser in the serum protein electrophoresis than in CSF electrophoresis because alpha-2-microglobulin and haptoglobin do not cross the blood−brain barrier due to relatively large molecular weight. However, if the brain barrier is damaged or if the CSF is collected from a traumatic tap, this band may be denser in CSF electrophoresis.
- In contrast to serum protein electrophoresis where two bands are present in the beta region, CSF electrophoresis usually shows three bands in the beta region, including the beta-1 band, another band for C3, and the beta-2 band due to the presence of desialated transferrin (also known as tau protein).

Another reason for ordering CSF electrophoresis is to establish or rule out leakage of CSF through the nose. The presence of prealbumin and tau protein helps identify the sample as CSF leaking through the nose. In addition, immunofixation studies can also be performed with antibodies directed against the tau protein.

KEY POINTS

- Serum protein electrophoresis, urine electrophoresis, serum immunofixation, and urine immunofixation are all performed primarily to investigate suspicion of monoclonal gammopathy in a patient. Monoclonal gammopathy is present in a patient when a monoclonal protein (also called a paraprotein or M protein) is identified in a patient's serum, urine, or both. The paraprotein can be an intact immunoglobulin, only light chains (light chain myeloma, light chain deposition disease, or amyloid light chain amyloidosis), or rarely, as heavy chains (heavy chain disease).

- Agarose gel electrophoresis and capillary electrophoresis are two principal methods employed in screening for paraproteins. Both methods are applicable for both serum and urine specimens. Paraproteins are usually seen in the gamma region of the electrophoresis, but also may be present in the beta or (rarely) alpha-2 regions.

- Once a paraprotein is detected, confirmation and isotyping of the paraprotein are essential. This is usually achieved by immunofixation. In about 5% of cases, two paraproteins may be detected. This is referred to as biclonal gammopathy. A patient may also have non-secretory myeloma, as in the case of a plasma cell neoplasm in which the clonal cells are neither producing nor secreting M proteins. The most commonly observed paraprotein is IgG, followed by IgA, light chain, and, rarely, IgD. A normal serum protein electrophoresis does not exclude diagnosis of myeloma because approximately 11% of myeloma patients can have normal serum protein electrophoresis. Therefore, serum and urine immunofixation studies should be performed regardless of serum electrophoresis results if clinical suspicion is high.

- Other components of serum protein electrophoresis include albumin, alpha-1 globulins (alpha-1 zone), alpha-2 globulins (alpha-2 zone), beta globulins (beta zone often splits into beta-1 and beta-2 band), and gamma globulins (gamma zone).

- Reduced intensity of the albumin band is observed in inflammation, liver dysfunction, uremia, nephrotic syndrome, and other conditions that lead to hypoalbuminemia. A smear observed in front of the albumin band may be due to hyperbilirubinemia or to the presence of certain drugs. A band in front of the albumin band may be due to prealbumin (a carrier for thyroxine and vitamin A), which is commonly seen in cerebrospinal fluid specimens or serum specimens in patients with malnutrition. Two (rather than one) albumin bands may represent bisalbuminemia. This is a familial abnormality with no clinical significance.

- The alpha-1 band mostly consists of alpha-1-antitrypsin (AT, 90%), alpha-1-chymotrypsin, and thyroid-binding globulin. Alpha-1-antitrypsin is an acute phase reactant and its concentration is increased in inflammation and other conditions. The alpha-1-antitrypsin band is decreased in patients with alpha-1-

antitrypsin deficiency or decreased production of globulin in patients with severe liver disease. At the leading edge of this band, a haze due to high density lipoprotein (HDL) may be observed. The alpha-2 band consists of alpha-2-macroglobulin, haptoglobin, and ceruloplasmin. Because both haptoglobin and ceruloplasmin are acute phase reactants, this band is increased in inflammatory states. Alpha-2-macroglobulin is increased in nephrotic syndrome and cirrhosis of the liver.

- The beta zone may consist of two bands, beta-1 and beta-2. Beta-1 is mostly composed of transferrin and low density lipoprotein. An increased beta-1 band is observed in iron deficiency anemia due to an increased level of free transferrin. This band may also be elevated in pregnant women. Very low density lipoprotein usually appears in the pre-beta zone. The beta-2 band is mostly composed of complement proteins. If two bands are observed in the beta-2 region, it implies either electrophoresis of plasma specimen (fibrinogen band) instead of serum specimen, or IgA paraprotein.

- There are certain situations where a band may be apparent, but in reality it is not a monoclonal band. For example, fibrinogen is seen as a discrete band between the beta and gamma regions when electrophoresis is performed on plasma instead of serum specimens. If the electrophoresis is repeated after the addition of thrombin, this band should disappear. In addition, an immunofixation study should be negative. Intravascular hemolysis results in the release of free hemoglobin in circulation (which binds to haptoglobin). The hemoglobin—haptoglobin complex may appear as a large band in the alpha-2 area. Serum immunofixation studies should be negative in such cases.

- In patients with iron deficiency anemia, concentrations of transferrin may be high, which can result in a band in the beta region. Again, immunofixation should be negative.

- Patients with nephrotic syndrome usually show low albumin and total protein, but this condition can also produce increased alpha-2 and beta fractions. Bands in either of these regions may mimic a monoclonal band.

- Hypogammaglobulinemia can be congenital or acquired. Amongst the acquired causes are multiple myeloma and primary amyloidosis. Panhypogammaglobulinemia can occur in about 10% of cases of multiple myeloma. Most of these patients have a Bence—Jones protein in their urine, but lack intact immunoglobulins in their serum.

- Common features of serum protein electrophoresis in various disease states other than monoclonal gammopathy include:
  - Inflammation: Increased intensity of alpha-1 and alpha-2 with a sharp leading edge of alpha-1 may be observed, but with chronic inflammation the albumin band may be decreased with increased gamma zone due to polyclonal gammopathy.

- Nephrotic syndrome: The albumin band is decreased due to hypoalbuminemia. In addition, the alpha-2 band may be more distinct.
- Cirrhosis or chronic liver disease: A low albumin band due to significant hypoalbuminemia with a prominent beta-2 band and beta—gamma bridging are characteristic features of liver cirrhosis or chronic liver disease. In addition, polyclonal hypergammaglobulinemia is observed.

- Proteinuria can be classified as glomerular, tubular, or combined proteinuria. Glomerular proteinuria can be sub-classified as selective glomerular proteinuria (urine has albumin and transferrin bands) or non-selective glomerular proteinuria (urine has all different types of proteins). In glomerular proteinuria, the dominant protein is always albumin. In tubular proteinuria, albumin is a minor component. The presence of alpha-1-microglobulin and beta-2-microglobulin are indicators of tubular damage.

- One source of possible error in urine immunofixation studies is the "step ladder" pattern. Here, multiple bands are seen in the kappa (more often) or lambda lanes, and are indicative of polyclonal spillage rather than monoclonal spillage into the urine. During urine immunofixation, five or six faint, regular, diffuse bands with hazy background staining between bands may be seen. This is more often seen in the kappa lane than the lambda lane. This is referred to as the step ladder pattern and is a feature of polyclonal hypergammaglobulinemia with spillage into the urine.

- The rapid clearance of light chains by the kidney is reduced in renal failure. Levels may be 20–30 times higher than normal in end-stage renal disease. In addition, the kappa:lambda ratio may be as high as 3:1 in renal failure (normal is 0.26/1.65). Therefore, patients with renal failure may be misdiagnosed as having kappa light chain monoclonal gammopathy. If a patient has lambda light chain monoclonal gammopathy, with the relative increase in kappa light chain in renal failure, the ratio may become normal. Thus, a case of lambda light chain monoclonal gammopathy may be missed.

- Characteristic features of cerebrospinal fluid (CSF) electrophoresis include:
  - Prealbumin band (transthyretin band) is seen anodal (positive electrode) to albumin (a band above the albumin band). This band is not present in serum electrophoresis.
  - Albumin band in CSF electrophoresis is slightly anodal to the albumin band present in the corresponding serum electrophoresis.
  - The alpha-2 band is significantly denser in the serum protein electrophoresis than CSF electrophoresis because alpha-2-microglobulin and haptoglobin do not cross the blood—brain barrier due to relatively large molecular weight. However, if the brain barrier is damaged or if the CSF is collected from a traumatic tap, this band may be denser in CSF electrophoresis.
  - In contrast to serum protein electrophoresis where two bands are present in the beta region, CSF electrophoresis usually shows three bands in the beta region, including the beta-1 band, another band for C3, and a beta-2 band due to the presence of desialated transferrin (also known as tau protein).

- CSF is used in the diagnosis of multiple sclerosis to identify intrathecal IgG synthesis as reflected qualitatively by the presence of oligoclonal bands in CSF electrophoresis and quantitatively by IgG index or IgG synthesis ratio.
- Oligoclonal bands are defined as at least two bands seen in the CSF lane with no corresponding band present in the serum lane. Thus, it is crucial to perform CSF and serum electrophoresis simultaneously.
- Another reason for ordering CSF electrophoresis is to establish or rule out leakage of CSF through the nose. The presence of prealbumin and tau protein helps to identify the sample as CSF leaking through the nose. In addition, immunofixation studies can also be performed with antibodies directed against the tau protein.

REFERENCES

[1] Agarwal A, Ghobrial IM. Monoclonal gammopathy of undetermined significance and smoldering multiple myeloma: a review of current understanding of epidemiology, biology, risk stratification, and management of myeloma precursor disease. Clin Cancer Res 2013;19:985−94.

[2] Jones CI, Zabolotskaya MV, King AJ, Stewart HJ, et al. Identification of circulating micro RNAs as diagnostic biomarkers for use in multiple myeloma. Br J Cancer 2012;107:1987−96.

[3] Gregersen H, Mellemkjaer L, Ibsen JS, Dahlerup JF, et al. The impact of M component type and immunoglobulin concentration on the risk of malignant transformation in patients with monoclonal gammopathy of undetermined significance. Haematologica 2001;86:1172−9.

[4] Vodopick H, Chaskes SJ, Solomon A, Stewart JA. Transient monoclonal gammopathy associated with cytomegalovirus infection. Blood 1974;44:189−95.

[5] Gerritsen E, Vossen J, van Tol M, Jol-van der Zijde C, et al. Monoclonal gammopathies in children. J Clin Immunol 1999;9:296−305.

[6] Gras J, Padros R, Marti I, Gomez-Acha JA. Pseudo-analbuminemia due to the presence of a slow albumin variant moving into the alpha 1 zone. Clin Chim Acta 1980;104:125−8.

[7] Mahato M, Mohapatra S, Sumitra G, Kaushik S, et al. A case of light chain deposition disease (LCDD) in a young patient. Ind J Clin Biochem 2011;26:420−2.

[8] Gay-Bellile C, Bengoufa D, Houze P, Le Carrer H, et al. Automated multi-capillary electrophoresis for analysis of human serum proteins. Clin Chem 2003;49(11):1909−15.

[9] Roy V. Artifactual laboratory abnormalities in patients with paraproteinemia. South Med J 2009;102:167−70.

[10] Cascavilla N, Falcone A, Sanpaolo G, D'Arena G. Increased serum bilirubin level without jaundice in patients with monoclonal gammopathy. Leuk Lymphoma 2009;50:1392−4.

CHAPTER 23

Human Immunodeficiency Virus (HIV) and Hepatitis Testing

23.1 HUMAN IMMUNODEFICIENCY VIRUS (HIV) TESTING

Human immunodeficiency virus (HIV) is a slowly replicating retrovirus that causes acquired immunodeficiency syndrome (AIDS). HIV infection is sexually transmitted, but is also transmitted due to sharing infected needles, blood transfer, and from mother to newborn. In body fluids HIV is present as free viral particles and also as a virus within infected immune cells. HIV infects vital cells involved in human immune functions such as helper T cells, especially CD4+ T cells, macrophages, and dendritic cells. HIV is transmitted as an enveloped RNA virus, and, upon entry into target cells, the viral RNA genome is converted into double-stranded DNA by a virally encoded reverse transcriptase that is transported along with the viral genome in virus particles during infection.

Two major types of HIV have been characterized: HIV-1 and HIV-2. HIV-1 was the first discovered and is the cause of the majority of HIV infections worldwide. HIV-2 has lower infectivity than HIV-1 and is largely confined to West Africa. Although HIV-2 is endemic in West Africa, infection in North America has also recently been described. HIV-2 infection progresses to a symptomatic disease at a much slower rate than HIV-1. While treatment of HIV-1 infection is well characterized, there is far less experience among physicians in treating persons infected with HIV-2, and there are also controversies as to when to initiate therapy, the goal being to reach immune restoration while minimizing drug toxicity [1]. Several groups are classified under HIV-1 (Table 23.1). The HIV-1 M group is the main type of HIV seen in clinical practice; the M type can be divided into several subtypes [2].

407

A. Dasgupta and A. Wahed: Clinical Chemistry, Immunology and Laboratory Quality Control
DOI: http://dx.doi.org/10.1016/B978-0-12-407821-5.00023-1

Table 23.1 Various HIV-1 Viral Types and Groups

Type	Group	Comments
HIV-1		Related to viruses found in chimpanzees and gorillas.
HIV-1	M	M denotes "major." Most common HIV. Responsible for AIDS pandemic. sub-divided into A–D, F–H, and J–K. Sub-type B remains the predominant viral infection throughout North America and many developed countries, but sub-type C has the highest prevalence worldwide among infected individuals.
HIV-1	N	N denotes "non-M, non-O." Seen in Cameroon.
HIV-1	O	O denotes "outlier." Common in Cameroon, but usually seen outside West–Central Africa.
HIV-1	P	P denotes "pending identification of further human cases." Virus was isolated from Cameroonian woman residing in France.

23.2 WINDOW PERIOD IN HIV INFECTION

The diagnosis of HIV infection is most commonly achieved by detecting antibodies against HIV in body fluid using a screening test, followed by a confirmatory test. In 1985, the U.S. Food and Drug Administration (FDA) approved the first enzyme-linked immunosorbent assay (ELISA) to detect the presence of antibodies against HIV in serum. The major purpose of HIV testing is:

- Diagnosis of HIV in individuals suspected of infection.
- Testing for an individual who wishes to know if they are infected.
- Since 1985 HIV screening of blood products has been used to provide protection to countless individuals from transmission of HIV through blood or blood products.
- Testing of potential donors before organ or tissue transplantation.
- Epidemiological surveillance so that health care officers can determine specific needs in a community.

After HIV infection, viral RNA can be detected within 10 to 12 days, and viral p24 antigen can be detected afterward. The time needed before appearance of HIV-specific antibodies in serum is known as the serological "window period." In this period only viral RNA and possibly p24 antigen can be detected, but if screening is performed using a method that detects HIV-specific antibodies, the test could be negative. In general, IgM may be the first antibody against HIV that can be detected in circulation, followed by IgG antibodies (which appear approximately 3–4 weeks after infection). Within 1–2 months HIV antibodies are usually present in almost all infected individuals, although for a few individuals it may take up to 6 months for antibodies to appear in circulation. In order to reduce the spread of HIV infection, The Centers for Disease Control and Prevention (CDC) recommends HIV testing as a part of routine health care to all patients living in an area with a high prevalence of HIV infection (>1%), to high-risk patients who reside in low HIV-infected areas, to pregnant women, as well as to anyone requesting HIV testing. Patients with HIV infection benefit from early detection because infections at late stages may have advanced immune suppression and may not get the full benefits of antiretroviral therapy [3]. In 2006 the CDC recommended expanded HIV screening in emergency departments.

False negative HIV-1 antibody testing is most commonly attributed to a "window period" prior to the development of HIV-1-specific antibodies. In several case reports, patients treated with highly active antiretroviral therapy (HAART) very early in the course of disease may not have HIV-1 antibody, possibly due to drug-induced viral load suppression. In very rare cases, a patient infected with HIV, as evidenced by high viral load, may not have HIV-specific antibodies for a long time (or persistent lack of antibodies), but these patients present with severe immunodeficiency [4].

Safety of blood products used for transfusion is important, and in the U.S. blood and plasma are tested for antibodies to HIV-1 and HIV-2. However, despite a dramatic reduction in transmitting HIV through transfusion, 1 in 450,000 to 1 in 600,000 U.S. blood donations may transmit HIV, and nearly all cases of transfusion-associated HIV infection are caused by donations made during the "window period" (i.e. prior to seroconversion) [5]. The introduction of nucleic acid amplification technology (NAT) for screening individual donations has remarkably improved the safety of blood products. Testing for HIV can be broadly divided into screening tests and confirmatory tests. Screening tests include standard testing, rapid HIV testing, and combination HIV antibody and antigen testing. Confirmatory testing is performed by Western blot and recombinant immunoblot or line immunoassay (LIA).

23.3 STANDARD HIV TESTING

Standard screening is performed with various immunoassays. The test is based on the detection of IgG antibody against HIV-1 antigens in the serum. HIV antigens include p24, gp 120, and gp 41, and antibodies to gp 41 and p24 are the first detectable serologic markers following HIV infection. IgG antibodies appear as early as 3 weeks, but most likely within 12 weeks following HIV infection in the majority of patients, and they generally persist for life. Assays for IgM antibodies are not used because they are relatively insensitive. HIV viruses are categorized into several groups (Table 23.1); M is considered to be the pandemic strain and accounts for the vast majority of infections. The two important issues regarding HIV screen tests are the ability of the test to detect non-M strains and the timing of the test post-exposure. If the patient has not yet seroconverted, the antibody is absent and the individual may have a false negative test. There are also rare patients with HIV infection who become seronegative, although these patients show earlier seropositive results after exposure to HIV. Some other causes of false negative results include:

- Fulminant HIV infection.
- Immunosuppression or immune dysfunction.
- Delay in seroconversion following early initiation of antiretroviral therapy.

The chance of a false positive serologic test for HIV is extremely low. However, false positive test results for HIV infection have been documented in individuals who have received HIV vaccines in vaccine trials. Some of the individuals who became HIV-positive on screening tests also showed positive Western blot results. Testing of viral RNA is an approach that should be used to resolve such issues. Enzyme immunoassays (EIA) are widely used in clinical laboratories for HIV testings, and various versions of EIA are used depending on the assay platform. Since its introduction in 1985, the ELISA method has evolved, and now third-generation ELISA assays (sandwich format) can detect IgG and IgM antibodies and antibodies to all of the M subtypes (as well as N and O groups). Antibody testing methods capable of detecting O groups are important for screening of blood products. More recently, fourth-generation assays have been introduced that simultaneously detect HIV antibodies and p24 antigen, thus reducing the window period to only 13–15 days. These assays are often referred to as HIV antigen/antibody combo assays. The first such assay was the Architect HIV combo assay, approved by the FDA in 2010. The sensitivity and specificity of these assays can reach over 99%. Enzyme-linked immunofluorescent assays (ELFA) are modified versions of the ELISA technique that utilize solid phases with a greater surface contact area in order to reduce incubation time. These assays use enzyme and fluorescent substances that are converted into fluorescent products by the action of enzymes and can be measured with fluorescent detectors. Assays based on this method can be adopted in automated analyzers. Chemiluminescence methods where chemiluminescent compounds are used to label antigen or antibody can also be adopted in automated high-throughput analyzers. Examples of some automated HIV testings are summarized in Table 23.2. Although various standard assays for HIV testing have excellent sensitivity and specificity, false positive test results may still be observed. Vardinon et al. studied 520 patients undergoing hemodialysis and observed 23 (4.4%) positive test results with EIA. However, results were indeterminant using confirmation Western blot. Five years of follow-up showed

Table 23.2 Examples of Automated HIV Tests

Analyzer	Diagnostic Company	Comment
Architect	Abbott Laboratories	Fourth-generation chemiluminescent sandwich assay. Chemiluminescence microparticle immunoassay (CMIA; detects antibodies and antigens).
Siemens	Centaur XP and CP	Third-generation chemiluminescent sandwich assay (detects antibodies). OR Fourth-generation chemiluminescent sandwich assay (detects antibodies and antigens).
Roche	Cobas	Fourth-generation chemiluminescent sandwich assay. Electrochemiluminescence (ECLIA; detects antibodies and antigens).
Diasorin	Liaison XL	Fourth-generation chemiluminescent sandwich assay (detects antibodies and antigens).
Izasa	Access 2 Unicel Dxl	Fourth-generation chemiluminescent sandwich assay (detects antibodies and antigens).

no seroconversion, indicating positive test results using EIA as false positive test results [6]. Therefore, patients undergoing hemodialysis may show false positive test results with EIA, indicating the need for confirmatory testing.

23.4 RAPID HIV ANTIBODY TESTING

Rapid HIV antibody tests can provide results in less than 30 min and can be adopted in point of care settings. These tests are also based on either immunochromatography (lateral flow) or immunoconcentration (flow through) techniques. Since 2002 the FDA has approved six rapid HIV tests (Table 23.3). Most of these tests can utilize whole blood, thus avoiding the need to centrifuge specimens to obtain serum. However, OraQuick Advanced HIV1/2 assay can use whole blood, serum, or oral fluid. The FDA has recently approved the OraQuick for Home HIV test, a rapid home-use HIV kit that uses oral fluid; test results can be obtained in 20−40 min. Most rapid HIV tests are based on the principles of enzyme immunoassays that are utilized in clinical laboratories and use automated or semi-automated analyzers. Most tests detect HIV antibodies by incorporating HIV envelop-region antigens in the test methodology. However, both false positive and false negative results can occur with rapid HIV tests, and it is important to confirm initial findings with a laboratory-based HIV assay. These tests are used for rapid screening only. Delaney et al. evaluated six rapid HIV antibody tests and observed sensitivities over 95% and specificity over 99% for all rapid tests.

Table 23.3 Examples of Available Rapid HIV Tests

Rapid Test	Specimen	Methodology
OraQuick Advance HIV-1/2	Whole blood, oral fluid, plasma	Lateral flow
MultiSpot HIV-1/2	serum, plasma	Flow through
Uni-Gold Recombigen HIV-1	Whole blood, serum, plasma	Lateral flow
Reveal G3 HIV-1	Serum, plasma	Flow through
Clearview Complete HIV-1/2	Whole blood, serum, plasma	Lateral flow
Clearview STAT-PAK HIV-1/2	Whole blood, serum, plasma	Lateral flow

However, false negative and false positive results were observed in all rapid HIV assays [7]. Facente et al. observed that false positive rates for oral fluid HIV tests increase near the expiration date of the kit [8].

23.5 CONFIRMATORY HIV TEST

Commonly used confirmatory tests for diagnosis of HIV infection include Western blot and line immune assay (LIA). A confirmatory test is conducted if a screening assay is positive. In Western blot, HIV denatured proteins are blotted on strips of a nitrocellulose membrane that is then incubated with serum obtained from the patient. If the serum contains antibodies against these viral proteins (most commercial assays use antigens from both HIV-1 and HIV-2), they will bind to these proteins; the antigen—antibody reaction is visualized using an enzyme-labeled secondary antibody and a matching substrate. A colorimetric reaction leads to the formation of bands that represent the antigen—antibody complex (indicating a positive result). The results of Western blot can be positive, negative, or indeterminant. The disadvantage of Western blot is high cost and subjective interpretation. However, LIA methods based on recombinant proteins and/or synthetic peptides capable of detecting antibodies to HIV-1 and HIV-2 are more specific and produce fewer indeterminant results than Western blot.

CASE REPORT

A 49-year-old man visited an outpatient clinic with fever, malaise, generalized rash, anal itching, and rectal discharge after repeated unprotected anal and oral sex with an anonymous man a week before presenting to the clinic. In the previous 15 months he had repeated protected and unprotected anal and oral sex with 15 different male partners. His last HIV test (performed 1 year prior) was negative. He was previously treated for syphilis infection. Upon arrival at the clinic his rapid HIV test was negative. Suspecting that the patient had a high risk for HIV infection, a p24 antigen test was conducted. It was positive, indicating a highly infectious stage of HIV infection, but the rapid HIV test was negative due to the window period when antibodies specific to HIV are absent. An immunoblot assay confirmed HIV infection 14 days later [9].

23.6 HIV VIRAL LOAD TEST AND RELATED ASSAYS

HIV viral load test determines the number of copies of HIV RNA present per milliliter of serum or plasma, and is expressed as copies/mL or in log scale. HIV viral load test indicates viral replication and is often conducted to monitor progress of antiretroviral therapy. In addition, CD4 lymphocyte counts are also measured to evaluate the immune system. The success of antiretroviral therapy is measured both clinically and by suppression of viremia below 50 copies/mL in two successive measurements (this cut-off value has been contested). Nucleic acid tests (NAT) are commercially available tests that can identify HIV nucleic acid (either RNA or proviral DNA). These tests are based on the principles of polymerase chain reaction (PCR), real-time PCR, nucleic acid sequence-based amplification, or ligase chain reaction. NAT assays are useful in special situations such as in the window period of infection when the antibody against HIV is absent in serum, and in newborns of HIV-infected mothers where maternal antibodies against HIV are present in the newborn's serum. Amplification of proviral DNA allows detection of immune cells that harbor quiescent provirus as well as cells infected with actively replicating HIV. This test is useful in diagnosis of HIV infection in infants and children up to 18 months of age born to HIV-infected mothers [10]. However, false negative DNA PCR test results can occur in children treated with antiretroviral therapy and may lead to inappropriate discontinuation of antiretroviral therapy [11].

23.7 INTRODUCTION TO HEPATITIS TESTING

Hepatitis infection is a worldwide problem. Hepatitis A is a problem in many developing countries because the hepatitis A virus can spread from contaminated water (fecal−oral route). However, hepatitis A testing is straightforward, where IgM anti-hepatitis A antibody denotes recent infection and IgG anti-hepatitis A antibody appears in the convalescent phase of acute hepatitis. Hepatitis E virus is also an enterically transmitted virus. Hepatitis E can also be transmitted by blood transfusion, particularly in endemic areas. Chronic hepatitis does not develop after acute hepatitis E infection, except in the transplant setting, and possibly in other settings of immunosuppression. Fulminant hepatitis can occur, resulting in an overall case fatality rate of 0.5–3%. For reasons as yet unclear, the mortality rate in pregnant women can be as high as 15−25%, especially in the third trimester. The diagnosis of hepatitis E is based upon the detection of hepatitis E virus in serum or stool by PCR or by the detection of IgM antibodies against hepatitis E. Antibody tests against hepatitis E alone are less than ideal since they have been associated with frequent false positive and negative results. The hepatitis D virus (also called the delta virus) is a defective

Table 23.4 Characteristics of Various Hepatitis Viruses

Hepatitis Virus	Type	Transmission	Comment
Hepatitis A	RNA virus	Fecal–oral	Low mortality, does not cause chronic liver disease.
Hepatitis B	DNA virus	Parenteral, vertical, sexual transmission	10% fail to clear virus and may have chronic liver disease/ hepatocellular carcinoma.
Hepatitis C	RNA virus	Parenteral, vertical, sexual transmission	50–70% fail to clear virus and may have chronic liver disease/ hepatocellular carcinoma.
Hepatitis D	RNA virus	Parenteral, vertical sexual transmission	Can only infect patients with hepatitis B infection.
Hepatitis E	RNA virus	Fecal–oral	Low mortality (1%) except pregnant women (10–20%), does not cause chronic liver disease.

pathogen that requires the presence of the hepatitis B virus for infection. Hepatitis D can elicit a specific immune response in the infected host, consisting of antibodies of the IgM and IgG classes (anti-hepatitis D antibodies). In summary, various tests for laboratory diagnosis of hepatitis include:

- Liver function tests, including bilirubin, aspartate aminotransferase (AST), alanine aminotransferase (ALT), alkaline phosphatase (ALP), bilirubin, albumin, and prothrombin time.
- Blood test to identify IgM antibody against hepatitis A.
- For hepatitis B, testing of hepatitis B surface antigen (HBsAg), hepatitis e antigen (HBeAg), as well as c, e, and s antibodies and viral load.
- For hepatitis C, testing of antibodies and hepatitis C viral load.
- For hepatitis D, IgG and IgM antibodies against hepatitis D.
- For hepatitis E, IgM and PCR.

Characteristics of various hepatitis viruses are summarized in Table 23.4. Because hepatitis B and C infections are the most important components of hepatitis testing, the following sections detail laboratory tests for hepatitis B and C.

23.8 TESTING FOR HEPATITIS B

An estimated 350 million people worldwide are chronically infected with hepatitis B, resulting in an estimated 600,000 deaths per year from cirrhosis, liver failure, and liver carcinoma. In the U.S. an estimated 0.8 to 1.4 million people suffer from chronic hepatitis B infection, but with the introduction of

Table 23.5 Hepatitis B Testing

Test	Comments
HBsAg	First detectable agent in acute infection.
HBcAg	Not tested as it is not detectable in blood.
HBeAg	Indicates virus is replicating and patient is highly infectious.
Anti-HBc	First antibody to appear in blood. This test is positive when other tests for hepatitis B are negative during the window period (HBsAg is negative and anti-HBs not yet detectable).
Anti-HBe	Indicates virus is not replicating.
Anti-HBs	Patient has achieved immunity.

the hepatitis B vaccine in 1991, new cases of hepatitis B infection number only 1.6 out of 1,000,000 people. Hepatitis B is a small diameter (42 nm), incompletely double-stranded, DNA hepadnavirus with eight distinguishable genotypes (A through H) [12].

Serologic markers available for hepatitis B infection are HBsAg (hepatitis B surface antigen), HBeAg (hepatitis B e antigen), anti-HBc (antibody against hepatitis B core antigen, both IgG and IgM), anti-HBs (antibody against hepatitis B surface antigen), anti-HBe (antibody against hepatitis B e antigen), and testing for viral DNA (Table 23.5). HBsAg is the first marker to be positive after exposure to hepatitis B virus. It can be detected even before the onset of symptoms. Most patients can clear the virus, and HBsAg typically becomes undetectable within 4−6 months. Persistence of HBsAg for more than 6 months implies chronic infection. The disappearance of HBsAg is followed by the presence of anti-HBs. During the window period (after the disappearance of HBsAg and before the appearance of anti-HBs), evidence of infection is documented by the presence of anti-HBc (IgM). The co-existence of HBsAg and anti-HBs has been documented in approximately 24% of HBsAg-positive individuals. It is thought that the antibodies fail to neutralize the virus particles. These individuals should be considered carriers of hepatitis B virus. There exists a subset of patients who have undetectable HBsAg but are positive for hepatitis B viral DNA. Most of these patients have very low viral load with undetectable levels of HBsAg. Uncommon situations are infection with hepatitis B variants that decrease HBsAg production, or mutant strains that have altered epitopes normally used for detection of HBsAg. The timeframe for release of various virological and serological markers in acute hepatitis B infection with recovery is presented in Figure 23.1.

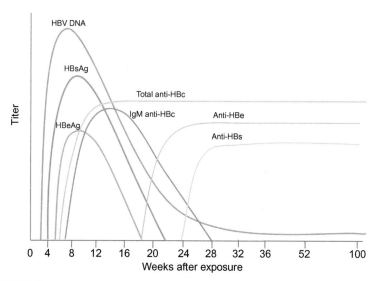

FIGURE 23.1

Virological and serological response to acute hepatitis B infection with recovery. This figure is reproduced in color in the color plate section. *(Courtesy of Andres Quesda, M.D., Department of Pathology and Laboratory Medicine, University of Texas, Houston Medical School.)*

Individuals with recent infection will develop anti-HBc IgM antibodies. Individuals with chronic infection and individuals who have recovered from an infective episode will develop anti-HBc IgG antibodies. However, anti-HBc IgM antibodies can remain positive for up to 2 years after an acute infection. Levels can also increase and be detected during exacerbations of chronic hepatitis B. This can lead to diagnostic confusion.

In certain individuals, isolated positive anti-HBc antibodies may be detected. The clinical significance of this finding is unclear. Some of these individuals have been found (by PCR) to have hepatitis B DNA. Transmission of hepatitis B has been reported from blood and organ donors who have isolated positive anti-HBc antibodies. On the other hand, there are reports that a certain percentage of these individuals are false positive. The presence of HBeAg usually indicates that the hepatitis B is replicating and the patient is infectious. Seroconversion to anti-HBe typically indicates that the virus is no longer replicating. This is associated with a decrease in serum hepatitis B DNA and clinical remission. In some patients seroconversion is still associated with active liver disease. This may be due to low levels of wild-type HBV or HBV variants that prevent or decrease HBeAg production.

Table 23.6 Interpretation of Hepatitis B Serology

Test	Result	Interpretation
HBsAg Anti-HBc Anti-HBs	Negative Negative Negative	Susceptible/no infection.
HBsAg Anti-HBc Anti-HBs	Negative Positive Positive	Immunity after natural infection.
HBsAg Anti-HBc Anti-HBs	Negative Negative Positive	Immunity due to hepatitis B vaccine.
HBsAg Anti-HBc IgM anti-HBc Anti-HBs	Positive Positive Positive Negative	Acute infection.
HBsAg Anti-HBc IgM anti-HBc Anti-HBs	Positive Positive Negative Negative	Chronically infected.
HBsAg Anti-HBc Anti-HBs	Negative Positive Negative	Unclear interpretation, most likely resolved infection. Other possibilities include false positive anti-HBc, low level chronic infection, or resolving acute infection.

Testing for HBsAg is most common for investigating a suspected hepatitis B infection, but this surface antigen is present for a short time after hepatitis B vaccination because hepatitis B vaccines contain the surface antigen. However, such a positive test is unlikely approximately 14 days after vaccination; a weakly positive test result may persist in a few individuals beyond 2 weeks [13]. Hepatitis B screening is also important in blood donors, and serological testing for HBsAg and anti-HBc are standard screening techniques. However, more recently, nucleic acid-based testing (NAT) for hepatitis viral DNA has been implemented to avoid transmission-related infection in order to identify the HBsAg-negative window period in a blood donor who may have early acute infection [14]. All major diagnostic companies market immunoassays for testing serological markers of hepatitis B. The immunoassays typically use the automated analyzers sold by these companies. Interpretation of hepatitis B serology is summarized in Table 23.6.

CASE REPORT

A 56-year-old male suffering from acute myeloid leukemia received 13 units of red cell concentrate and 21 adult doses of platelet during a 2-month period from May to July 1999. Routine follow-up of the patient showed an abnormal liver function test and he was diagnosed with acute hepatitis B infection. However, his blood prior to transplant was negative for any serological marker of hepatitis B. The archived sample of the donated blood in May 1999 showed negative HBsAg and anti-HBc. In addition, PCR analysis of viral DNA was also negative. A follow-up blood specimen was collected from the original donor in February 2000 and the specimen showed negative HBsAg but was positive for anti-HBc and anti-hepatitis B surface antigen, indicating that the donor suffered from a hepatitis B infection between his blood donation and recent investigation of his hepatitis status. The authors speculated that there were insufficient viral copies in the donated blood and that the level was below the detection limit of the assay. In addition, because the recipient received red blood cells, the possible titer of virus was high enough to infect him. The case indicated that the goal of "zero risk" may not be achievable even with use of anti-HBc screening of possible blood donors [15].

23.9 TESTING FOR HEPATITIS C

Hepatitis C virus (HCV) is a single-stranded RNA virus. An estimated 170 million people worldwide are chronically infected with hepatitis C. HCV is categorized into nine genetically distinct genotypes. Unlike hepatitis B, no vaccine is available for hepatitis C. Most patients with hepatitis C infection (60−70%) may be asymptomatic, but chronic HCV infection can lead to liver cirrhosis in some patients. Persons with HCV infection who drink alcohol on a regular basis have a higher risk of liver diseases, including fibrosis and cirrhosis [16]. The diagnostics tests available for HCV can be classified into four broad categories:

- Immunoassays for detecting anti-HCV IgG antibodies.
- Recombinant immunoblot assays for detecting anti-HCV IgG antibodies: these tests are more specific than immunoassays for detecting antibodies against HCV.
- Rapid assays for detecting anti-HCV IgG antibodies: these assays may be less specific than laboratory-based immunoassays.
- Polymerase chain reaction or related techniques: the gold standard for detection of viral RNA in HCV testing. These tests are in general called nucleic acid testing (NAT).

Immunoassays are commonly used screening assays for detecting anti-HCV antibodies. The latest third-generation assays have better sensitivity and specificity compared to first- and second-generation assays. In addition, the mean time to detect seroconversion is shortened by 2−3 weeks, and now third-generation assays are capable of detecting HCV infection as early as 10 weeks after exposure; these assays have over 99% specificity. Nevertheless, even third-generation immunoassays can produce false negative results in patients undergoing hemodialysis and in immunocompromised patients. False positive results can also occur. In the case of false positive tests, recombinant immunoblot assays and strip immunoblot assays can be used to detect antibodies against HCV. These tests are more specific than immunoassays. For establishing the diagnosis of past infection, recombinant immunoblot assay (RIBA) can also be used. If the anti-HCV test yields a false positive, then the RIBA should be negative. If it is a case of past infection, an RIBA should be positive. There are some individuals who have HCV RNA, but their anti-HCV antibody tests are negative. This can be seen in immunocompromised individuals or in early acute infections. HCV RNA tests are positive earlier than anti-HCV antibody tests. Rapid tests for detection of anti-HCV also exist, but these tests are more expensive than automated immunoassays. Recently the FDA approved the OraQuick rapid HCV test for use with the fingerstick of venous blood specimens. If an individual is positive for anti-HCV, then the logical next step is to test for viral RNA. Viral RNA is detectable in serum or plasma as early as 1 week after exposure to HCV. NAT assays for detecting HCV RNA include PCR-based assays, branched DNA signal amplification, and transcription-mediated amplification. Although both qualitative and quantitative methods are available, quantitative methods are gaining more acceptance [17]. If there is a true infection, both anti-HCV antibody as well as tests for HCV RNA should be positive. If the HCV RNA is negative, then the possibilities include:

- A false positive anti-HCV antibody test.
- In newborns the anti-HCV may be that of the mother, with transfer of antibodies across the placenta.
- Intermittent viremia.
- Past infection.

Interpretations of HCV test results are summarized in Table 23.7. Blood products are routinely screened for hepatitis C using both immunoassays and NAT

Table 23.7 Interpretation of HCV Test Results

Immunoassay (HCV Antibody)	Recombinant Immunoblot (HCV Antibody)	HCV RNA (Nucleic Acid Testing)	Interpretation
Negative	Negative	Negative	Most likely no infection.
Negative	Negative	Positive	Recent infection (window period of seroconversion).
Positive	Positive	Positive	Current infection.
Positive	Positive	Negative	Past infection or in newborn, antibody from mother.
Positive	Negative	Negative	False positive screen.

methods to identify donors with early-stage infection without seroconversion. False positive anti-HCV test results can also occur with immunoassays, and in such cases viral RNA must be tested. False positive hepatitis C antibodies may be observed in a patient with autoimmune hepatitis, but after responding to the therapy, antibodies against HCV should be negative. Autoimmune hepatitis is a periportal hepatitis where the human immune system attacks liver cells, mistaking them as foreign bodies and causing increased immunoglobulins and auto-antibodies. This disease affects more women than men (70% females), especially young girls, and may respond to immunosuppression therapy. It is postulated that a drug, virus, or an environmental agent can trigger a T-cell-mediated cascade directed against liver antigens in genetically susceptible individuals who are prone to autoimmune disease.

CASE REPORT

A 13-year-old boy with a diagnosis of autoimmune hepatitis based on liver biopsy showed positive anti-HCV using enzyme-linked immunosorbent assay (ELISA, Ortho Diagnostics). He also showed elevated AST (1,110 U/L), ALT (1,506 U/L), hyperglobulinemia (4.2 g/dL, normal < 3.0 g/dL), and serum titer for antinuclear antibody and anti-smooth muscle antibody. He responded well to prednisone therapy, and a repeat test of his anti-HCV antibody showed a negative result. The authors speculated that his false positive result in anti-HCV testing was due to hyperglobulinemia, and they recommended repeat testing for anti-HCV when the patient responded to therapy. The additional anti-HCV test should be negative [18].

23.10 IMMUNIZATION AND FALSE POSITIVE HIV AND HEPATITIS TESTING

There are reports of false positive HIV and hepatitis C serological test results following influenza vaccination in some subjects. Although for HIV serology this phenomenon is transient, for hepatitis C serology it may persist for a longer time. Erickson et al. observed false positive HIV antibody test results in an individual when using an enzyme immunoassay 11 days after the individual received an influenza vaccine; however, the individual's antibody against hepatitis C was negative. A Western blot test was indeterminate and no viral load was detected in his blood. The authors considered that his HIV serology test was false positive, and when he was tested 1 month later his viral load was again undetectable and his Western blot reverted to non-reactive [19]. Rubella vaccination and transitory false positive results for HIV Type I in blood donors has also been reported [20].

Table 23.8 Serological Profile of Epstein—Barr Virus

Test	Acute infection	Past Infection
IgM antibody against viral capsid antigen (VCA)	Positive	Negative
IgG antibody against viral capsid antigen (VCA)	Positive	Positive
Antibody to early antigen (EA)	Positive/negative	Negative
Antibody to Epstein—Barr nuclear antigen (EBNA)	Negative	Positive
Heterophilic antibody	Positive	Negative

23.11 TESTING FOR EPSTEIN—BARR VIRUS (EBV)

Epstein—Barr virus targets B lymphocytes, and the infected B lymphocytes disseminate the virus throughout the reticuloendothelial system. The T cells mount an immune response against the infected B cells and these T cells, which appear as morphologically reactive cells, respond to EBV infection by producing two types/classes of antibodies:

- Antibodies specific to EBV.
- Polyclonal antibodies such as heterophil antibodies, cold agglutinins, rheumatoid factors, and antinuclear antibodies (ANA).

EBV-specific antibodies are antibodies to viral capsid antigen (VCA), antibodies to early antigen (EA), and antibodies to Epstein—Barr nuclear antigen (EBNA). Antibodies to VCA can be IgG or IgM. Patients with acute infection are positive for VCA IgM. Antibodies to EA may be positive or negative, and antibodies to EBNA are usually negative. With past infection, VCA IgM is negative but VCA IgG is positive. In addition, antibodies to EA are negative but antibodies to EBNA are positive. Heterophil antibodies are cross-reacting antibodies to antigens that occur in several species that are not phylogenetically related. Heterophil antibodies usually include two different types:

- Antibodies capable of agglutinating horse red blood cells (basis of the Monospot test).
- Antibodies that do not react with any EBV antigens.

Serological markers for diagnosis of EPV infection are summarized in Table 23.8.

KEY POINTS

- Human immunodeficiency virus (HIV) is a slowly replicating retrovirus responsible for acquired immunodeficiency syndrome (AIDS). HIV infection is sexually transmitted, but is also transmitted by sharing infected needles, via blood transfer, and from mother to newborn. HIV infects vital cells involved in human immune function, such as helper T cells (especially CD4+ T cells), macrophages, and dendritic cells. Two major types of HIV virus have been characterized: HIV-1 and HIV-2. HIV-1 was discovered first and is the cause of the majority of HIV infections worldwide. HIV-2 has lower infectivity than HIV-1, is largely confined to West Africa, but is spreading slowly in North America. Several groups are classified within HIV-1. HIV-1 M group is the main type of HIV virus seen in clinical practice. The M type can be divided into several sub-types.

- After HIV infection, viral RNA can be measured within 10−12 days and viral p24 antigen can be measured a little bit later. In general, IgM is the first antibody against HIV, followed by IgG antibodies that appear approximately 3−4 weeks after infection. In general, within 1−2 months after infection, HIV antibodies are usually present in almost all infected individuals, although for a few individuals it may take up to 6 months for antibodies to appear in circulation.

- Testing for HIV can be broadly divided into screening tests and confirmatory tests. Screening tests include standard testing, rapid HIV testing, and combination HIV antibody−antigen testing. Confirmatory testing is performed by Western blot, recombinant immunoblot, or line immunoassay (LIA).

- Standard screen is performed by various immunoassays. The test is based on the detection of IgG antibody against HIV-1 antigens in the serum. HIV antigens include p24, gp 120, and gp 41; antibodies to gp 41 and p24 are the first detectable serological markers following HIV infection. IgG antibodies appear as early as 3 weeks, but most likely within 12 weeks in the majority of patients, and they generally persist for life.

- Causes of false negative HIV screening tests include testing during the window period, fulminant HIV infection, immunosuppression, or immune dysfunction and delay in seroconversion due to early initiation of antiretroviral therapy.

- Since its introduction in 1985 the ELISA method has evolved, and third-generation ELISA assays (sandwich format) can now detect IgG and IgM antibodies and antibodies to all of the M subtypes (as well as N and O groups). Antibody testing methods capable of detecting the O group are important for screening blood products. More recently, fourth-generation assays have been introduced that simultaneously detect HIV antibodies and p24 antigen, thus reducing the window period to only 13−15 days.

- Rapid HIV antibody tests can provide test results in less than 30 min and can be adopted in point-of-care settings. However, both false positive and false negative results can occur with rapid HIV tests and it is important to confirm initial findings with a laboratory-based HIV assay.

- HIV viral load tests detect the number of copies of RNA of HIV present per milliliter of serum or plasma; they are expressed as copies/mL or in log scale. HIV viral load tests indicate viral replication and are often conducted to monitor the progress of antiretroviral therapy. In addition, CD4 lymphocyte counts are also measured to evaluate the immune system.

- The nucleic acid tests (NAT) are commercially available tests that can identify HIV nucleic acid (either RNA or proviral DNA). These tests are based on the principles of polymerase chain reaction (PCR), real time PCR, nucleic acid sequence-based amplification, or ligase chain reaction. NAT assays are useful in special situations, such as in the window period of infection when an antibody against HIV is absent in serum, and in newborns of HIV-infected mothers where maternal antibodies against HIV are present in the newborn's serum.

- Hepatitis A testing is straightforward: IgM anti-hepatitis A antibody denotes recent infection and IgG anti-hepatitis A antibody appears in the convalescent phase of acute hepatitis.

- Hepatitis E virus is also an enterically transmitted virus. Hepatitis E can also be transmitted by blood transfusion, particularly in endemic areas. Chronic hepatitis does not develop after acute hepatitis E infection, except in the transplant setting and possibly in other settings of immunosuppression. The diagnosis of hepatitis E is based upon the detection of hepatitis E virus in serum or stool via PCR or the detection of IgM antibodies against hepatitis.

- The hepatitis D virus (also called the delta virus) is a defective pathogen that requires the presence of the hepatitis B virus for infection. Hepatitis D can elicit a specific immune response in the infected host, consisting of antibodies of the IgM and IgG classes (anti-hepatitis D antibodies).

- Serologic markers available for hepatitis B infection are HBsAg (hepatitis B surface antigen), HBeAg (hepatitis B e antigen), anti-HBc (antibody against hepatitis B core antigen; both IgG and IgM), anti-HBs (antibody against hepatitis B surface antigen), anti-HBe (antibody against hepatitis B e antigen), and viral DNA. HBsAg is the first marker to be positive after exposure to hepatitis B virus. It can be detected even before the onset of symptoms. Most patients may clear the virus, and HBsAg typically becomes undetectable within 4–6 months. Persistence of HBsAg for more than 6 months implies chronic infection. The disappearance of HBsAg is followed by the presence of anti-HBs. During the window period (after the disappearance of HBsAg and before the appearance of anti-HBs) evidence of infection is documented by the presence of anti-HBc (IgM). The co-existence of HBsAg and anti-HBs has been documented in approximately 24% of HBsAg-positive individuals. It is thought that the antibodies fail to neutralize the virus particles. These individuals should be considered carriers of the hepatitis B virus. There exists a subset of patients who have undetectable HBsAg but are positive for hepatitis B viral DNA. Most of these patients have very low viral load with undetectable levels of HBsAg. Uncommon situations are infection with hepatitis B variants that decrease HBsAg production, or mutant strains that have altered epitopes normally used for detection of HBsAg.

- Individuals with recent infection will develop anti-HBc IgM antibodies. Individuals with chronic infection and individuals who have recovered from an infective episode will develop anti-HBc IgG antibodies. However, anti-HBc IgM antibodies may remain positive for up to 2 years after an acute infection. Levels may also increase and be detected during exacerbations of chronic hepatitis B. This can lead to diagnostic confusion.
- The diagnostic tests available for hepatitis C virus (HCV) can be classified under four broad categories: immunoassays for detecting anti-HCV IgG antibodies, recombinant immunoblot assays to detect anti-HCV IgG antibodies (these tests are more specific than immunoassays for detecting antibodies against HCV), rapid assays for detecting anti-HCV IgG antibodies (these assays may be less specific than laboratory-based immunoassays), and PCR or related techniques for detection of viral RNA (the gold standard for HCV testing, these are in general called nucleic acid tests, NATs).
- Epstein—Barr virus (EBV) targets B lymphocytes. The infected B lymphocytes disseminate the virus throughout the reticuloendothelial system. The T cells mount an immune response against the infected B cells and these T cells, which appear as morphologically reactive cells, respond to EBV infection by producing two types/classes of antibodies, including antibodies specific to EBV and polyclonal antibodies such as heterophil antibodies, cold agglutinins, rheumatoid factors, and antinuclear antibodies (ANA).
- EBV-specific antibodies are antibodies to viral capsid antigen (VCA), antibodies to early antigen (EA), and antibodies to Epstein—Barr nuclear antigen (EBNA). Antibodies to VCA can be IgG or IgM. In acute infection, patients are positive for VCA IgM. Antibodies to EA may be positive or negative, and antibodies to EBNA are usually negative. With past infection VCA IgM is negative but VCA IgG is positive. In addition, antibodies to EA are negative but antibodies to EBNA are positive.

REFERENCES

[1] Hollenbeck BL, Beckwith CG. HIV-2 infection in Providence, Rhode Island from 2001 to 2011. HIV Med 2013;14:115—9.

[2] Luft L, Gill MJ, Church DL. HIV-1 viral diversity and its implications for viral load testing: review of current platforms. Int J Infect Dis 2011;15:e661—70.

[3] Gallant JE. HIV counseling, testing and referral. Am Fam Physician 2004;70:295—302.

[4] Spivak A, Sydnor E, Blankson JN, Gallant JE. Seronegative HIV-1 infection: a review of the literature. AIDS 2010;24:1407—14.

[5] Ling AE, Robbins KE, Brown BS, Dunmire V, et al. Failure of routine HIV-1 tests in a case involving transmission with pre seroconversion blood components during the infection window period. JAMA 2000;284:210—4.

[6] Vardinon N, Yust I, Katz O, Laina A, et al. Anti-HIV indeterminate Western blot in dialysis patients: a long term follow up. Am J Kidney Dis 1999;34:146—9.

[7] Delaney KP, Branson BM, Uniyal A, Phillips S, et al. Evaluation of the performance characteristics of 6 rapid HIV antibody tests. Clin Infect Dis 2011;52:257−63.

[8] Facente SN, Dowling T, Vittinghoff E, Sykes DL, et al. False positive rate of rapid oral fluid HIV test increases as kits near expiration date. PLoS One 2009;4:e8217.

[9] Van Oosten HE, Damen M, de Vries HJ. Symptomatic primary HIV infection in a 49 year old man who has sex with men: beware of the window phase. Euro Surveill 2009;14:iii.

[10] Butto S, Suligoi B, Fanales-Belasio E, Raimondo M. Laboratory diagnostics for HIV infection. Ann 1st Super Sanita 2010;46:24−33.

[11] Garcia-Prats AJ, Draper HR, Sanders JE, Agarwal AK, et al. False negative post 18 month confirmatory DNA PCR-positive children: a retrospective analysis. AIDS 2012;26:1927−34.

[12] Wilkins T, Zimmerman D, Schade RR. Hepatitis B diagnosis and treatment. Am Fam Physician 2010;81:965−72.

[13] Rysgaard C, Morris CS, Dress D, Bebber T, et al. Positive hepatitis B surface antigen test due to recent vaccination: a persistent problem. BMC Clin Pathol 2012;12:15.

[14] Kuhns MC, Busch MP. New strategies for blood donor screening for hepatitis B virus: nucleic acid testing versus immunoassay methods. Mol Diag Ther 2006;10:877−91.

[15] Dow BC, Peterkin MA, Green RH, Cameron O. Hepatitis B virus transmission by blood donation negative for hepatitis B surface antigen, antibody to HBsAg, antibody to hepatitis B core antigen and HBV DNA. Vox Sang 2001;81:140.

[16] Wilkins T, Malcolm J, Raina D, Sachade R. Hepatitis C: diagnosis and treatment. Am Fam Physician 2010;81:1351−7.

[17] Kamili S, Drobeniuc J, Araujo A, Hayden TM. Laboratory diagnostics for hepatitis C virus. Clin Infect Dis 2012;55(Suppl. 1):S43−8.

[18] Rosenthal P. False positive results of tests for hepatitis C in autoimmune liver disease. J Pediatr 1992;120:160−1.

[19] Erickson CP, McNiff T, Klausner JD. Influenza vaccination and false positive HIV result. N Eng J Med 2006;354:1422−3.

[20] Araujo PB, Albertoni G, Arnoni C, Ribeiro J, et al. Rubella vaccination and transitory false positive results for human immunodeficiency virus Type I in blood donors. Transfusion 2009;49:2516−7.

CHAPTER 24

Autoimmunity, Complement, and Immunodeficiency

24.1 INTRODUCTION TO THE IMMUNE SYSTEM AND COMPLEMENT

The immune system is a complex biological structure in an organism that is capable of protecting the organism from disease by neutralizing the invading entity. The human immune system can detect a wide variety of pathogens, including viruses, bacteria, parasitic worms, and even certain cancer cells. It is capable of distinguishing such objects from an organ's healthy tissue. In autoimmune diseases, this distinction is lost, and the body's own immune system can target certain cells in an organ or a variety of organs and cause disease. Disorders of the immune system can also cause inflammatory diseases and cancer. Immunity can be broadly divided into innate (natural) and adaptive (specific) immunity. Innate immunity does not provide defense against any particular pathogen, but can be considered as all-purpose immunity and can attack a number of pathogens in a short period of time when challenged. Therefore, innate immunity is the first line of defense. A characteristic feature of innate immunity is that no prior memory of infection by an agent is needed for its activation. The components of innate immunity include certain cells and chemicals. These cells include granulocytes (neutrophils, eosinophils, and basophils), mast cells, monocytes, dendritic cells, and natural killer lymphocytes. Chemical molecules involved in innate immunity include:

- Complement
- Enzymes (e.g. lysozyme)
- Collectins
- Pentraxin (e.g. C reactive protein)

The largest group of cells involved in innate immunity is granulocytes. Neutrophils phagocytose microorganisms and possess enzymes such as myeloperoxidase, cathepsins, proteinase 3, elastase, and defensins for killing

A. Dasgupta and A. Wahed: Clinical Chemistry, Immunology and Laboratory Quality Control
DOI: http://dx.doi.org/10.1016/B978-0-12-407821-5.00024-3

these invaders. Eosinophils play a major role against multicellular parasites (e.g. worms). Eosinophils release toxic proteins against parasites, including major basic protein, eosinophilic cationic protein, and eosinophil neuro-toxin. Basophils and mast cells release chemicals such as histamine, which is an inflammatory mediator. Mast cells also release tumor necrosis factor-alpha (TNF-alpha), which recruits and activates neutrophils. Monocytes and macrophages have phagocytic and microbicidal properties, and also release cytokines as well as clear cellular debris. Monocytes and dendritic cells present antigens to T lymphocytes, which can trigger adaptive immunity. Therefore, major steps in achieving innate immunity are phagocytosis, enzyme activation, liberation of cytokines, and activation of complement. Cells use receptors to recognize microbes. The most important of these receptors are known as Toll-like receptors. The complement system consists of plasma proteins, which have an important role in immunity and inflammation. Complement proteins are present in plasma in inactive states. These proteins are numbered C1 to C9. Many complement proteins, once activated, can in turn activate other complement proteins.

Specific or adaptive immunity is characterized by greater specificity than innate immunity, but response is slower than innate immunity. Lymphocytes play an important role in adaptive immunity. Three types of lymphocytes (T helper cells, cytotoxic T cells, and B cells) mediate specific immunity.

24.1.1 T Lymphocytes and Cell-Mediated Immunity

T cells constitute about two-thirds of lymphocytes in the peripheral blood. In general, T cells are not activated by free or circulating antigens, but they can recognize antigens bound to major histocompatibility complex (MHC) molecules. The T cell receptor complex consists of heterodimers of alpha and beta chains, CD3 proteins (gamma, delta, and epsilon), as well as two zeta chains. The antigen associated with the MHC molecule of the cell binds with the alpha and beta chains. The other proteins interact with the constant portion of the alpha and beta chains of the T cell receptors to generate intracellular signals after the alpha and beta chains have recognized the antigen. If the T cell is a CD4+ cell it will recognize antigens presented in association with Class II MHC molecules. If the T cell is a CD8+ cell, it will recognize antigens presented in association with Class I MHC molecules. There is also a minority of T cells that have gamma and delta chains as the receptors instead of alpha and beta chains. These gamma delta cells do not express either as CD4 or CD8. Typically, more than one signal is required for activation of these T cells. CD8+ lymphocytes differentiate into cytotoxic T lymphocytes, which kill cells harboring microbes in their cytoplasm. CD4+ T cells, once activated, also secrete IL-2, which in turn causes proliferation of T lymphocytes. Interferon gamma (IFN-gamma) activates macrophages and also stimulates B cells to produce antibodies. Interleukin-4 (IL-4) stimulates B cells to differentiate into IgE-producing plasma cells. Some T cells also differentiate into long-living memory cells.

24.1.2 B Cells and Humoral Immunity

B cells can be directly stimulated by antigens. The antigens are recognized by immunoglobulin molecules on the surface of B cells. These immunoglobulin molecules are either IgM or IgD. A second signal also comes from activated CD4+ T helper cells. Once activated, the B cells differentiate into plasma cells. The first exposure to antigen results in production of the IgM class (primary response), but, with subsequent exposure to the same antigen, IgG antibodies are produced (secondary response). This ability of changing the antibody class is called "class switching." Immunoglobulins contain both heavy chains and light chains. The names of various antibodies are derived from the names of the heavy chain; for example, IgG contains gamma heavy chain, hence the name. Various features of immunoglobulins are summarized in Table 24.1. Characteristics of various immunoglobulins include:

- IgG is the dominant class of immunoglobulins that are produced as a secondary response in adaptive immunity.
- IgM is the primary response of adaptive immunity.
- IgA is found in mucous membrane secretions.
- IgE is required for defense against parasites (especially worms).
- IgD is present as a B cell surface receptor.

Many factors can affect immune response. Interestingly, acute stress can be associated with adaptive upregulation of some parameters of natural immunity and downregulation of some functions of specific immunity, while chronic stress can cause suppression of both cellular and humoral response [1].

Table 24.1 Various Classes of Immunoglobulins

Specific Feature	IgG	IgM	IgA	IgE	IgD
Heavy chain	Gamma (γ)	Mu (μ)	Alpha (α)	Epsilon (ε)	Delta (δ)
Molecular type	Monomer	Pentamer	Dimer	Monomer	Monomer
Sub-class	IgG1–4	None	IgA1–2	None	None
Interact with complement	Yes (Classical)	Yes (Classical)	Yes (Alternate)	No	No
Bind to mast cells	No	No	No	Yes	No
Crosses placenta	Yes	No	No	No	No

24.2 PATHWAYS OF COMPLEMENT ACTIVATION

The complement system was discovered more than 100 years ago by Jules Bordet, and since then its importance in protecting humans and animals against infection has been well recognized. The name "complement system" is derived from the ability of this system to complement the action of antibodies and phagocytic cells to destroy pathogens from an organism. At present, more than 30 complement proteins have been discovered [2]. Complement proteins can interact with each other and are also capable of interacting with cell surface proteins. A key step in the activation of complements is the activation of C3, which can be accomplished through three different pathways:

- Classical pathway: Initiated by antigen−antibody complexes fixating with C1.
- Alternate pathway: Spontaneous but bacterial polysaccharides and proteins such as properdin and factors B and D are also involved in activating this pathway.
- Lectin pathway: Plasma lectin binds to mannose on microbes and activates the pathway.

The classical pathway is triggered by activation of C1 complement. There are three components of C1: C1q, C1r, and C1s. IgG or IgM antibodies already bound with antigen can also bind to C1q, thus activating complements through the classical pathway. One IgM molecule is capable of activating this pathway, but multiple IgG molecules are needed for activation. Subsequently, C1r and then C1s are activated, which in turn cleaves C4 and C2 to yield C4a, C4b, C2a, and C2b. Then C4b and C2a form C4b2a (C3 convertase of the classical pathway) that splits C3 into C3a and C3b. At that point C3b combines with C4b2a to form C5 convertase that cleaves C5 into C5a and C5b, which eventually leads to the formation of C5b, C6, C7, C8, and C9 membrane attack complex (MAC). C1 is inhibited by C1 inhibitor. C3 convertase is inhibited by decay accelerating factor, a protein absent in paroxysmal nocturnal hemoglobinuria (PNH).

The alternative pathway is a low-level activation pathway that is spontaneous but can also be activated by insoluble bacterial polysaccharides, yeast cell walls, etc., in the absence of antibody. This is due to spontaneous hydrolysis of C3 to C3a and C3b. In the presence of factors B and D, this eventually forms C3bBb complex (which in turn acts as C3 convertase). Properdin stabilizes the C3 convertase by binding to the complex, which in turn cleaves other C3 molecules to continue the cascade. Factors H and I can inhibit alternative pathways.

In the lectin pathway, plasma lectins bind to mannose on microbes and activation of C4 and C2 takes place as in the classical pathway. Regardless of pathway, the end result of activation of C3 leads to the formation of C3 convertase that breaks down C3 to yield C3a and C3b. C3b can act as an opsonin or bind to C3 convertase to form C5 convertase, which breaks down C5 to yield C5a and C5b and subsequently activates C6, C7, C8, and C9. Thus, complement, once activated, promotes inflammation, recruits cells, and kills targeted cells. Various activities of complement include:

- Opsonins: C3b and C4b can promote phagocytosis by phagocytic cells.
- Anaphylatoxins: C3a, C4a, and C5a.
- Leukocyte activation and chemotaxis: C5a.
- Cell lysis: Membrane attack complex (C5 to C9, activated).
- Removal of circulating antigen−antibody complexes.

Complement deficiency may result in impaired innate immunity (a well-known example is increased susceptibility to *Neisseria meningitidis* infection) and immune complex-mediated inflammation: glomerulonephritis, vasculitis, and systematic lupus erythematosus. Deficiency of C1 inhibitor may be inherited, and is known as hereditary angioedema.

24.3 IMMUNODEFICIENCY

Immunodeficiency may be primary or secondary. Primary immunodeficiency includes a broad category of diseases, including B cell defect, T cell defect, both B and T cell defects, complement deficiency, and defective phagocytosis.

24.3.1 B Cell Defects

Patients with B cell defects are typically susceptible to recurrent bacterial infections, especially respiratory tract infections involving influenzae virus, Strep. Pneumonia, and *Staphylococcus aureus*. Diarrhea may also be present due to infection from enterovirus and/or *Giardia lamblia*. However, most viral, fungal, and protozoal infections are cleared due to intact cell-mediated immunity.

Burton's disease (X-linked agammaglobulinemia) is due to a mutation on chromosome Xq22 that affects the gene for a tyrosine kinase known as Bruton's tyrosine kinase (BTK) or B cell tyrosine kinase. This mutation results in arrest in B cell maturation, from pre-B cells to B cells. Clinical features are seen soon after birth once maternal immunoglobulins (that have crossed the placenta) start to decline. B cells, plasma cells, and immunoglobulin levels are all decreased. Lymphoid tissue lacks germinal centers. Subsequently there is an increased incidence of leukemia, lymphoma, and autoimmune diseases. This is a rare disease affecting 1 in 200,000 live births. If a patient is diagnosed at an early age and treated with regular intravenous gammaglobulin therapy before the sequelae of recurrent infection, prognosis for this disease is relatively good [3].

The genetic basis of common variable immunodeficiency disease is not clear, but in this condition B cells are present but fail to differentiate into plasma cells. Immunoglobulin levels are usually low. Clinical features are seen during the second or third decade of life. However, selective IgA deficiency is the most common primary immunodeficiency, where patients develop recurrent infections due to pyogenic organisms affecting mucosal sites. Individuals with IgA deficiency are susceptible to developing anaphylactic reactions with blood products.

24.3.2 T Cell Defects

Individuals with T cell defects have recurrent and persistent viral, fungal, and protozoal infections. Individuals are also at risk for transfusion-associated graft vs. host disease. DiGeorge syndrome is characterized by failure of development of the thymus and parathyroids (due to failure of development of the third and fourth pharyngeal pouches). Features of T cell immunodeficiency include hypoparathyroidism, dysmorphic facies, and cardiac defects.

24.3.3 Both B and T Cell Defects

In severe combined immunodeficiency (SCID), underlying genetic defects may be diverse. About 50% of cases are transmitted as X-linked disorders. These are due to mutations in the gene coding for the gamma chain for receptors for various interleukins (IL), including IL-2, IL-4, IL-7, IL-9, and IL-15. In addition, 40−50% of cases of SCID are transmitted as autosomal recessive. The most common example of this type of SCID is due to a mutation in the gene encoding the adenosine deaminase (ADA) enzyme. ADA deficiency causes accumulation of adenosine and deoxyadenosine triphosphate metabolites (which are lymphotoxic). In SCID patients the thymus is hypoplastic, and lymph nodes and lymphoid tissue lack germinal centers as well as paracortical T cells. Lack of help from T cells prohibits B cells from being functional, and clinical features seen in both B and T cell defects are also present in patients with combined B and T cell defects.

As mentioned earlier in this chapter, IgM is produced first as a primary immune response, with subsequent production of other classes of immunoglobulins. This is called class switching. One factor involved in class switching is interaction of CD40 molecules on B cells and CD40 ligands (CD40L or CD154) on T helper cells. The most common cause of hyper-IgM syndrome is mutation in the gene encoding for CD40L, which is located in the X chromosome. The interaction between CD40 and CD40L is also required for T helper-mediated activation of macrophages. Thus, both humoral immunity and cell-mediated immunity are affected in hyper-IgM syndrome, and IgM levels are normal or high with low levels of IgG, IgA, and IgE. Wiskott−Aldrich Syndrome (WAS) is an X-linked disease characterized by immunodeficiency, eczema, and thrombocytopenia. In this syndrome, platelets are small in size. The *WAS* gene codes for the Wiskott−Aldrich Syndrome Protein (WASP). Lymphoreticular malignancies and autoimmune diseases complicate this syndrome.

Ataxia telangiectasia is an autosomal recessive condition due to mutation of the ataxia telangiectasia mutated gene (*ATM* gene) that encodes for the ATM protein kinase involved in DNA repair. Clinical features include cerebellar ataxia, oculocutaneous telangiectasia, and lymphoreticular malignancy.

Epstein−Barr virus (EBV)-associated immunodeficiency (Duncan's syndrome or X-linked lymphoproliferative disease) is a disease where individuals develop overwhelming EBV infections, immunodeficiency, aplastic anemia, and lymphomas. Complement deficiency can be associated with deficiency of C3, which leads to increased infections by pyogenic organisms; deficiency of C5−C9 leads to increased infections by *Neisseria* (both *gonococcus* and *meningococcus*).

Defective phagocytosis can be classified under three broad categories:

- Chronic granulomatous disease (CGD): This disease is characterized by a deficiency of NADPH oxidase, resulting in lack of oxidative burst and defective killing of bacteria and fungus that are catalase-positive (e.g. *Staphylococcus* and *Aspergillus*). This disorder may be inherited as X-linked recessive or autosomal recessive fashion. The gene encoding for the Kx antigen of the Kell blood group system is very close to the gene encoding for NADPH oxidase, and, if both are affected, Kx antigen may also be lacking. This is called the McLeod phenotype, which is associated with the presence of acanthocytes. A screening test used for CGD is the nitroblue tetrazolium test (NBT) where a yellow dye is converted into a blue dye if NADPH oxidase function is intact.
- Chédiak−Higashi syndrome is an autosomal recessive condition where defective trafficking of intracellular organelles leads to defective fusion of lysosomes with phagosomes. This syndrome is related to a mutation in the lysosomal trafficking regulator gene. Granulocytes, lymphocytes, and monocytes exhibit giant lysosomes. Neutropenia, thrombocytopenia, and oculocutaneous albinism are seen in this syndrome along with immunodeficiency.
- Leukocyte adhesion deficiency (LAD): LAD Type 1 (LAD-1) is characterized by defective synthesis of LFA-1 and Mac-1, which are integrins. This results in defective leukocyte adhesion to the endothelium, impaired leukocyte migration, and defective leukocyte phagocytosis. LAD Type 2 (LAD-2) is due to the absence of sialyl-Lewis X in leukocytes, which binds to selectin on the endothelium.

24.4 MAJOR HISTOCOMPATIBILITY COMPLEX (MHC)

Histocompatibility molecules are important for immune response, but these molecules are also responsible for evoking transplant rejections. Histocompatibility molecules bind to peptide fragments of foreign proteins and render them susceptible to attack by specific T cells. The genes encoding the histocompatibility molecules are clustered on a small segment (small arm) of chromosome 6. The cluster of genes is known as major histocompatibility complex (MHC) or human leukocyte antigen (HLA) complex. The HLA system is highly polymorphic. The proteins encoded by certain HLA genes are also called antigens, which are essential elements of immune function and play major roles in histocompatibility during an organ transplant.

Class I MHC molecules are present on all nucleated cells and platelets. There are three different Class I MHC molecules: A, B, and C. Class I molecules are heterodimers of an alpha (or heavy) chain and a smaller beta-2-microglobulin. The beta-2-microglobulin molecule is extracellular and the alpha chain has extracellular components as well as parts that traverse the cell membrane into the cell. The extracellular part of the alpha chain has three domains, alpha1, alpha2 and alpha3. Peptides are able to bind within a groove formed by the alpha1 and alpha2 domains.

Class II MHC molecules are present on B lymphocytes and monocytes. There are three different Class II MHC molecules: DP, DQ, and DR. Class II molecules are also heterodimers of one alpha chain and one beta chain. Both chains have extracellular components with parts that traverse the cell membrane into the cell. The extracellular portions of both chains have domains alpha1, alpha2, beta1, and beta2. The peptide or antigen-binding site is formed between the alpha1 and beta1 domains. Class III MHC molecules are components of the complement system.

Antigens within a cell may bind with a Class I MHC molecule, which is produced within the cell. This binding takes place in the endoplasmic reticulum. The complex of Class I molecule and antigen is transported to the cell surface for presentation to CD8+ cytotoxic T lymphocytes. The T cell receptor (TCR) recognizes and binds with the MHC–peptide complex (MHC molecule–antigen complex). The CD8 molecule also binds with the alpha3 domain of the Class I MHC molecule. The T cell is thus activated. CD8+ T cells are Class I MHC-restricted because they can only be activated with antigens that are bound to MHC Class I molecules. Similarly, CD4+ T cells are Class II-restricted. A variety of diseases are associated with certain HLA alleles (Table 24.2). A link between HLA-B27 and ankylosing spondylitis has been well established. Recently, Li et al. reported an association between HLA-46 and Graves' disease [4].

Table 24.2 Association Between Various HLA Alleles and Diseases

HLA Allele	Increased Risk of Disease
HLA-B27	Ankylosing spondylitis
	Reactive arthritis
HLA-B46	Graves' disease (Asian population)*
HLA-B47	21-Hydroxylase deficiency
HLA-DR2	Systemic lupus erythematosus
HLA-DR3	Type 1 diabetes mellitus
	Systemic lupus erythematosus
HLA-DR4	Myasthenia gravis
	Rheumatoid arthritis

*Other HLA types (such as HLA-B8) may also be associated with Graves' disease.

24.5 HUMAN LEUKOCYTE ANTIGEN TESTING

Human leukocyte antigen (HLA) testing, also known as HLA typing or tissue typing, is used to identify antigens on blood cells to determine the compatibility between an organ recipient and a donor organ. If HLA antigens of the recipient are well matched with a donor organ, the possibility of organ rejection is minimized. However, HLA matching is more complex than blood group matching because there are six loci on chromosome 6 where the genes that code HLA antigens are inherited (HLA-A, HLA-B, HLA-C, HLA-DR, HLA-DQ, and HLA-DP). In classical serological HLA testing, antibodies are used to distinguish between different variants of HLA antigens. Each antibody is specific for a particular antigen, and by using different antibodies HLA serotyping is performed to determine if the donor serotype is a good match for the recipient. However, serological testing is limited by the number of antibodies available against specific HLA antigens. More recently, molecular techniques have been used for HLA DNA typing; these are superior to classical serological tests. The HLA Class I genes are by far the most polymorphic genes in the human genome. Current molecular techniques for HLA DNA typing include recombinant DNA technology, chain-termination Sanger sequencing, and polymerase chain reaction (PCR)-based amplification. These molecular tests can recognize more alleles than traditional serological testing [5].

HLA typing along with ABO (blood type) grouping is used to evaluate tissue compatibility between a donor and a potential transplant recipient. HLA typing is performed before various transplant procedures, including those for kidneys, liver, heart, pancreas, and bone marrow. The success of a transplant increases with the number of identical HLA antigens between a recipient and a potential donor. Major types of HLA testing include:

- HLA antigen typing between donor and recipient: Classically this type of testing is done using serological markers, but more recently molecular (DNA) typing (which provides more information) is replacing classical serological testing.
- HLA antibody screening: Performed on the recipient in order to determine if there is any antibody present that might target donor organs, which would trigger organ rejection. HLA antibody is not always present in an individual unless the person has received a blood transfusion or is a woman post-pregnancy.
- Lymphocyte cross-matching: This step takes place when a donor is identified; the objective is to identify any antibody that, if present in the recipient, might be directed against antigens present on the donor's lymphocyte. In this test, serum from the intended recipient is mixed with T and B lymphocytes (white blood cells) from the donor to investigate potential reactions (a positive test result) that might destroy white blood cells of the recipient.

24.6 TRANSPLANT REJECTION

Graft rejection is due to recognition by the host that the graft is a foreign entity. The antigens responsible for graft rejection belong to the HLA system. Rejection involves both cell-mediated immunity and humoral immunity. In T cell-mediated rejection, individual T cells recognize a single peptide antigen in the graft by two distinct pathways, direct and indirect. In the direct pathway, host T cells encounter donor MHC molecules through interstitial dendritic cells that are present in the donor organ and function as antigen-presenting cells (APC). These dendritic cells have a high density of MHC molecules capable of directly stimulating the host T cells. The encounter of the host T cells and donor dendritic cells can take place in the graft or when the dendritic cells move out of the graft and migrate to regional lymph nodes. Both CD4+ and CD8+ cells are activated. CD8+ cells are responsible for cell-mediated cytotoxicity. CD4+ cells secrete cytokines, which results in accumulation of lymphocytes and macrophages. In the indirect pathway, T cells recognize peptides (as antigens) presented by host APCs but not by donor APCs. The peptides (antigens) are, however, derived from the graft tissue. The direct pathway is the major pathway in acute cellular rejection, and the indirect pathway is thought to be responsible for chronic rejection.

Antibody-mediated rejection is also an important aspect of graft rejection where a host may have developed preformed antibodies to donor antigens even before the transplant. Prior blood transfusions can lead to the development of anti-HLA antibodies because platelets and white blood cells are rich in HLA antigens. Multiparous women can also develop anti-HLA antibodies. These antibodies are directed against paternal antigens that are shed from the fetus. Presence of these preformed antibodies can be detected and they are referred to as panel reactive antibodies (PRA). High titers of such antibodies will likely cause development of hyperacute rejection. In those individuals who are not pre-sensitized, exposure to donor antigens may result in formation of antibodies. These antibodies can cause graft damage by antibody-dependent, cell-mediated cytotoxicity, complement-mediated cytotoxicity, and inflammation. The primary target of such antibodies is the vessels of the graft. After transplant, a patient can be assessed for donor-specific antibodies (DSA). During antibody-mediated rejection, the complement is activated and C4 is converted to C4a and C4b. C4b is converted to C4d, which can bind to the endothelial and collagen basement membranes. C4d can be detected by monoclonal antibodies on graft biopsies, thus establishing the process of antibody mediated rejection (AMR). Patterns of rejection can be hyperacute, acute (acute cellular rejection, acute humoral rejection), or chronic.

24.6.1 Graft vs Host Disease

Graft vs host disease (GVHD) is typically observed in bone marrow transplant recipients where the host is severely immunocompromised (due to underlying disease, drugs, or irradiation) and the donor tissue has fully immunocompetent cells. The donor T cells recognize the host HLA antigens as foreign entities and become activated. HLA matching in bone marrow transplantation reduces chances of GVHD, but subtle differences may be enough to trigger such a response. GVHD can be acute or chronic. Acute GVHD, which occurs within days to weeks after transplant, is usually accompanied by skin rash, liver dysfunction, and diarrhea. Features of chronic GVHD include dermal fibrosis, cholestatic jaundice, and immunodeficiency.

CASE REPORT

A 27-year-old male on hemodialysis for 7 years underwent HLA minor mismatch renal transplantation after receiving a kidney from his 52-year-old father. On postoperative Day 12, the patient developed fever, skin rash, and watery diarrhea. Although an allograft kidney biopsy did not show any sign of rejection, antibody was detected in his plasma and his hemoglobin value was decreased to 5.6 g/dL, indicating hemolytic anemia due to passenger lymphocyte syndrome. His clinical symptoms were probably related to graft vs. host disease. An endoscopic biopsy of his colon revealed apoptotic cells consistent with graft vs. host disease. The patient responded to Solu-Medrol® pulse therapy and his symptoms were resolved. On Day 36, mycophenolic acid was added to his immunosuppression therapy with tacrolimus and prednisolone. The patient was discharged in stable condition with stabilized renal function on post-operative Day 53 [6].

24.7 AUTOIMMUNE SEROLOGY

Antinuclear antibodies (ANA) are antibodies directed against various components of the nucleus. An ANA test is ordered in patients suspected of autoimmune diseases, most commonly in patients suspected of suffering from systemic lupus erythematosus (SLE). An ANA test can be performed by indirect immunofluorescence (IIF) assay on Hep-2 cells or by using enzyme-linked immunosorbent assay (ELISA); the result is reported as a titer. In an IIF assay the patient's serum is incubated with Hep-2 cells (a line of human epithelial cells), followed by addition of fluorescein-labeled anti-human globulin (AHG). The serum is serially diluted until the test becomes negative, which provides an estimation of the strength of positivity. Low titers (1:40 to 1:160) are typically observed, but titers higher than 1:160 are likely to be significant; titers greater than 1:320 are likely indicators of true positive results.

ANA shows up on IIF assays as a fluorescent pattern in cells that are fixed to a slide. Therefore, the pattern can be further investigated under a microscope. Although there are some overlaps, different patterns can be associated with certain autoimmune diseases. These various patterns include speckled, homogenous, anti-centromeric, and peripheral (Table 24.3).

For the diagnosis of SLE, once ANA is positive, further testing for antibodies must be considered. In general, ANA testing using Hep-2 cells is very effective in identifying patients with SLE because almost all patients show ANA positivity. In addition, an ANA test can also be positive in other diseases [7]. Unfortunately, a false positive ANA test is common in many conditions, and also in the elderly (Table 24.4). If an ANA test is positive, anti-dsDNA (antibody against double-stranded DNA) and anti-Smith antibody testing may be undertaken. Tests for anti-dsDNA can be done using the Farr assay or an IIF using *Crithidia luciliae*. The Farr assay is used to quantify the amount of anti-dsDNA antibodies in serum. Ammonium sulfate is used to precipitate an antigen−antibody complex that is formed if the serum contains antibodies to dsDNA. The quantity of these antibodies is determined by using radioactively labelled dsDNA. *Crithidia luciliae* is a protozoon that contains a kinetoplast, which is a mitochondrion rich in dsDNA. A patient's serum reacts (if positive) with the kinetoplast; binding is identified with a fluorescent antibody.

If an ANA test is positive, testing for various other antibodies may be undertaken because, in addition to SLE, an ANA test may also be positive for other disorders. Anti-Smith antibodies were first discovered in the 1960s when a patient named Stephanie Smith was treated for SLE and a unique set of antibodies against nuclear proteins was detected in her blood. These antibodies were called anti-Smith ("anti-Sm") antibodies. They are specific in patients

Table 24.3 Various Patterns of ANA in Indirect Immunofluorescence Assay

Pattern	Disease	Further Testing/Autoantibody
(Peripheral)	Systemic lupus erythematosus (SLE)	Anti-dsDNA
Speckled	SLE, scleroderma, Sjögren's syndrome	Smith antibody
	Mixed connective tissue disease	Anti-SAA (Anti-Ro)
		Anti-SSB (Anti-La)
		Anti-topoisomerase I (Scl-70)
		U1-RNP antibody
		PCNA antibody
Homogenous	SLE, drug-induced SLE	Anti-dsDNA
		Anti-histone
Nucleolar	Scleroderma, CREST syndrome	RNA-polymerase I
		U3-RNP antibody
		PM-Scl antibody
Centromere	CREST syndrome, Raynaud's	Anti-centromere
		Antibody
Diffuse	Non-specific for any disease	

Abbreviations: U1-RNP antibody, U1-ribonuclear protein antibody; PCNA antibody, Proliferating cell nuclear antigen antibody; PM-Scl antibody, Polymyositis-associated antibody; CREST syndrome, Limited cutaneous form of systemic scleroderma is often referred as CREST syndrome (acronym of calcinosis, Raynaud's syndrome, esophageal dysmotility, sclerodactyly, and telangiectasia).

Table 24.4 Positive and False Positive ANA Tests in Diseases

Positive ANA Test:	Systemic lupus erythematosus, drug-induced lupus, scleroderma, Sjögren's syndrome, rheumatoid arthritis, mixed connective tissue disease, polymyositis, dermatomyositis, systemic vasculitis
False Positive ANA:	Elderly, liver disease, hIV infection, multiple sclerosis, diabetes, pulmonary fibrosis, pregnancy, and patients with silicone implants

with SLE. Anti-histone antibody testing is useful for patients with a positive ANA test and a history of exposure to medications (e.g. procainamide and isoniazid) associated with drug-induced lupus. ELISA assays are available for detecting anti-histone antibodies or sub-fractions (H1, H2a, H2b, H3, and H4). However, such antibodies may also be detected in patients with rheumatoid arthritis, localized scleroderma, and other diseases characterized by the presence of autoantibodies [8]. Anti-Ro (anti-Sjögren syndrome A,

Table 24.5 Association Between Various Autoantibodies in ANA Testing with Diseases

Antibody	Association	Antigen	Appearance on IIF* Using Hep-2 Cells
Anti-dsDNA	SLE (specific)	DNA backbone	Homogenous
Anti-Smith	SLE (highly specific)	Non-histone nuclear protein complexed with U1-RNP, involved in mRNA splicing	Speckled
Anti-histone	Drug-induced SLE, SLE	Histone H1, H2A, H2B, H3 or H4	Homogenous
Anti-SSA (Ro)	Sjögren (70%), SLE (30%)	Small ribonuclear protein	Speckled
Anti-SSB (La)	Sjögren (50%), SLE (15%)	Small ribonuclear protein and without RNA polymerase III	Speckled
Anti-RNP	Mixed connective tissue disease	U1-RNP associated protein	Speckled
Anti-Scl-70	Scleroderma	DNA topoisomerase I	Finely speckled
Anti-centromere	CREST scleroderma	CENP B	Anti-centromere

Indirect Immunofluorescence Assay (IIF).

anti-SSA) and anti-La (anti-Sjögren syndrome B, anti-SSB) autoantibodies are usually associated with Sjögren syndrome. Anti-ribonuclear protein antibody (anti-RNP) can be observed in mixed connective tissue disease. Anti-topoisomerase I is also called anti-scl-70. This autoantibody is present in patients with scleroderma. Anti-centromere antibody is found in patients with CREST syndrome and scleroderma. Associations of various antibodies in ANA testing with diseases are listed in Table 24.5. Various cytoplasmic antibodies are also associated with different autoimmune diseases. These are summarized in Table 24.6.

CASE REPORT

An 18-year-old woman presented with lower limb edema, abdominal pain, and diarrhea that had started 8 months prior to admission. An abdominal ultrasound was normal except for ascites. Further investigation revealed low serum albumin (1.6 g/dL), an ANA titer of 1:2,560 using indirect immunofluorescence assay, and also showed a speckled pattern along with the presence of anti-Smith antibody. She also showed a low C3 serum level of 35 mg/dL. Scintigraphy using 99mTc-labeled albumin (99mTc: metastable nuclear isomer of technetium-99; used as a radioactive tracer) was positive for abdominal protein loss. A diagnosis of systemic lupus erythematosus related to protein-losing enteropathy (a rare manifestation of lupus) was made, and the patient was treated with prednisolone (40 mg/day). A month later azathioprine (100 mg/day) was added to her drug regime [9].

Table 24.6 Association of Cytoplasmic Antibodies with Various Autoimmune Diseases

Antibody	Disease
Anti-smooth muscle (SMA) against actin	Autoimmune hepatitis
Anti-mitochondrial against various mitochondrial antigens (M2 most specific)	Primary biliary cirrhosis antigens
Anti-Jo-1 against histidyl tRNA synthase	Polymyositis, dermatomyositis (interstitial lung disease)
Anti-parietal cell	Pernicious anemia
Anti-endomysial	Celiac sprue, dermatitis herpetiformis (specific)
Anti-microsomal	Hashimoto's disease
Anti-thyroglobulin	Hashimoto's disease

24.7.1 Anti-Neutrophil Cytoplasmic Antibodies

Anti-neutrophil cytoplasmic antibodies (ANCA) are autoantibodies mainly of IgG type that are directed against antigens present in cytoplasmic granules of neutrophils and monocytes. ANCA may recognize multiple antigens, but antibodies against only two antigens (proteinase 3 and myeloperoxidase) have clinical significance. There are two main types of ANCA, cytoplasmic-ANCA (c-ANCA) and perinuclear-ANCA (p-ANCA). Immunofluorescence on ethanol-fixed neutrophils is used for detection of ANCA. When serum is incubated with alcohol-fixed neutrophils, two different types of reactivity may be observed in individuals with ANCA. If c-ANCA is present, cytoplasmic granular immunofluorescence activity is observed, where c-ANCA has specificity against proteinase-3 and is seen in Wegener's granulomatosis. The other type of reactivity is where a perinuclear immunofluorescence pattern is observed if p-ANCA is present (which has specificity against myeloperoxidase). This pattern is observed in patients with microscopic polyarteritis nodosa, polyarteritis nodosa, and Churg–Strauss syndrome.

24.8 HYPERSENSITIVITY REACTION-MEDIATED DISEASES

Hypersensitivity reactions are generally of four types: immediate (type I), antibody-mediated (type II), immune complex-mediated (type III), and T cell-mediated (type IV).

In immediate (type I) hypersensitivity reactions, when a host is exposed to an antigen, IgE antibodies are produced and bind to the surface of mast cells, which triggers mast cell degranulation. Mast cell products are responsible for subsequent clinical manifestations such as allergic rhinitis, bronchial asthma, and even anaphylactic reactions.

Antibody-mediated (type II) hypersensitivity disorder is due to antibodies directed against antigens, which are components of cells. Sometimes the antigen is exogenous in nature and is adsorbed onto the cell surface (e.g. a drug or its metabolites). Examples of this type of hypersensitivity disorder include myasthenia gravis, Goodpasture syndrome, autoimmune hemolytic anemia, and autoimmune thrombocytopenia. In Graves' disease, the antibody binds to the thyroid-stimulating hormone (TSH) receptor and stimulates it, resulting in hyperthyroidism. Therefore, Graves' disease is an example of a type II hypersensitivity reaction; however, some authors prefer to put this disease in a different category, type V.

In immune complex-mediated (type III) hypersensitivity disorders, large amounts of antigen—antibody complexes are formed which, especially if they persist in circulation, can deposit in various tissues and cause an inflammatory response. Common sites of immune complex deposition are kidneys, joints, and skin. Examples of disease states due to this mechanism are SLE, polyarteritis nodosa, post-streptococcal glomerulonephritis, and serum sickness.

T cell-mediated (type IV) reactions can be sub-classified into delayed-type hypersensitivity (DTH) and T cell-mediated cytotoxicity reactions. The classic example of delayed-type hypersensitivity is the tuberculin reaction. Upon first exposure to tubercle bacilli, macrophages take up bacteria, process their antigens, and present them on their surface. This antigen, in association with Class II MHC molecules, is recognized by CD4+ T lymphocytes, which can remain as memory CD4+ lymphocytes. However, during subsequent exposure (tuberculin testing) these CD4+ memory T cells gather at the site of inoculation. Gamma interferon is secreted and recruits macrophages; this is the major mediator of DTH. Prolonged DTH reactions yield a granulomatous inflammation that causes accumulation of macrophages. Some of these macrophages may be converted into epithelioid cells and some into giant cells. Macrophages can be surrounded by lymphocytes, and even by a rim of fibrous tissue. In T cell-mediated cytotoxicity, CD8+ T cells are responsible for killing antigen-bearing target cells. This type of cytotoxicity is important against viral infections and tumor cells.

KEY POINTS

- Immunity can be broadly divided into innate (natural) and adaptive (specific) immunity. Innate immunity is non-specific and all-purpose. Adaptive immunity is characterized by greater specificity but slower response than innate immunity.

- T cells constitute about two-thirds of lymphocytes in the peripheral blood. A T cell receptor complex consists of heterodimers of alpha and beta chains, CD3 proteins (gamma, delta, epsilon), and two zeta chains. The antigen associated with the major histocompatibility complex (MHC) molecule of the cell binds with the alpha and beta chains. If the T cell is a CD4+ cell it will recognize antigens presented in association with Class II MHC molecules. If the T cell is a CD8+ cell it will recognize antigens presented in association with Class I MHC molecules. There is also a minority of T cells that, instead of alpha and beta chains, possess gamma and delta chains as the receptors. These gamma delta cells do not express as either CD4 or CD8. Typically, more than one signal is required for activation of these T cells.
- B cells can be directly stimulated by antigens recognized by immunoglobulin molecules on their surface. These immunoglobulin molecules are either IgM or IgD. A second signal also comes from activated CD4+ T helper cells. Once activated, B cells differentiate into plasma cells. The first exposure to antigen results in production of the IgM class (primary response), but subsequently IgG antibodies are produced (secondary response). This ability to change the antibody class is called class switching.
- Activation of complements involves activation of C3, which can be accomplished through various pathways: (1) classical pathway (initiated by antigen−antibody complexes fixating with C1); (2) alternate pathway (spontaneous, but bacterial polysaccharides and proteins such as properdin and factors B and D are also involved in activation); and (3) lectin pathway (plasma lectin binds to mannose on microbes for activation).
- Various activities of the complement include opsonins (C3b and C4b can promote phagocytosis by phagocytic cells), anaphylatoxins (C3a, C4a, and C5a), leukocyte activation and chemotaxis (C5a), cell lysis (membrane attack complex C5 to C9, activated), and removal of circulating antigen−antibody complexes.
- Complement deficiency can result in impaired innate immunity (a well known example is increased susceptibility to *N. meningitidis* infection) and immune complex-mediated inflammation: glomerulonephritis, vasculitis, and systemic lupus erythematosus. In addition, a deficiency of C1 inhibitor may be inherited (hereditary angioedema).
- Patients with B cell defects are typically susceptible to recurrent bacterial infections, especially respiratory tract infections involving influenzae virus, strep. pneumonia, and *Staphylococcus aureus*. Diarrhea may also be present due to infection caused by enterovirus and/or *Giardia lamblia*. However, most viral, fungal, and protozoal infections are cleared due to intact cell-mediated immunity.
- Burton's disease (X-linked agammaglobulinemia) is due to a mutation on chromosome Xq22 that affects the gene for a tyrosine kinase known as Bruton tyrosine kinase (BTK) or B cell tyrosine kinase. This mutation results in arrest in B cell maturation, from pre-B cells to B cells.

- Individuals with T cell defects have recurrent and persistent viral, fungal, and protozoal infections. Individuals are also at risk for transfusion-associated graft vs. host disease.
- DiGeorge syndrome is characterized by failure of development of the thymus and parathyroids (due to failed development of the third and fourth pharyngeal pouches). Features of T cell immunodeficiency include hypoparathyroidism, dysmorphic facies, and cardiac defects.
- In severe combined immunodeficiency (SCID), underlying genetic defects may be diverse. About 50% of cases are transmitted as X-linked disorders. These are due to mutations in the gene coding for the gamma chain for receptors for various interleukins (IL), including IL-2, IL-4, IL-7, IL-9, and IL-15. In addition, 40–50% of SCID cases are transmitted as autosomal recessive, the most common example being due to mutations in the gene encoding for adenosine deaminase (ADA) enzyme. ADA deficiency causes accumulation of adenosine and deoxyadenosine triphosphate metabolites, which are lymphotoxic. In SCID patients the thymus is hypoplastic and lymph nodes and lymphoid tissue lack germinal centers as well as paracortical T cells. Lack of help from T cells prohibits B cells from being functional, and clinical features related to both B and T cell defects are present in the patient.
- One factor involved in class switching is interaction of CD40 molecules on B cells and CD40 ligands (CD40L or CD154) on T helper cells. The most common cause of hyper-IgM syndrome is a mutation in the gene encoding for CD40L, which is located in the X chromosome. The interaction between CD40 and CD40L is also required for T helper-mediated activation of macrophages. Thus, both humoral immunity and cell-mediated immunity are affected in hyper-IgM syndrome and IgM levels are normal or high with low levels of IgG, IgA, and IgE.
- Wiskott–Aldrich syndrome (WAS) is an X-linked disease characterized by immunodeficiency, eczema, and thrombocytopenia. In this syndrome the platelets are small in size. The *WAS* gene codes for the Wiskott–Aldrich syndrome protein (WASP). Lymphoreticular malignancies and autoimmune diseases complicate this syndrome.
- Ataxia telangiectasia is an autosomal recessive condition due to mutation of the ataxia telangiectasia gene (*ATM* gene), which encodes for the ATM protein kinase involved in DNA repair.
- Defective phagocytosis can be classified under three broad categories:
  - Chronic granulomatous disease (CGD): This disease is caused by a deficiency of NADPH oxidase that results in lack of oxidative burst and defective killing of bacteria and fungi that are catalase-positive (e.g. *Staphylococcus* and *Aspergillus*). This disorder may be inherited as X-linked recessive or autosomal recessive. The gene encoding for the Kx antigen of the Kell blood group system is very close to the gene encoding for NADPH oxidase, and if both are affected, Kx antigen may also be lacking; it is the called McLeod phenotype, which is associated with the presence of acanthocytes.

- Chédiak—Higashi syndrome: An autosomal recessive condition where defective trafficking of intracellular organelles leads to defective fusion of lysosomes with phagosomes. This syndrome is related to a mutation in the lysosomal trafficking regulator gene. Granulocytes, lymphocytes, and monocytes exhibit giant lysosomes. Neutropenia, thrombocytopenia, and oculocutaneous albinism are seen in this syndrome along with immunodeficiency.
- Leukocyte adhesion deficiency (LAD): LAD Type 1 (LAD-1) is due to defective synthesis of LFA-1 and Mac-1, which are integrins. This results in defective leukocyte adhesion to the endothelium, impaired leukocyte migration, and defective leukocyte phagocytosis. LAD Type 2 (LAD-2) is due to the absence of sialyl-Lewis X in leukocytes, which binds to selectin on the endothelium.

- The genes encoding for the histocompatibility molecules are clustered on a small segment (small arm) of chromosome 6. The cluster of genes is known as major histocompatibility complex (MHC) or human leukocyte antigen (HLA) complex. The HLA system is highly polymorphic. Class I MHC molecules are present on all nucleated cells and platelets. There are three different Class I MHC molecules: A, B, and C. Class I molecules are heterodimers of an alpha (or heavy) chain and a smaller beta-2-microglobulin. The beta-2-microglobulin molecule is extracellular and the alpha chain has extracellular components as well as parts that traverse the cell membrane into the cell. The extracellular part of the alpha chain has three domains: alpha1, alpha2, and alpha3. Class II MHC molecules are present on B lymphocytes and monocytes. There are three different Class II MHC molecules (DP, DQ, and DR). Class II molecules are also heterodimers of one alpha chain and one beta chain. Both chains have extracellular components with parts that traverse the cell membrane and into the cell. The extracellular portions of both chains have domains alpha1 and alpha2, and beta1 and beta 2.
- Antigens within a cell may bind with a Class I MHC molecule, which is produced within the cell. This binding takes place in the endoplasmic reticulum. The complex of Class I molecule and the antigen is transported to the cell surface for presentation to CD8+ cytotoxic T lymphocytes. The T cell receptor (TCR) recognizes and binds with the MHC—peptide complex (MHC molecule—antigen complex). The CD8 molecule also binds with the alpha3 domain of the Class I MHC molecule. The T cell is thus activated. CD8+ T cells are Class I MHC-restricted because they can only be activated with antigens, which are bound to MHC Class I molecules. Similarly CD4+ T cells are Class II-restricted.
- Patterns of transplant rejection can be hyperacute, acute (acute cellular rejection, acute humoral rejection), or chronic.
- Graft vs. host disease is typically observed in bone marrow transplant recipients where the host is severely immunocompromised (due to underlying disease, drugs, or irradiation) and the donor tissue has fully immunocompetent cells. The donor T cells recognize the host HLA antigens as foreign entities and become activated.

- Antinuclear antibody (ANA) tests can be performed by indirect immunofluorescence (IIF) assay on Hep-2 cells or by using enzyme-linked immunosorbent assay (ELISA); results are reported as a titer. In IIF assay, a patient's serum is incubated with Hep-2 cells (a line of human epithelial cells), followed by addition of fluorescein-labeled anti-human globulin (AHG). The serum is serially diluted until the test becomes negative, which provides an estimation of the strength of positivity. Low titers (1:40 to 1:160) are observed in general, but titers higher than 1:160 are likely to be significant; titers greater than 1:320 are likely indicative of true positive results.
- ANA shows up on indirect immunofluorescence assay as a fluorescent pattern in cells that are fixed to a slide. Therefore, the pattern can be further investigated under a microscope. Although there are some overlaps, different patterns can be associated with certain autoimmune diseases. These various patterns include speckled, homogenous, anti-centromeric, and peripheral. If an ANA test is positive, anti-dsDNA (antibody against double-stranded DNA) and anti-Smith antibody testing may be undertaken. Tests for anti-dsDNA may be done using the Farr assay or an IIF using *Crithidia luciliae*.
- Anti-neutrophil cytoplasmic antibodies (ANCA) are autoantibodies mainly of IgG type that are directed against antigens present in cytoplasmic granules of neutrophils and monocytes. ANCA may recognize multiple antigens, but antibodies against only two antigens (proteinase 3 and myeloperoxidase) have clinical significance. There are two main types of ANCA: cytoplasmic-ANCA (c-ANCA) and perinuclear-ANCA (p-ANCA). Immunofluorescence on ethanol-fixed neutrophils is used for detection of ANCA.
- If c-ANCA is present, cytoplasmic granular immunofluorescence activity is observed where c-ANCA has specificity against proteinase-3, and this is seen in Wegener's granulomatosis. The other type of reactivity is where a perinuclear immunofluorescence pattern is observed if p-ANCA is present (which has specificity against myeloperoxidase). This pattern is observed in patients with microscopic polyarteritis nodosa, polyarteritis nodosa, and Churg–Strauss syndrome.
- Hypersensitivity reactions are generally of four types: immediate (type I), antibody-mediated (type II), immune complex-mediated (type III), and T cell-mediated (type IV).
- In immediate (type I) hypersensitivity reaction, when a host is exposed to an antigen, IgE antibodies are produced, which are bound to the surface of mast cells, that trigger mast cell degranulation.
- Antibody-mediated (type II) hypersensitivity disorder is due to antibodies directed against antigens, which are components of cells.
- In immune complex-mediated (type III) hypersensitivity disorder, large amounts of antigen–antibody complexes are formed which, especially if they persist in the circulation, may deposit in various tissues and cause an inflammatory response. Common sites of immune complex deposition are kidneys, joints, and skin.

■ T cell-mediated (type IV) reactions can be sub-classified under two categories: delayed-type hypersensitivity (DTH) and T cell-mediated cytotoxicity. The classic example of DTH is the tuberculin reaction. In T cell-mediated cytotoxicity, CD8+ T cells are responsible for killing antigen-bearing target cells. This type of cytotoxicity is important against viral infections and tumor cells. Associations of cytoplasmic antibodies with various autoimmune diseases are listed in Table 24.6.

REFERENCES

[1] Segerstrom SC, Miller GE. Psychological stress and the human immune system: a meta-analytical study of 30 years of inquiry. Psychol Bull 2004;130:601–30.

[2] Glovsky MM, Ward PA, Johnson KJ. Complement determinations in human disease. Ann Allergy Asthma Immunol 2004;93:513–23.

[3] Chun JK, Lee TJ, Song JW, Linton JA, et al. Analysis of clinical presentation of Burton disease: a review of 20 years of accumulated data from pediatric patients at Severance Hospital. Yonsei Med 2008;49:28–36.

[4] Li Y, Yao Y, Yang M, Shi L, et al. Association between HLB-B*46 allele and Graves disease in Asian population: A meta-analysis. Int J Med Sci 2013;10:164–70.

[5] Erlich H. HLA DNA typing: past, present and future. Tissue Antigens 2012;80:1–11.

[6] Kato T, Yazawa K, Madono J, Saito J, et al. Acute graft versus host disease in kidney transplantation: case report and review of literature. Transplant Proc 2009;41:3949–52.

[7] Lane SK, Gravel JW. Clinical utility of common serum rheumatologic test. Am Fam Physician 2002;65:1073–80.

[8] Hasegawa M, Sato S, Kikuchi K, Takehara K. Antigen specificity of antihistone antibodies in systemic sclerosis. Ann Rheum Dis 1998;57:470–5.

[9] Carneiro FO, Sampaio LR, Brandao LA, Braga LL, et al. Protein losing enteropathy as initial manifestation of systemic lupus erythematosus. Lupus 2012;21:445–8.

CHAPTER 25

Effect of Herbal Supplements on Clinical Laboratory Test Results

25.1 USE OF HERBAL REMEDIES IN THE UNITED STATES

Throughout the history of mankind, herbal remedies were the only medicines available. However, when a pharmaceutical is prepared from a plant source, the active ingredient is sold in the pure form following extensive steps of extraction, purificaTest Resultstion, and standardization. In contrast, herbal remedies are crude extracts of plant products and may contain active ingredients along with other active components that can cause toxicity. In contrast to the popular belief that herbal medicines are safe and effective, many herbal medicines have known toxicity and may interact with Western drugs to cause treatment failure.

The popularity of herbal supplements is steadily increasing among the general population in the United States. According to one survey, approximately 1 out of 5 adults reported using an herbal supplement within the past year. In general, more women use herbal supplements than men, and patients suffering from HIV infection, cancer, and various chronic illnesses use more herbal supplements than the healthy population. The ten most commonly used herbal supplements are echinacea, ginseng, ginkgo biloba, garlic, St. John's wort, peppermint, ginger, soy, chamomile, and kava [1]. Cavaliere et al. reported that in 2009 the sale of herbal supplements in the U.S. alone was estimated to be $5.03 billion [2]. Unfortunately, the sale of herbal supplements is not regulated by the FDA since they are classified as food supplements by the 1994 Dietary Supplement Health and Education Act. However, in Germany, the German E commission publishes monographs regarding safety and efficacy of various herbal supplements.

CONTENTS

A. Dasgupta and A. Wahed: Clinical Chemistry, Immunology and Laboratory Quality Control
DOI: http://dx.doi.org/10.1016/B978-0-12-407821-5.00025-5

25.2 HOW HERBAL REMEDIES AFFECT CLINICAL LABORATORY TEST RESULTS

An herbal remedy may affect clinical laboratory test results by one of the following mechanisms:

- Herbal remedies may produce an unexpected test result by a direct physiological effect on the human body. For example, use of a hepatotoxic herb such as kava can cause elevated liver function tests due to hepatotoxicity.
- Herbal supplements may interact with a therapeutic drug to cause clinically significant drug–herb interactions. For example, St. John's wort induces liver enzymes that metabolize cyclosporine, thus reducing its blood level. Reduced blood cyclosporine levels may cause treatment failure or even the possibility of organ rejection.
- An herbal supplement ingredient can cross-react with assay antibodies to cause interference. This has been reported only in the therapeutic drug monitoring of digoxin using immunoassays where Chinese medicine such as Chan Su can cause a falsely elevated digoxin concentration (see Chapter 15).
- Herbal products can contain undisclosed drugs as adulterants. An unexpected drug level (such as phenytoin in a patient who never took phenytoin but took a Chinese herb) may confuse the laboratory staff and the clinician.
- Herbal supplements can be contaminated with a heavy metal or a heavy metal may be an active ingredient in the herbal supplement, such as in Indian Ayurvedic medicine. Heavy metal toxicity can occur after use of such herbal supplements.

25.3 LIVER DAMAGE AS REFLECTED BY ABNORMAL LIVER FUNCTION TEST AFTER USING CERTAIN HERBALS

The best documented organ toxicity due to use of certain herbal supplements is liver toxicity, and abnormal liver function tests are the first indication of such toxicity. Measurements of the serum or plasma activities of the enzymes aspartate aminotransferase (AST), alanine aminotransferase (ALT), γ-glutamyl transferase (GGT), and alkaline phosphatase (ALP) are routinely performed to assess any liver injury. In general, an abnormal liver function test in the absence of any hepatitis or other infections is a strong indication of liver

Table 25.1 Herbal Supplements That May Cause Liver Damage

Herbal Supplements	Indication for Use	Death Associated with Use
Kava	Herbal sedative/anxiolytic agent	Yes
Chaparral	Antioxidant, anticancer, anti-HIV	No
Comfrey	Repairing broken bone, gout, arthritis	Yes
Germander	Herbal weight loss	No
Mistletoe	Digestive aid, heart tonic	No
LipoKinetix	Herbal weight loss product	No
Pennyroyal	Aromatherapy, inducing abortion	Yes
Noni juice	Stimulating immune system	No

damage due to use of an herbal supplement. Key points regarding herb-induced liver injury include:

- Abnormal liver function tests (such as elevated liver enzymes and possibly bilirubin) along with negative serological tests (for hepatitis or related viral infection) are a strong indication of liver toxicity due to use of an herbal supplement.
- The most common herbal supplement associated with liver damage is kava, an herbal sedative and anxiolytic agent.
- Other hepatotoxic herbals are chaparral, comfrey, germander, and pennyroyal oil. Prolonged use of a certain hepatotoxic herb (3 months or more of continuous use) may cause irreversible liver damage and even death. Various hepatotoxic herbs are summarized in Table 25.1.

CASE REPORT

A 42-year-old healthy Caucasian male presented to the clinic with general weakness, loss of appetite, and jaundice. Three weeks prior he had gone to a Samoan island with his wife for 20 days on their honeymoon. His physical examination was unremarkable except for scleral and skin jaundice and some pain in his liver. Laboratory tests revealed markedly elevated liver enzymes, including AST (1602 U/L), ALT (2841 U/L), GGT (121 U/L), ALP (285 U/L), and lactate dehydrogenase (460 U/L). His total bilirubin was also significantly elevated to 9.3 mg/dL. However, coagulation tests, blood cell counts, and serum protein electrophoresis results were all normal. In addition, serological tests for viral hepatitis (A, B, and C), Epstein–Barr virus, and cytomegalovirus were negative. Genetic testing for hemochromatosis was also negative. Because available information did not explain the abnormal liver function test, the patient was interviewed extensively, and he admitted that during his honeymoon he repeatedly participated in kava ceremonies, consuming a total of 2 to 3 liters of traditional kava preparation. The patient was discharged 19 days after admission. He recovered fully after 36 days. His liver toxicity was related to use of kava [3].

To date, more than 100 cases of hepatotoxicity have been linked to kava exposure. Co-ingestion of alcohol may potentiate the hepatotoxicity. In one case that resulted in death, the individual was reported to have consumed a standardized extract containing 30–70% kava lactones [4]. In addition to kava, use of chaparral, comfrey, and germander can also cause severe hepatotoxicity, and even death, but these herbs are encountered less frequently than kava in clinical practice. Key points regarding hepatotoxicity of comfrey include:

- Pyrrolizidine alkaloids found in comfrey are responsible for liver damage.
- Russian comfrey is more toxic than European or Asian comfrey.

LipoKinetix has been promoted as a weight loss aid and an alternative to exercise that increases metabolism. This product contains phenylpropanolamine, caffeine, yohimbine, diiodothyronine, and sodium usniate. Both phenylpropanolamine (a banned drug) and sodium usniate may be responsible for liver damage after use of LipoKinetix. Sodium usniate is derived from usnic acid, which is also present in Kombucha Tea (also known as Manchurian Mushroom or Manchurian Fungus tea), prepared by brewing Kombucha mushroom in sweet black tea. Acute liver damage due to drinking Kombucha tea has been reported. Herbalene®, also promoted for weight reduction, may also cause liver injury.

Pennyroyal (*Mentha pulegium*) is a plant in the mint genus whose leaves release a spearmint-like fragrance when crushed. Portions of the plant, as well as the essential oil, are used for a variety of purposes, including as an additive to bath products and in aromatherapy. Traditionally, pennyroyal has been brewed as a tea to be ingested in small amounts as an abortifacient and emmenagogue. Ingestion of as little as 10 mL of pennyroyal oil can cause severe toxicity. Death has been reported from ingestion of pennyroyal oil. Interestingly, the antidote used in acetaminophen overdose, N-acetylcysteine, has been used successfully in treating pennyroyal toxicity. Noni juice, which is prepared from noni fruits that grow in Tahiti, is indicated for stimulating the heart and is also used as a digestive aid. There are case reports that noni juice may cause hepatotoxicity, but such effects are usually reversed after discontinuation.

25.4 KIDNEY DAMAGE AND HERBAL SUPPLEMENTS

In 1993, rapidly progressing kidney damage was reported in a group of young women who were taking pills containing Chinese herbs while attending a weight loss clinic in Belgium. It was discovered that one prescription Chinese herb had been replaced by another Chinese herb containing aristolochic acid, a known toxin to the kidney [5]. Later there were many reports of kidney

> ## Box 25.1 COMMON HERBS ASSOCIATED WITH KIDNEY DAMAGE
>
> - Aristolochic acid-containing chinese herbs
> - Wormwood plant
> - Sassafras
> - Horse chestnut
> - Kava
> - Calamus
> - Chaparral
> - Wormwood oil
> - White sandalwood oil

damage due to use of herbal supplements contaminated with aristolochic acid in the medical literature. There are several herbal supplements which are known to cause hematuria and proteinuria. Examples of these herbs are kava, calamus, chaparral, horse chestnut seed, and wormwood oil. Common herbs associated with kidney damage are listed in Box 25.1.

25.5 KELP AND THYROID FUNCTION

Kelp (seaweed) is a part of the natural diet in many Asian countries. The popular Japanese food sushi is wrapped with seaweed. In addition, kelp extracts are available in the form of tablets in health food stores and are used as a thyroid tonic, anti-inflammatory, and metabolic tonic, as well as dietary supplement. Kelp tablets are rich in vitamins and minerals but also contain substantial amounts of iodine. Usually eating sushi or Japanese food should not cause any problem with the thyroid, although some Asian seaweed dishes may exceed the tolerable upper iodine intake of 1,100 microgram/day [6]. However, taking kelp supplements on a regular basis for a prolonged time may cause thyroid dysfunction, especially hyperthyroidism, due to the high iodine content of kelp supplements. Some kelp preparations may also contain arsenic.

25.6 MISCELLANEOUS ABNORMAL TEST RESULTS DUE TO USE OF CERTAIN HERBALS

Various abnormal test results can also be encountered due to use of certain herbal supplements. Although measuring hypertension is not a clinical laboratory test, blood pressure is one of the first few parameters measured when a person is presented to a clinic or emergency department. Although use of ephedra in weight loss products is banned in the United States, infrequently ephedra is encountered in weight loss products imported to the United States from various Asian countries. A popular example is ma huang. Hypertension is common after use of ephedra-containing products.

CASE REPORT

A 39-year-old woman had an enlarged thyroid but she had no sign of hyperthyroidism or hypothyroidism. Ultrasonography demonstrated a multi-nodular goiter with a total volume of 62 mL. The patient was presented with treatment options that included either thyroid resection or routine follow-up. The patient decided not to have the surgery, and at that point she was scheduled for a routine follow-up. She was advised to avoid excess iodine, including iodine-containing drugs and radiographic contrast agents. Two months later the patient was in good health and her thyroid hormones and thyroid-stimulating hormone (TSH) were all within normal limits. Four months later, the patient presented with typical symptoms of hyperthyroidism, including tachycardia, palpitation, tremor, increased sweating, and weight loss. Laboratory test results indicated elevated levels of free T3 (781 pg/dL; upper limit of normal: 420 pg/dL), free T4

(3.2 ng/dL; upper limit of normal: 3.2 ng/dL), and suppressed levels of TSH (<0.01 mU/L). Ultrasonography showed a multi-nodular goiter with a total volume of 67 mL. The patient did not report any exposure to iodine or iodine-containing medications, but admitted that for the last 4 weeks she had been taking a Chinese herbal tea prescribed by a Chinese herbal specialist. The tea preparation contained large amounts of kelp. The patient was advised to discontinue the tea and was treated with an anti-thyroid drug (40 mg thiamazole) and 40 mg propranolol daily. After 7 months, her free T4 and T3 returned to normal values, but her TSH was still slightly decreased (0.14 mU/L). Because her hyperthyroidism was resolved clinically, her thiamazole dosage was reduced to 20 mg per day. The iodine-induced thyrotoxicosis in this patient was due to ingestion of kelp-containing herbal tea [7].

Another relatively safe herbal product, licorice, which is also used in candies as a flavoring agent, may further increase blood pressure in a person suffering from hypertension. In addition, these patients are also vulnerable to developing hypokalemia and possibly pseudo-hyperaldosteronism from regular use of licorice. Glycyrrhizic acid found in licorice is possibly responsible for increasing blood pressure after licorice use.

Many herbal supplements such as ginseng, fenugreek seed, garlic, bitter melon, bilberry, dandelion, burdock, and prickly pear cactus are indicated for lowering blood glucose. In addition, dietary supplement of chromium is also capable of lowering serum glucose levels. Patients suffering from diabetes mellitus and taking oral hypoglycemic agents should not use any such herbals without the approval of their physicians because severe hypoglycemia may occur due to interaction of these herbals with oral hypoglycemic agents. Patients suffering from insulin-dependent diabetes should also refrain from using such herbal supplements.

25.7 DRUG−HERB INTERACTIONS INVOLVING ST. JOHN'S WORT AND WARFARIN−HERB INTERACTIONS

Although many drug−herb interactions have been reported in the literature, clinically significant drug−herb interactions more commonly encountered in clinical situations involve St. John's wort, an herbal antidepressant and Western drug. In addition, warfarin also interacts with many herbal supplements. In general, it has been recommended that the following groups of patients not take any herbal supplements because they are susceptible to drug–herb interactions:

- Organ transplant recipients must not take any herbal supplements because immunosuppressants, especially cyclosporine and tacrolimus, are susceptible to various interactions with the supplements. Clinically significant interaction between St. John's wort and cyclosporine or tacrolimus may cause potential rejection of the transplanted organ due to increased clearance of both drugs as a result of pharmacokinetic interaction with St. John's wort.
- Patients taking warfarin should avoid herbal supplements because many clinically significant interactions have been reported between warfarin and various supplements.
- Patients suffering from HIV infection and being treated with HAART (highly active antiretroviral therapy) should avoid all herbal supplements due to potential treatment failure as a result of interaction between antiretroviral agents and certain herbs.

St. John's wort is a popular herbal antidepressant that is composed of dried alcoholic extract or the alcohol/water extract of hypericum, a perennial aromatic shrub with bright yellow flowers that bloom from June to September. The flowers are believed to be most abundant and brightest around June 24, the day traditionally believed to be the birthday of John the Baptist. Therefore, the name St. John's wort became popular for this herbal product. Active components of St. John's wort, hypericin and hyperforin, are responsible for pharmacokinetic interactions between many Western medications and St. John's wort. Although St. John's wort interacts with most drugs pharmacokinetically, pharmacodynamic interaction of St. John's wort with several drugs has also been reported. Key points involving interaction of St. John's wort with various drugs are as follows:

- Hyperforin, an active component of St. John's wort, induces cytochrome P-450-mixed function oxidase, the major liver enzyme responsible for metabolism of many drugs; thus, it increases clearance of many drugs, which can result in treatment failure.
- Hypericin, another active component of St. John's wort, modulates the P-glycoprotein pathway, thus affecting clearance of drugs that are not metabolized by liver enzymes, such as digoxin.
- Pharmacodynamic interaction of St. John's wort with various selective serotonin reuptake inhibitors (SSRIs) such as paroxetine, sertraline, or venlafaxine may produce life-threatening serotonin syndrome.

Table 25.2 Pharmacokinetic Interactions Between Various Drugs and St. John's Wort*

Drug Class	Comments
Immunosuppressant agents	Reduced levels of cyclosporine and tacrolimus. No interaction with mycophenolic acid.
Antiretroviral agents	Reduced levels of indinavir, saquinavir, atazanavir, lamivudine, and nevirapine.
Anticancer agents	Reduced levels of imatinib and irinotecan.
Cardiovascular drugs	Reduced levels of digoxin, verapamil, and nifedipine.
Benzodiazepines	Reduced levels of alprazolam and midazolam.
Hypoglycemic agents	Reduced levels of gliclazide.
Anti-asthmatic agents	Reduced levels of theophylline.
Statins	Reduced efficacy of simvastatin and atorvastatin.
Oral contraceptives	Failure of contraception by ethinyl estradiol and related compounds.
Antidepressants	Reduced level of amitriptyline.
Synthetic opioid	Reduced levels of methadone and oxycodone.

*May cause treatment failure.

The most important pharmacokinetic interaction of St. John's wort with various drugs includes its interaction with immunosuppressants (reduced efficacy of cyclosporine and tacrolimus, but no interaction with mycophenolic acid), warfarin (reduced efficacy), and various antiretroviral agents (reduced efficacy) [8]. Pharmacokinetically important drug interactions with St. John's wort are summarized in Table 25.2.

CASE REPORT

A 65-year-old patient who received a renal transplant in November 1998 had a trough whole blood level tacrolimus concentration between 6 and 10 ng/mL, which was within therapeutic range. The patient experienced depression in July 2000 and started self-medication with St. John's wort (600 mg per day). In August 2000, the patient showed an unexpectedly low tacrolimus concentration of 1.6 ng/mL. Interestingly, her serum creatinine was also decreased to 0.8 mg/dL from an initial value of between 1.6 and 1.7 mg/dL. Whe▮ patient stopped taking St. John's wort, the tacrolimus returned to the previous range of 6–10 ng/mL. After 1 m▮ the creatinine value was also gradually increased to 1.3 dL. Because the patient showed no rejection episode▮ new tacrolimus target level was set to 4–6 ng/mL by ad▮ ing the tacrolimus dosage in order to lower potential ne▮ toxicity of tacrolimus [9].

Many herbal supplements are known to potentiate the effect of warfarin and may produce excessive anticoagulation, causing bleeding problems. In such cases, increased INR with no change of dosage may be an early indication of such warfarin—herb interactions. In general it is assumed that angelica root, anise, asafoetida, bogbean, borage seed oil, bromelain, capsicum, chamomile, clove, fenugreek, feverfew, garlic, ginger, ginkgo biloba, horse chestnut, licorice root, meadowsweet, passionflower herb, red clover, turmeric extract, and willow bark potentially increase the effectiveness of warfarin, thus increasing the risk of bleeding in a patient taking warfarin and one of these supplements. In contrast, green tea extract and St. John's wort reduce the efficacy of warfarin.

25.8 HERBS ADULTERATED WITH WESTERN DRUGS AND CONTAMINATED WITH HEAVY METALS

Sometimes herbal medicines manufactured in various Asian countries are contaminated with Western drugs but the product labels do not mention the presence of such drugs. Of 2,069 samples of traditional Chinese medicines collected from eight hospitals in Taiwan, 23.7% contained pharmaceuticals, most commonly caffeine, acetaminophen, indomethacin, hydrochlorothiazide, and prednisolone [10]. Lau et al. reported a case of phenytoin poisoning in a patient after using Chinese medicines. This patient was treated with valproic acid, carbamazepine, and phenobarbital for epilepsy, but was never prescribed phenytoin [11]. Heavy metal contamination is another major problem with Asian medicines. Ko reported that 24 of 254 Asian patent medicines collected from herbal stores in California contained lead, 36 products contained arsenic, and 35 products contained mercury [12]. Lead and other heavy metal contaminations (cadmium and mercury) are common in Indian Ayurvedic medicines. Unfortunately, some Ayurvedic medicines contain heavy metals as a part of the active ingredient.

KEY POINTS

- Abnormal liver function tests (such as elevated liver enzymes and possibly bilirubin) along with negative serological tests (for hepatitis or related viral infection) are an indication of liver damage due to use of herbal remedies (most commonly kava). Other hepatotoxic herbals are chaparral, comfrey, germander, and pennyroyal oil.
- Chinese herbs used for weight loss may contain aristolochic acid, a known toxin to the kidney that causes nephrotoxicity.

- Kelp (seaweed) is rich in iodine, and taking kelp supplements on a regular basis may cause thyroid dysfunction.
- Weight loss products such as ma huang may contain ephedra, which can cause hypertension and even damage to the heart.
- St. John's wort, an herbal antidepressant, interacts with many drugs and can cause treatment failure due to reduced concentration of a particular drug in the blood. Hyperforin, an active component of St. John's wort, induces cytochrome P-450-mixed function oxidase, causing increased clearance of many drugs. Hypericin, another active component of St. John's wort, modulates the P-glycoprotein pathway, thus affecting clearance of drugs that are not metabolized by liver enzymes.
- Clinically significant interaction between St. John's wort and cyclosporine or tacrolimus may cause potential rejection of transplanted organs due to increased clearance of both drugs as a result of pharmacokinetic interaction with St. John's wort. Patients taking warfarin should avoid St. John's wort because St. John's wort significantly reduces efficacy of warfarin by increasing its clearance. Patients suffering from HIV infection and being treated with HAART (highly active antiretroviral therapy) should avoid St. John's wort because it reduces the efficacy of many protease inhibitors.
- Pharmacodynamic interaction of St. John's wort with various SSRIs such as paroxetine, sertraline, or venlafaxine may produce life-threatening serotonin syndrome.
- Indian Ayurvedic medicines and herbal supplements manufactured in Asia may be contaminated with heavy metals, most commonly lead, mercury, and arsenic. In addition, certain herbal supplements manufactured in Asian countries may be contaminated with Western drugs.

REFERENCES

[1] Bent S. Herbal medicine in the United States: review of efficacy, safety and regulation. J Gen Intern Med 2008;23:854–9.

[2] Cavaliere C, Rea P, Lynch M, Blumenthal M. Herbal supplement sales rise in all channels in 2009. HerbalGram 2010;86:62–65.

[3] Christl SU, Seifert A, Seeler D. Toxic hepatitis after consumption of traditional kava preparation. J Travel Med 2009;16:55–6.

[4] Denham A, McIntyre MA, Whitehouse J. Kava-the unfolding story: report on a work-in-progress. J Alternative Complementary Med 2002;8:237–63.

[5] Vanhaelen M, Vanhaelen-Fastre R, Nut P, Abramowicz D, et al. Rapidly progressive interstitial renal fibrosis in young women: association with slimming regimen including Chinese herb. Lancet 1993;341:387–91.

[6] Teas J, Pino S, Critchley A, Braverman LE. Variability of iodine content in common commercially available edible seaweeds. Thyroid 2004;14:836–41.

[7] Mussig K, Thamer C, Bares R, Lipp HP, et al. Iodine induced thyrotoxicosis after ingestion of kelp containing tea. J Gen Intern Med 2006;21:C11–4.

[8] Di YM, Li CG, Xue CC, Zhou SF. Clinical drugs that interact with St. John's wort and implications in drug development. Curr Pharm Des 2008;14:1723–42.

[9] Bolley R, Zulke C, Kammerl M, Fischereder M, Kramer BK. Tacrolimus induced nephrotoxicity unmasked by induction of CYP3A4 system with St. John's wort. [Letter] Transplantation 2002;73:1009.

[10] Huang WF, Wen KC, Hsiao ML. Adulteration by synthetic therapeutic substances of traditional Chinese medicine in Taiwan. J Clin Pharmacol 1997;37:344–50.

[11] Lau KK, Lai CK, Chan AYW. Phenytoin poisoning after using Chinese proprietary medicines. Hum Exp Toxicol 2000;19:385–6.

[12] Ko RJ. Adulterants in Asian patent medicines. N Eng J Med 1998;339:847.

Index

Note: Page numbers followed by *"f"*, *"t"* and *"b"* refers to figures, tables and boxes respectively.

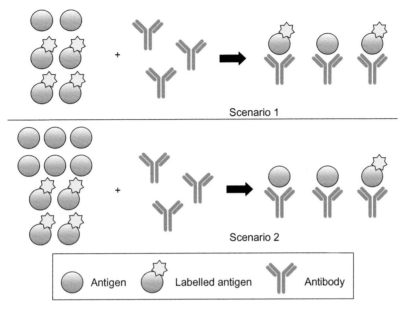

FIGURE 2.1
Competitive immunoassay. *(Courtesy of Stephen R. Master, MD, PhD, Perelman School of Medicine, University of Pennsylvania)*

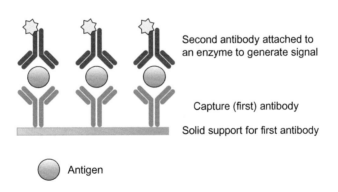

FIGURE 2.2
Sandwich immunoassay. *(Courtesy of Stephen R. Master, MD, PhD, Perelman School of Medicine, University of Pennsylvania)*

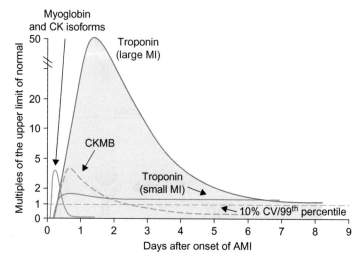

FIGURE 8.1

Timing of release of various cardiac biomarkers after myocardial injury. © *American Heart Association. Reprinted with permission.[3]*

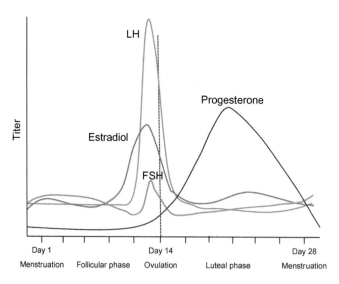

FIGURE 9.1

Titers of various hormones during menstrual cycle. *(Courtesy of Andres Quesda, M.D, Department of Pathology and Laboratory medicine, University of Teaxs-Houston Medical School.)*

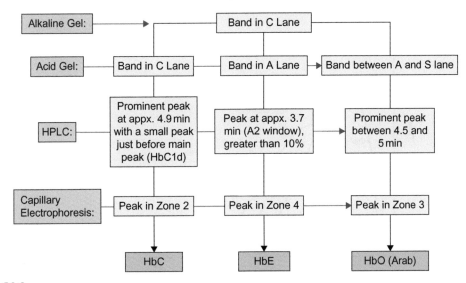

FIGURE 21.1

Interpretation of hemoglobinopathy when a band is present in the C lane in the alkaline gel. *(Courtesy of Andres Quesda, M.D, Department of Pathology and Laboratory Medicine, University of Texas, Houston Medical School.)*

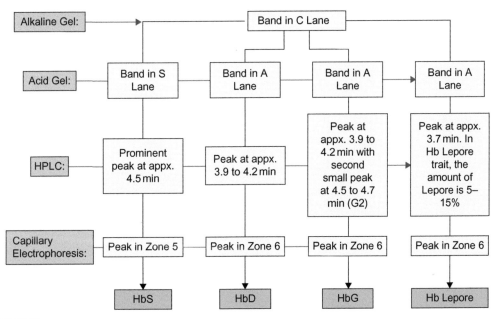

FIGURE 21.2

Interpretation of hemoglobinopathy when a band is present in the E lane in the alkaline gel. *(Courtesy of Andres Quesda, M.D, Department of Pathology and Laboratory Medicine, University of Texas, Houston Medical School.)*

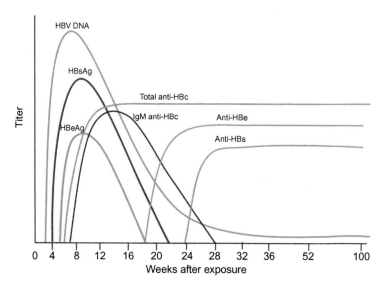

FIGURE 23.1

Virological and serological response to acute hepatitis B infection with recovery. *(Courtesy of Andres Quesda, M.D, Department of Pathology and Laboratory Medicine, University of Texas, Houston Medical School.)*

Printed and bound by CPI Group (UK) Ltd, Croydon, CR0 4YY

03/10/2024

01040321-0005